Lithium-Ion Batteries: Basics and Applications

Produced with the kind support of:

VDE INSTITUTE

The independent and neutral VDE Testing and Certification Institute is a national and internationally accredited institution in the field of testing and certification of electrotechnical devices, components and systems. Also in the battery industry, we have been a recognized partner for testing, certification and standard development for years.

www.vde.com/institute

Reiner Korthauer
Editor

Lithium-Ion Batteries: Basics and Applications

Translator Michael Wuest, alphabet & more

Springer

Editor
Reiner Korthauer
LIS-TEC GmbH
Kriftel
Germany

Translator Michael Wuest (alphabet & more, Landau, Germany)

ISBN 978-3-662-53069-6 ISBN 978-3-662-53071-9 (eBook)
https://doi.org/10.1007/978-3-662-53071-9

Library of Congress Control Number: 2017936665

Translation of the German book Korthauer: Handbuch Lithium-Ionen-Batterien, Springer 2013, 978-3-642-30652-5

Printed on acid-free paper

This Springer imprint is published by the registered company Springer-Verlag GmbH, DE part of Springer Nature.
The registered company address is: Heidelberger Platz 3, 14197 Berlin, Germany

Foreword

Life without batteries is inconceivable. Stored energy has become an integral part of our everyday lives. Without this over 100-year-old technology, the success story of laptops, cell phones, and tablets would not have been possible. Although there are many ways of storing power, there is only one system that enables the functions that meet consumers' expectations of a storage medium – the rechargeable battery. A battery that can be discharged and charged at the push of a button. Strictly speaking, the battery is not a storage system for electric power but an electrochemical energy converter. And in recent decades its development has followed many convoluted paths.

The history of the battery, both as a primary and secondary element, has not yet been fully elucidated today. We know that the voltaic pile was introduced by A. Volta (1745–1827) around 1800. Some 65 years later, around 1866, G. Leclanché (1839–1882) was awarded a patent for a primary element, the so-called Leclanché element. The element consisted of a zinc anode, a graphite cathode, and an electrolyte made of ammonium chloride. The cathode had a manganese dioxide coating on the boundary surface with the electrolyte. C. Gassner (1855–1942) further developed this system, and in 1901 P. Schmidt (1868–1948) succeeded in inventing the first galvanic dry element based on zinc and carbon.

The further development of batteries – both as primary and secondary elements – can be described as tentative. There were not any major breakthroughs with regard to an increase in specific energy or specific power. Nevertheless, the technical and chemical properties of the elements were improved on an ongoing basis. Today, nearly all battery systems have high cycling stability and safety and are completely maintenance-free.

It was not until the beginning of the 1970s that a new era began. The first ideas for a new system were born at the Technical University of Munich, Germany: lithium batteries with reversible alkaline-metal-ion intercalation in the carbon anode and an oxidic cathode. It was some years before the first commercial lithium battery was launched on the market by Sony in 1991. Constant development – which also involved implementing new materials – resulted in this unparalleled success.

Today we are faced with new challenges. The change in paradigms in mobility and energy supply (the shift away from fossil fuels) requires new, low-cost, low-maintenance, and lightweight energy storage systems. These requirements are, to a certain extent, contradictory and therefore not fully realizable. As a result, there is tremendous pressure on research and development as well as on the industrial sector to come up with innovations that bring us closer to this goal. Although R&D activities have increased in recent years, partly because new institutes have been set up in universities and research centers, only time will tell whether they are sufficient.

The aim of *Lithium-Ion Batteries: Basics and Applications* is to make a small contribution toward successfully managing the pending change in paradigms. 32 articles by 54 authors provide a broad overview of all of the relevant areas of the lithium-ion battery: the chemistry and design of a battery cell, production of batteries, deployment of the battery system in its two most important applications as well as issues concerning safety, transport, and recycling.

The book is divided into five sections. At the beginning, an overview of the different storage systems implementing the electrochemical conversion of energy is provided. The second section is devoted to all of the facets of the lithium-ion battery. Important materials and components of the cell are presented in detail. These components include the cathode's and anode's chemical materials as well as the conducting salts and the electrolyte. Several chapters are dedicated to the battery system's modular design; the modules are in turn made up of a large number of cells and necessary mechanical components. Next, the electric components are explained. This section closes with details on thermal management and the battery management system in addition to an outlook.

The third section focuses on the production resources required for manufacturing batteries, followed by the necessary test procedures. Before the battery is deployed, a series of questions regarding transport, safety, and recycling – and more – need to be addressed. The fourth section is devoted to these issues. Last but not least, the applications – in the area of electric mobility and stationary uses – are described in the fifth, and last, section.

The main aim of this manual is to provide help to all people who want to acquire an understanding of state-of-the-art battery technology. It describes the lithium-ion battery in great detail in order to show the difficulties that manufacturers are still battling with today with 20 years of experience under their belts. It also strives to demonstrate the tremendous potential of this technology and the possibilities it holds for users and newcomers in research and development. The book does not, however, provide the same degree of depth as a scientific paper on one of the many issues related to the lithium-ion battery. It is intended as a reference book at a high technical level.

I would like to thank all of those who contributed to the success of this book. First and foremost, my thanks go to the authors of the individual chapters as well as to our translator Mr. Wuest from alphabet & more and – last but not least – to Ms. Hestermann-Beyerle and Ms. Kollmar-Thoni from Springer Verlag.

The data in this version of *Lithium-Ion Batteries: Basics and Applications* were retrieved from current data sources.

I hope that all of the readers of *Lithium-Ion Batteries: Basics and Applications* acquire important information for their day-to-day work and wish them an enjoyable read.

Frankfurt am Main, Germany, May 2017 Reiner Korthauer

Preface

In 1780, the Italian physicist Alessandro Volta produced electricity for the very first time with the "Voltaic pile" – a battery made of copper, zinc, and an electrolyte. He was thus the first person to succeed in generating electricity from electrochemical energy stored in an electrolyte, rather than from friction. Already in 1802, William Cruickshank invented the trough battery, the first mass-produced battery. Since then, the use of electricity has been inextricably linked to the development and use of electrochemical energy storage systems. Nowadays we are accustomed to finding batteries in different shapes and sizes almost everywhere – in small electronic appliances and industrial-scale applications alike.

Nevertheless, storage technologies have recently become the focus of public interest in a very specific field. The transition of the energy supply to renewable energies is becoming increasingly important worldwide. In Germany, the government made the decision to abandon the use of nuclear energy by 2022 and, instead, to feed large quantities of renewable energies into our energy grid. Ever since, it has become clear that the large yet fluctuating amounts of energy generated by renewable energies can only be efficiently used if at the same time we are able to provide sufficient capacities for storing energy until it is needed. Integrated energy storage systems and their integration into decentralized, intelligent networks play a key role. Worldwide investment needs are therefore expected to significantly exceed EUR 300 billion by 2030.

This book focuses on the lithium-ion battery, a very important storage medium in this context, and examines all of its facets. Lithium-ion batteries have a vital role to play in several respects because they are able to react rapidly, can be installed locally, are easily scalable, and have a broad field of applications both in mobile and stationary operations.

They are considered to be the most important "door opener" to the future of battery-electric vehicles. Due to their high energy density, they appear to be the only technology that has the potential to enable sufficiently high ranges for electric vehicles. In addition, their value-added share for the entire vehicle is as high as 40 percent. These are already two very good reasons for focusing intensively on lithium-ion batteries because high added value also secures jobs. In a report drawn up for the German Chancellor in 2011, the experts of the German National Platform for Electric Mobility stated that Germany has a lot of catching up to do in the field of battery technology. They also concluded that German companies are capable of taking the technology lead in the field of cells and batteries and of developing added value across the battery process chain within Germany. They recommended a dual strategy: optimization of today's solutions and, at the same time, research on successor generations.

Electric vehicles of all types are an essential milestone on the path toward emission-free mobility. Parallel to consuming "green power", they also make it possible for "green power" to be fed into the grid because the traction battery generates an operating reserve. Thus, in their mobile mode, they serve as a means of transport. And when stationary – operating in bi-directional mode – they can provide part of the urgently needed operating reserve for the power grid.

Fully stationary lithium-ion batteries are also a key component for successfully converting the power grid. One of research and development's primary aims is to make Germany a leading center for research in electrochemistry and leader in the mass production of safe, affordable battery systems.

This book constitutes an important step forward along the challenging yet rewarding path toward a new energy system. In addition to presenting all of the technical aspects of lithium-ion batteries in detail, it also sets out equally important topics such as production, recycling, standardization, and electrical and chemical safety.

Industry Chairman of the Steering Committee of the German National Platform for Electric Mobility. Henning Kagermann

Contents

List of Authors

Prof. Dr. Philipp Adelhelm Institut fuer Technische Chemie und Umweltchemie Center for Energy and Environmental Chemistry (CEEC Jena) Friedrich-Schiller-Universitaet Jena Philosophenweg, Jena, Germany, philipp.adelhelm@uni-jena.de

Dr.-Ing. Ralf Bindel Robert Bosch GmbH, Wernerstrasse 51, 70469 Stuttgart, Germany, ralf.bindel@zeiss.com

Dr. Klaus Brandt Lithium Battery Consultant, Taunusstrasse 43, 65183 Wiesbaden, Germany, akkubrandt@aol.com

Dipl.-Ing. Frank Dallinger Robert Bosch GmbH, Wernerstrasse 51, 70469 Stuttgart, Germany, Frank.Dallinger@de.bosch.com

Dipl.-Ing. Christoph Deutskens PEM Aachen GmbH, Karl-Friedrich-Straße 60, 52074 Aachen, Germany, c.deutskens@pem-aachen.de

Roland Dorn Texas Instruments Deutschland GmbH, Haggertystr. 1, 85356 Freising, Germany, Roland.Dorn@ti.com

Dr. rer. nat. Harry Doering ZSW, Helmholtzstrasse 8, 89081 Ulm, Germany, harry.doering@zsw-bw.de

Markus Eckel TE Connectivity Germany GmbH, Ampèrestrasse 12-14, 64625 Bensheim, Germany, meckel@te.com

Dr. Frank Edler elbon GmbH, Freibadstrasse 30, 81543 Muenchen, Germany, frank.edler@elbon.de

Hans-Joachim Faul TE Connectivity Germany GmbH, Tempelhofer Weg 62, 12347 Berlin, Germany, joachim.faul@te.com

Meike Fleischhammer ZSW, Lise-Meitner-Strasse 24, 89081 Ulm, Germany, mfleischhammer@gmx.de

Dr. Christian Graf Chemische Fabrik Budenheim KG, Rheinstrasse 27, 55257 Budenheim, Germany, c_graf@gmx.net; batteries@budenheim.com

Dr. Christoph Hartnig Heraeus Deutschland GmbH & Co. KG, Heraeusstraße 12-14, 63450 Hanau, Germany, christoph.hartnig@heraeus.com

Heiner Heimes Chair of Production Engineering of E-Mobility Components, Campus Boulevard 30, 52074 Aachen, Germany, H.Heimes@pem.rwth-aachen.de

Ansgar von Hemdt Chair of Production Engineering of E-Mobility Components, Campus Boulevard 30, 52074 Aachen, Germany, A.Hemdt@pem.rwth-aachen.de

Dr.-Ing. Claus-Rupert Hohenthanner Evonik Technology & Infrastructure GmbH, Rodenbacher Chaussee 4, 63457 Hanau, Germany, claus-rupert.hohenthanner@evonik.com

Prof. Dr. Juergen Janek Physikalisch Chemisches Institut & Laboratory for Materials Research (LaMa) Justus-Liebig-Universitaet Giessen, Heinrich-Buff-Ring 17, 35392 Giessen, Germany, juergen.janek@phys.chemie.uni-giessen.de

Prof. Dr.-Ing. Achim Kampker Chair of Production Engineering of E-Mobility Components, Campus Boulevard 30, 52074 Aachen, Germany, A.Kampker@pem.rwth-aachen.de

Dr. Uwe Koehler Conwitex GmbH, Am Leineufer 51, 30419 Hannover, uwe.koehler@conwitex.com

Dr. Reiner Korthauer LIS-TEC GmbH, Kriftel, Germany, korthauer@lis-tec.de.

Dr. rer. nat. Peter Kritzer Freudenberg Sealing Technologies GmbH & Co. KG, 69465 Weinheim, Germany, peter.kritzer@fst.com

Dr. Peter Lamp BMW AG, 80788 Munich, Germany, peter.lamp@bmw.de

Dr.-Ing. Stephan Leuthner Robert Bosch Battery Systems GmbH, Kruppstrasse 20, 70469 Stuttgart, Germany, stephan.leuthner@bosch-battery.de

Dr. Jan Marien Isabellenhuette Heusler GmbH & Co. KG, Postfach 1453, 35664 Dillenburg, Germany, jan.marien@isabellenhuette.de

Dr. Kai-Christian Moeller Fraunhofer Alliance Batteries, Fraunhofer-Gesellschaft, Corporate Strategy, Hansastrasse 27c, 80686 Munich, Germany, kai-christian.moeller@ict.fraunhofer.de

Dipl.-Wirtsch.-Ing. Karlheinz Mueller EABB Consulting, Berufsbildungsausschuss, ZVEI – Zentralverband Elektrotechnik- und Elektronikindustrie e. V., Merckstrasse 7, 64342 Seeheim-Jugenheim, Germany, mueller.zwingenberg@t-online.de

Dr. Olaf Nahrwold Freudenberg Sealing Technologies GmbH & Co. KG, 69465 Weinheim, Germany, olaf.nahrwold@fst.com

Dr. Oswin Oettinger SGL Carbon GmbH, Werner-von-Siemens-Strasse 18, 86405 Meitingen, Germany, oswin.oettinger@sglgroup.com

Prof. Dr. Karl-Heinz Pettinger Technologiezentrum Energie, Hochschule Landshut, Am Lurzenhof 1, 84036 Landshut, Germany, karl-heinz.pettinger@haw-landshut.de

Ehsan Rahimzei VDMA Verband Deutscher Maschinen- und Anlagebau e.V., Lyoner Str. 18, 60528 Frankfurt am Main, Germany, ehsan.rahimzei@vdma.org

Dipl.-Ing. (Univ.) Simon Ramer LEONI Silitherm S.r.l., S.S 10 – Via Breda, 29010 Monticelli d'Ongina (PC), Italy, simon.ramer@leoni.com

Dr. rer. nat. Bernhard Riegel HOPPECKE Batterien GmbH & Co. KG, Bontkirchener Strasse 1, 59929 Brilon, Germany, bernhard.riegel@hoppecke.com

Dr. Michael Roth Freudenberg Forschungsdienste KG, Hoehnerweg 2-4, 69465 Weinheim, Germany, michael.roth@freudenberg.de

Dr. Kerstin Sann-Ferro DKE, Stresemannallee 15, 60596 Frankfurt, Germany, kerstin.sann@vde.com

Heiko Sattler VDE-Pruef- und Zertifizierungsinstitut, Merianstrasse 28, 63069 Offenbach, Germany, heiko.sattler@vde.com

Dr.-Ing. Peter Schmid Robert Bosch GmbH, Wernerstrasse 51, 70469 Stuttgart, Germany, PeterK.Schmid@de.bosch.com

Dr. Michael Schmidt BASF SE, GCN/EE – M311, 67056 Ludwigshafen, Germany, michael.e.schmidt@basf.com

Dipl.-Ing. Timo Schuff ITK Engineering AG, Im Speyerer Tal 6, 76761 Ruelzheim, Germany, timo.schuff@itk-engineering.de

Reiner Schwartz STMicroelectronics Application GmbH, Bahnhofstrasse 18, 85609 Aschheim-Dornach, Germany, reiner.schwartz@st.com

Dr. Rudolf Simon M+W Group GmbH, Löwentorbogen 9b, 70376 Stuttgart, Germany, rudolf.simon@mwgroup.net

Dipl.-Ing. Harald Staeb Seuffer GmbH & Co. KG, Baerental 26, 75365 Calw, Germany, harald.staeb@seuffer.de

Bjoern Steurich Infineon Technologies AG, Am Campeon 1-12, 85579 Neubiberg, Germany, bjoern.steurich@infineon.com

Frank Treffer Umicore AG & Co. KG, Rodenbacher Chaussee 4, 63457 Hanau-Wolfgang, Germany, frank.treffer@eu.umicore.com

Dipl.-Ing. Michael Vogt SGS-TÜV GmbH, Hofmannstrasse 51, 81379 Munich, Germany, michael.vogt@sgs.com

Dipl.-Ing. Hermann von Schoenau Schoenau-Consulting, Hauptstrasse 1 a (Schlosshof), 79739 Schwoerstadt, Germany, hermann.schoenau@t-online.de

Dr. Kai Vuorilehto Aalto University Helsinki, Kemistintie 1, 02150 Espoo, Finland, kai.vuorilehto@helsinki.fi

Dr. Christoph J. Weber Freudenberg Vliesstoffe KG, Hoehnerweg 2-4, 69465 Weinheim, Germany, christoph.weber@freudenberg-nw.com

Dr.-Ing. Achim Wiebelt Behr GmbH & Co. KG, Heilbronner Strasse 393, 70469 Stuttgart, Germany, achim.wiebelt@behrgroup.com

Stephan Wittkaemper GTS Flexible Materials GmbH, Hagener Strasse 113, 57072 Siegen, Germany, stephanwittkaemper@gts-flexible.de

Dr. Thomas Woehrle BMW AG, Petuelring 130, 80788 Munich, Germany, thomas.woehrle@bmw.de

Dr. Călin Wurm Robert Bosch Battery Systems GmbH, Heilbronner Strasse 358-360, 70469 Stuttgart, Germany, calin.wurm@de.bosch.com

Dr.-Ing. Robert Zauter Wieland-Werke AG, Graf-Arco-Strasse 36, 89079 Ulm, Germany, robert.zauter@wieland.de

Dipl.-Ing. Michael Guenther Zeyen vancom GmbH & Co. KG, Marie-Curie-Strasse 5, 76829 Landau, Germany, m.zeyen@vancom.de

Part I

Electrochemical Storage Systems – An Overview

Overview of battery systems

1

Kai-Christian Moeller

Contents

1.1 Introduction

Electrochemical storage systems will increasingly gain in importance in the future. This is true for the energy supply of computers and mobile phones that are becoming more and more sophisticated and smaller. It is also true for power tools and electric vehicles as well as, on a larger scale, for stationary storage of renewable energy. This Chapter will provide an overview of today's most common electrochemical storage systems. It will discuss two primary systems, which in general cannot be recharged, or only in limited fashion. Among other things, problems

K.-C. Moeller (✉)
Fraunhofer Alliance Batteries, Fraunhofer-Gesellschaft, Corporate Business Development and Marketing, Hansastrasse 27c, 80686 Munich, Germany
e-mail: kai-christian.moeller@zv.fraunhofer.de

R. Korthauer (ed.), *Lithium-Ion Batteries: Basics and Applications*,
https://doi.org/10.1007/978-3-662-53071-9_1

of rechargeability are discussed, using the example of the anode materials zinc (for aqueous electrolytes) and lithium (for non-aqueous electrolytes). In terms of rechargeable systems, the whole spectrum from lead-acid batteries to rechargeable nickel-based or sodium-based batteries to lithium-ion batteries is covered. Redox flow-batteries also are discussed, as are electric double-layer capacitors. This will enable the reader to gain an insight into lithium-ion technology's competing and complementary technologies. The latter will be presented in other chapters of this book.

1.2 Primary systems

1.2.1 Cells with zinc anodes

One of the first cells of technical importance was the Leclanché cell (1866), which supplied railroad telegraphs and electric bells with electricity.

As with the current advanced zinc-carbon and alkaline cells, its anode material was metallic zinc. One reason for the employment of zinc is its high specific charge of 820 Ah/kg and, for employment in aqueous electrolytes, the high negative voltage of – 0.76 V vs. a standard hydrogen electrode (SHE). If combined with a manganese dioxide (MnO_2) cathode, a cell voltage of 1.5 V is achieved. The internal resistance of these cells, which are mainly used as device batteries, causes a low current capability.

The high specific charge of zinc is also advantageous in zinc-air cells, usually employed in hearing aids. In combination with diffusing oxygen from the air it enables the production of cells with high energy densities of more than 450 Wh/kg.

Unfortunately, these cells exhibit a limited electrochemical rechargeability. The reason is the morphologically poor plating ability of zinc. In spite of intensive research it was not possible to improve the dendritic precipitations of zinc. The Electric Fuel Corp. tried a different approach, namely to substitute the used anodes with new ones. These cells were employed in the nineties during a fleet test of the Deutsche Post (German postal service).

1.2.2 Cells with lithium anodes

Lithium is the perfect material for anodes: It is a very light element and has a specific charge of 3,862 Ah/kg. In addition to that, it features an extremely negative redox potential of – 3.05 V vs. SHE. Specific energies of more than 600 Wh/kg are achievable. However, aqueous electrolytes cannot be used due to the high reducing power of lithium. The electrolytes must be based on organic solvents. In most commercial lithium-metal batteries the cathode consists of manganese dioxide. This enables voltages of more than 3 V. Such cells are used in cameras and watches, for example. Cells with other cathode materials, e.g., thionyl chloride or sulfur dioxide, are employed in electronic energy meters and heat cost allocators as well as in medicine and the military. Since a few years, a new system has found its way into photography applications as a high-quality and powerful replacement for alkaline manganese cells. This system features cathodes made of iron sulfide (FeS_2) and a voltage of 1.5 V, which is similar to that of regular alkaline batteries.

Usually, metallic lithium cells are considered to be non-rechargeable, since the morphology of the electrochemically plated lithium is unsuitable for charging and discharging processes. Dendritic growth of the lithium precipitations through the separator might be induced, causing short circuits with the cathode and subsequent fires. At the end of the eighties, a recall of such problematic rechargeable lithium-metal batteries had to be undertaken by Moli Energy. Since then, the professional world has been skeptical toward this technology.

In spite of that, the French company Bolloré successfully uses lithium-metal polymer systems in more than 3,500 vehicles on the street. Their rechargeable batteries with capacities of 30 kWh exhibit a metallic lithium anode, a polymer electrolyte made of polyethylene oxide (PEO) that prevents dendritic growth.

1.3 Secondary systems

1.3.1 Lead-acid battery

The lead-acid battery is the oldest rechargeable storage system among the systems that are technically relevant today. First investigated in the middle of the 19th century, it has been continually developed all the way to today's valve-regulated lead-acid batteries (VRLA). The active materials of the lead-acid battery are lead and lead oxide (PbO_2) on parallel grid plates. The electrolyte is aqueous sulfuric acid. Its cell voltage of more than 2 V is quite high for an aqueous system. In the newest developments the battery is a closed-system, maintenance-free battery with a fixed electrolyte. In lead-acid gel batteries the electrolyte is gelled by means of adding silica (SiO_2). In absorbent glass mat (AMG) batteries it is physically bound in glass mats.

Due to the high weight of lead (equivalent to 259 Ah/kg), only 30 to 40 Wh/kg are achievable. Although the cycling stability for complete cycles (state of charge 0 to 100 %) is low, it is still possible to charge the lead-acid battery with high currents for short periods of time. This is used in the application as a starter battery in automobiles. The sulfation of the lead into electrically non-conductive lead sulfate ($PbSO_4$) with large particles that occurs as a reaction product on both the anode and the cathode raises the internal resistance. This leads to a deterioration of the battery. The lead-acid battery still has a share of more than 90 % in the battery market. This is due to the low production costs (material, technology) and the high recyclability.

1.3.2 Nickel-cadmium and nickel metal hydride batteries

Nickel-based rechargeable batteries were first developed around 1900: nickel-iron batteries by T. Edison and nickel-cadmium batteries by W. Jungner. The cathode material of both types of batteries is nickel oxide hydroxide (NiO[OH]). The electrolyte is 20 % caustic potash. While the nickel-iron battery never really took off, the rechargeable nickel-cadmium battery was developed into an extremely powerful system. Cadmium demonstrates a high specific charge of 477 Ah/kg and a cell voltage of 1.2 V. Based on these values, specific energies of 60 Wh/kg are reachable. Recent rechargeable batteries are produced using winding technology with active materials on thin current collector foils or lattices. They show a very high current capability and outstanding low-temperature characteristics, even up to – 40 °C.

Due to EU regulations for the usage of cadmium, it is only allowed in medical and safety-relevant applications, also in power tools with high power requirements.

In 1990, Sanyo commercialized the nickel metal hydride batteries. The cadmium was substituted by a hydrogen storage alloy made of nickel and noble earths. It has been possible to triple the specific energy of the cells since their introduction, to 80 Wh/kg. Both nickel-based battery systems feature an internal chemical overcharging and overdischarging protection. They are therefore suitable for battery packs without sophisticated electronics. Consumer electronics has since seen a squeezing out of the nickel metal hydride batteries by the lithium-ion batteries. Now, most nickel metal hydride batteries are used in hybrid vehicles.

1.3.3 Sodium-sulfur and sodium nickel chloride batteries

Both are battery systems for application at high temperatures of 250 to 300 °C. Sodium has a very high specific charge of 1,168 Ah/kg and a very negative voltage curve (– 2.71 V vs. SHE), which is perfect for an anode.

The cathode material of the sodium-sulfur battery is sulfur. Therefore, both electrode materials are liquid at operating temperature. The separator is a solid ceramic made of aluminum oxide (sodium-β-aluminate), which is sodium ion-conductive. At 300 °C, this ceramic exhibits a conductivity for sodium ions that is similar to that of aqueous electrolytes. The nominal voltage of the cells differs in relation to the state of charge due to the formation of various sodium sulfides as reaction products. It lies between 1.78 and 2.08 V, while the specific energy reaches 200 Wh/kg. One advantage of the production of this battery type is the low price of the materials. The high operating temperatures and the subsequent thermal losses correspond to a self-discharging of the battery. This is why it ideally is used as large stationary energy storage system in the MW range. This technology was used in cars in the nineties, e.g., in the BMW E1 and the Ford Ecostar EV.

The sodium nickel chloride battery ("ZEBRA battery") is a safer variant of sodium batteries since it demonstrates a (limited) tolerance toward overcharging and overdischarging, among other things. Its design is similar to that of the sodium-sulfur battery with an aluminum-oxide ceramic that is sodium ion-conductive. The cathode however consists of a porous nickel matrix as current collector with nickel chloride ($NiCl_2$). The nickel chloride is impregnated with sodium chloroaluminate ($NaAlCl_4$), which functions as a second electrolyte in the form of a molten salt at 250 °C. The specific energy of the cells is around 120 Wh/kg for a nominal voltage between 2.3 and 2.6 V. Its inverse structure (liquid sodium on the outside) is advantageous when compared to the sodium-sulfur battery. It enables the usage of low-cost rectangular steel housings instead of nickel housings. Assembly is facilitated by the fact that it is possible to insert the battery materials in an uncharged state as sodium chloride and nickel. Therefore, the charged active materials are not generated until the first charging cycle.

The sodium nickel chloride batteries are used in short runs of electric vehicles and in special applications. Examples: The first specimens of the Smart ForTwo electric drive were equipped with batteries made by FIAMM SoNick.

1.3.4 Redox-flow batteries

Redox-flow batteries are related to fuel cells in that both electroactive components (for the anode and cathode reaction) are fed into an electrochemical vessel (cell stack) for the reaction from two outside holding tanks. This results in an enormous advantage of redox-flow batteries: It is possible to scale energy content (size of the tanks) and power (size of the vessel) independently of each other. Vanadium-redox batteries (VRB) are of practical use. This technology uses the dissolved vanadium salts as active materials in varying oxidation states. The anodic and cathodic regions are isolated from each other by a separator. This separator is impermeable to electrolytes and is made of a proton-conducting plastic foil, e.g., Nafion®. Contrary to fuel cells it is possible to electrochemically regenerate the "consumed" active material solutions in the vessel. The specific energies are relatively low at around 10 Wh/kg. This is caused by the aqueous, diluted vanadium salt solutions and the elaborate system technology. Therefore, these batteries are currently restricted to usage in the stationary energy storage area.

1.3.5 Electric double-layer capacitors

Electric double-layer capacitors (or "Supercaps", according to their NEC brand name) are similar in structure to standard batteries: The electrodes are made of metallic current collector foils that are coated with particles and separated by a thin electrolyte-soaked separator. However, the charge storage is not achieved by chemical redox reactions like in batteries. It is rather accomplished by means of an electrostatic separation of charge at the electrochemical double a layer between the particles and the electrolyte. Highly porous active carbons with a large specific surface are employed to attain larger surfaces. Also, organic electrolytes such as acetonitrile and suitable conducting salts are employed. They enable higher voltages with subsequent higher specific energies than those of aqueous electrolytes, up to the 5 Wh/kg range. Due to their manner of storage the electric double-layer capacitors achieve a high cycle number of around 1 million. This is the main advantage of this system. Another one is the very high power densities of more than 20 kW/kg. This enables charging and discharge times of less than 20 seconds. They are used for instance in wind turbines, where they power the off-grid blade angle control, and for the boosting/recuperation of railway vehicles, saving around 30 % energy.

1.3.6 Lithium-ion batteries

Above, I mentioned the critical deposition of metallic lithium. In order to prevent this, intercalation compounds of lithium as anode materials were developed in the eighties. Using this technology should result in zero metallic lithium. Instead, lithium ions are charge carriers in the electrolyte and act as counterions to compensate the electrical current flow through the consumer load. The lithium ion is very small, allowing for a multitude of intercalation compounds, which exhibit electrochemical potentials from almost – 3 V to more than 2 V vs. SHE. Intercalation compounds in the lower voltage range are suitable for anodes, graphite or lithium alloys with silicon and tin, for instance. Compounds of lithium with carbon, idealized as lithium graphite (LiC_6), have a specific charge of 372 Ah/kg at a voltage of – 2.9 V vs. SHE. However, due to the fact that these are employed in unlithiated form (as opposed to the lithium-metal battery, the lithium-ion battery is assembled uncharged), the challenge arose to find a cathode material that already contained the necessary lithium: $LiCoO_2$ with 137 Ah/kg and 0.8 V vs. SHE. This material and a compatible electrolyte made of organic carbonates and lithium hexafluorophosphate ($LiPF_6$) were the components for the first lithium-ion battery with an average voltage of around 3.6 V. SONY commercialized the lithium-ion battery in 1991. This was the advent of that system for consumer applications and portable PCs and it took a decade to squeeze out the standard nickel metal hydride technology. The energy density up to 250 Wh/kg for high-energy consumer cells and the possible several hundred cycles have made the lithium-ion battery the undisputed champion. Only with this technology was it possible to achieve the current proliferation of smartphones and tablets. In the beginning of their development, the power density of these high-energy cells was considerably inferior to that of, particularly, nickel-cadmium batteries. This is the reason why they only have been employed in high-power devices such as power tools from 2005 on. Hybrid vehicles, especially plug-in hybrid vehicles, have seen a considerable increase in the use of these systems recently, as well. In the meantime, prototypes are being developed for their employment as stationary energy storage systems to stabilize the grid voltage or to store the fluctuating renewable energies' electricity.

1.4 Outlook

This overview of the varying current technically relevant storage systems has examined the different characteristics of rechargeable batteries based on lead, nickel, and sodium, of redox-flow batteries as well as electric double-layer capacitors. It has also given a short introduction to lithium-ion batteries.

Lithium-ion batteries will eventually satisfy most requirements thanks to their versatility and will, in part, replace some of the established battery systems. Apart from evolutionary improvements will new developments, especially those based on lithium-sulfur batteries and possibly (in the long run) those based on lithium-air batteries, fulfill the increasingly demanding consumer wishes with their very high energy densities.

Bibliography

1. Daniel C, Besenhard JO (ed.) (2011) Handbook of battery materials, 2. edition Wiley-VCH
2. Reddy TB (2010) Linden's handbook of batteries, 4. edition McGraw-Hill Professional
3. Yoshio M, Brodd RJ, Kozawa A (eds.) (2009) Lithium-ion batteries science and technologies, 1. edition Springer
4. Huggins RA (2009) Advanced batteries: materials science aspects, 1. edition Springer
5. Nazri G-A, Balaya P, Manthiram A, Yang Y (eds.) (2014) Advanced lithium-ion batteries. New materials for sustainable energy and development, 1. edition Wiley-VCH
6. Park J-K (ed.) (2012) Principles and applications of lithium secondary batteries, 1. edition Wiley-VCH

BOSCH
Invented for life
First choice for efficient, fuel-saving driving
Powertrain systems and electrified mobility by Bosch
With complete powertrain systems for all vehicle classes, Bosch combines mobility with energy efficiency. This not only means making continuous improvements to internal combustion engines, but also developing powertrain systems for low- and high-voltage hybrids as well as electric vehicles.
Connecting components and systems intelligently makes powertrains more efficient, and thus economical to drive. With this aim in mind, Bosch combines hardware, software, and services to create top-to-bottom powertrain solutions.
www.bosch-mobility-solutions.com

Part II

Lithium-ion Batteries – Materials and Components

Lithium-ion battery overview

2

Stephan Leuthner

Contents

2.1 Introduction

The history of lithium-ion batteries started in 1962. The first battery was a battery that could not be recharged after the initial discharging (primary battery). The materials were lithium for the negative electrode and manganese dioxide for the positive electrode. This battery was introduced on the market by Sanyo in 1972. Moli Energy developed the first rechargeable battery (secondary battery) in 1985. This battery was based on lithium (negative electrode) and molybdenum sulfide (positive electrode). However, its design exhibited safety problems due to the lithium on the negative electrode.

The next step toward a lithium-ion battery was the use of materials for both electrodes that enable an intercalation and deintercalation of lithium and also have a high voltage potential. Sony developed the first rechargeable lithium-ion battery and

S. Leuthner (✉)
Robert Bosch Battery Systems GmbH, Kruppstrasse 20, 70469 Stuttgart, Germany
e-mail: stephan.leuthner@bosch-battery.de

R. Korthauer (ed.), *Lithium-Ion Batteries: Basics and Applications*,
https://doi.org/10.1007/978-3-662-53071-9_2

introduced it on the market in 1991. The negative electrode's active material was carbon, that of the positive electrode lithium cobalt oxide [1]. Later on, lithium-ion batteries were developed especially in countries such as South Korea and Japan and were introduced in many applications.

2.2 Applications

Lithium-ion batteries have been used in mobile consumer devices in great numbers since 1991. This is due to their low weight and high energy content. They are mainly used in cell phones, followed by notebooks.

Almost all notebooks already were equipped with lithium-ion batteries in 2000 [2]. The battery packs for these devices usually consist of 3 to 12 cells, in parallel or serial connection. Another application of lithium-ion batteries are power tools, with a voltage of 3.6 to 36 V, depending on the usage.

In electric mobility, lithium-ion batteries play an increasingly important role. They are used in pedelecs (electrically assisted bicycles), electric bicycles, and electric scooters. The automotive industry employs lithium-ion batteries in different kinds of hybrid vehicles as well as in so-called plug-in hybrid vehicles and electric vehicles. Hybrid buses and trucks and electric busses are also equipped with lithium-ion batteries. In stationary applications, lithium-ion batteries are available as mini storage devices with around 2 kWh up to 40 MWh in larger plants.

2.3 Components, functions, and advantages of lithium-ion batteries

Fig. 2.1 shows the basic principle and function of a rechargeable lithium-ion battery. An ion-conducting electrolyte (containing a dissociated lithium conducting salt) is situated between the two electrodes. The separator, a porous membrane to electrically isolate the two electrodes from each other, is also in that position. Single lithium ions migrate back and forth between the electrodes of lithium-ion batteries during charging and discharging and are intercalated into the active materials. During discharging, when lithium is deintercalated from the negative electrode (copper functions as current collector), electrons are released, for example. The active materials of the positive electrode are, for example, mixed oxides. Those of the negative electrode mainly are graphite and amorphous carbon compounds. The positive electrode contains active materials such as mixed oxides. The active materials of the negative electrode mainly are graphite and amorphous carbon compounds. These are the materials into which the lithium is intercalated. As shown in Fig. 2.1, the lithium ions migrate from the negative electrode through the electrolyte and the separator to the positive electrode during discharging. At the same time, the electrons as electricity carriers migrate from the negative electrode via an outer electrical connection (cable) to the positive electrode (aluminum as current collector). During charging, this process is reversed: Lithium ions migrate from the positive electrode through the electrolyte and the separator to the negative electrode.

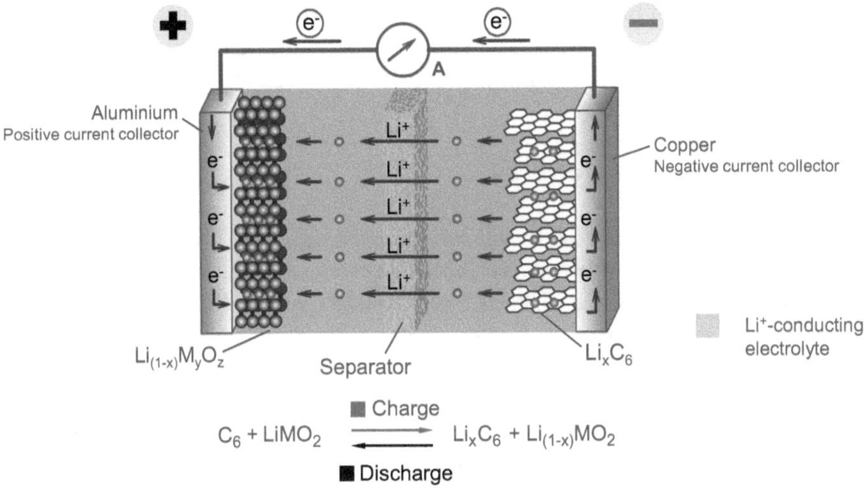

Fig. 2.1 Set-up of a lithium-ion battery (shown is the discharging process)

These cell materials are used to produce cylindrical, prismatic and pouch cells, the design of these cells is described in detail in Chapter 9.

Depending on the application, a single battery cell is used or several cells are connected in series in a module. Also, a parallel connection is possible, dependent on the required capacity. Several connected modules form a battery system for automotive applications (as an example, see Fig. 2.2).

For controlling purposes, automotive battery systems are equipped with a battery management system. This system performs cell monitoring functions and uses sensor technology to monitor cell voltages and temperatures. It also monitors the current and enables the switching on and off of the battery system. The battery management system is furthermore used to control the temperature management (cooling or heating) of the battery system.

Fig. 2.2 Set-up of a battery system for automotive applications (*left* battery module; *right* battery system) [3]

The advantages of lithium-ion batteries and the systems derived thereof are: high specific energy, high specific power, high efficiency during charging and discharging as well as low self-discharge rates.

2.4 Charging procedures

The standard charging process for lithium-ion batteries is CC-CV (constant current/ constant voltage): First, the battery is charged to a certain maximum voltage with a constant current (CC). Then, it is charged with a constant voltage (CV) and a decreasing current. The charging process ends after a predetermined time has elapsed or when a certain current value has been reached. Depending on the materials used, lithium-ion batteries can be charged up to different determined maximum voltages, but not any further.

Overcharging the batteries causes deterioration reactions from a certain voltage on. These deterioration reactions might differ in their intensity, depending on the employed safety measures. The charging currents with which a battery can be maximally charged are also dependent on the design and the temperature.

2.5 Definitions (capacity, electric energy, power, and efficiency)

Typical parameters for batteries are nominal capacity, electric energy and power. They are used to characterize a battery cell or system and are therefore discussed here.

Capacity describes the amount of electric charge a power source can deliver under specific discharge conditions. It depends on the discharging current, the cut-off voltage, the temperature, and the type and amount of active materials. The unit is Ah.

The energy of a battery or a rechargeable battery is calculated as the product of capacity and average discharge voltage. The unit is Wh. Specific energy refers to the mass of the rechargeable battery and its unit is Wh/kg. Energy density refers to the volume of the rechargeable battery and its unit is Wh/l.

Power is calculated as the product of current and voltage, for instance during discharging. The unit is W.

The efficiency of lithium-ion batteries is very high, usually above 95 %. Efficiency is the energy released during discharging divided by the energy stored during charging.

2.6 Safety of lithium-ion batteries

Fig. 2.3 shows, for an example of an automotive lithium-ion battery system, that the chemical, electrical, mechanical, and functional safety characteristics play an important role in product safety. The chemical safety is defined by the battery cell's design, for instance by the choice of active materials and the set-up. The electrical safety is achieved by the isolation of the battery system's cables, housing, and subcomponents. The mechanical safety depends on the respective design, for example the use of a special crash box. The functional safety is guaranteed by monitoring

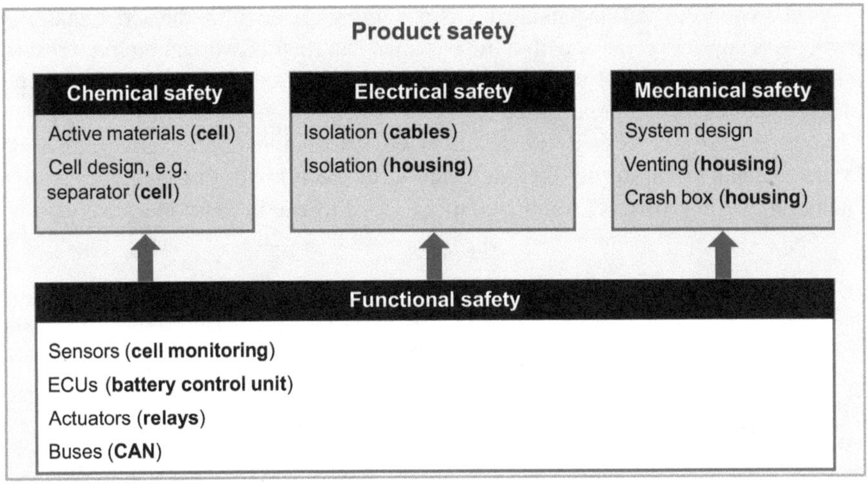

Fig. 2.3 Product safety of lithium-ion batteries (example: automotive lithium-ion battery system) [4]

the cells by means of sensors, by the battery control unit, by the actuators, e.g., the relays to connect or disconnect the battery system, and by the respective communication interfaces.

2.7 Lifetime

The characteristics of a battery system will change over time. Three different aging effects of lithium-ion battery cells are discussed below. The battery cells consist of different materials that are in contact and might react with each other. High temperatures accelerate these reactions. Therefore, the battery capacity decreases over time. Additionally, the internal resistance of the battery cell increases, causing the power of the cell to decrease as well. Battery cells are dimensioned in such a way that the defined capacity and/or the defined internal resistance of the cell is guaranteed until the end of the lifetime.

During the production processes, a durable layer, the so-called "solid electrolyte interface" (SEI) is formed on the active material of the negative electrode. This layer protects the active material from direct contact with the electrolyte. If that were not the case, the electrolyte would partially decomposed. During lifetime, chemical processes form additional layers on top of the SEI. This leads to a decrease in battery capacity, because some of the dissolved lithium ions in the electrolyte are transformed into compounds that are no longer available for the electrochemical reactions. Also, the thickness of the layer that the lithium ions in the electrolyte need to migrate through increases. This increase causes a bigger mass transfer resistance which results in a higher electrical resistance.

Mechanical load also causes aging. When the lithium ions are intercalated into the active materials, mechanical tension arises.

As shown in Fig. 2.4, mechanical tensions might arise within the active material particles. They form cracks within the particles and their pulverization. As a result, individual active material particles are no longer electrically connected. This type of stress and its effects are detailed in [5].

Another aging process is the result of the expansion of the active materials by mechanical strain during the intercalation of the lithium ions and it leads to a change in particle volume. As shown in Fig. 2.5, this might cause the separation of

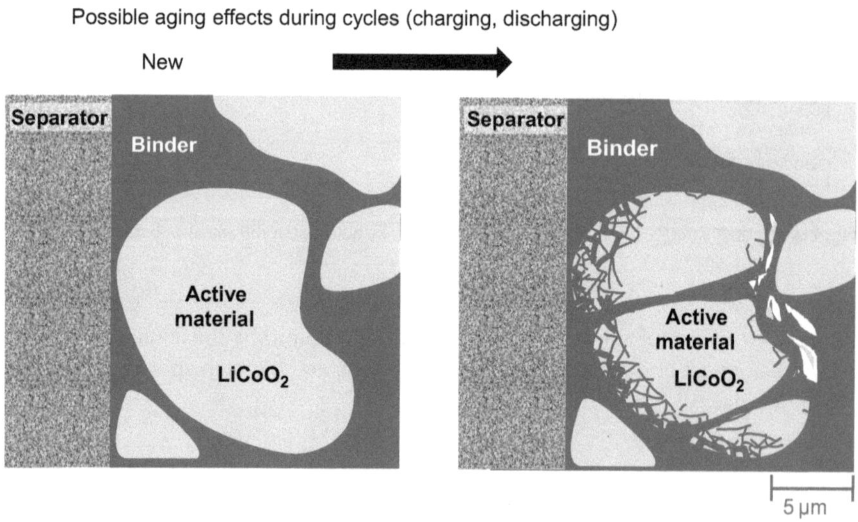

Fig. 2.4 Aging processes in the active material of the positive electrode during cycling [7]

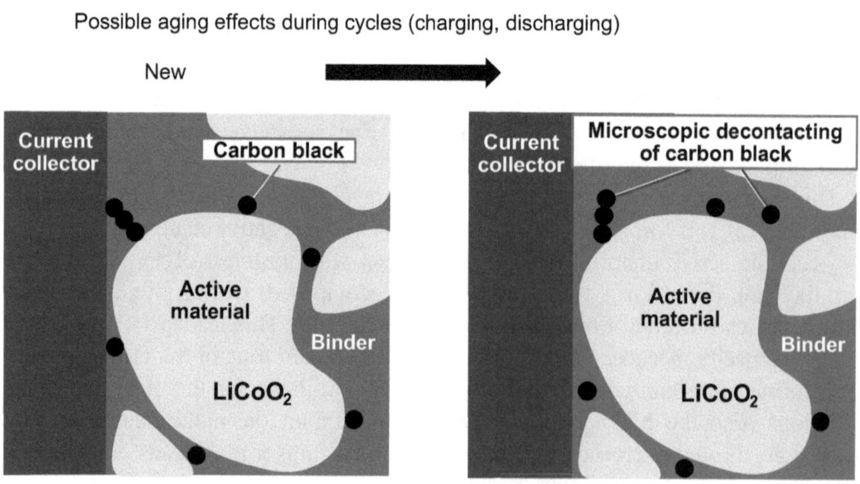

Fig. 2.5 Aging processes in the active material of the positive electrode during cycling. Separation of electrical conduction paths [7]

the electrical conduction paths (Specific electrical conduction paths are supplied between the particles and the current collector. This is done by means of carbon black, a special carbon conductor.) This entails that the active material particles are no longer electrically connected to the current collectors.

This aging process can become manifest at both the positive and the negative electrodes. Further aging processes are discussed in detail in [6]. The lifetime of the battery cells depends on the operating conditions, the materials applied, the electrolyte composition, and the quality of the production process. It differs in relation to the application, the design of the lithium-ion battery cell, and the operating conditions.

Bibliography

1. Ozawa K (2009) Lithium ion rechargeable batteries – materials, technology, and new applications. Wiley-VCH Verlag GmbH & Co. KGaA, Weinheim, Germany
2. Garche J (2009) Encyclopedia of electrochemical power sources, Vol. 6. Elsevier B. V.
3. Robert Bosch Battery Systems GmbH, Stuttgart, Germany
4. Reitzle A, Fetzer J, Fink H, Kern R (2011) Safety of lithium-ion batteries for automotive applications. AABC Europe, Mainz, Germany
5. Aifantis KE, Hackney SA, Kumar RV (2010) High energy density lithium batteries. Wiley-VCH Verlag GmbH & Co. KGaA, Weinheim, Germany
6. Garche J (2009) Encyclopedia of electrochemical power sources. Secondary batteries – lithium rechargeable systems – lithium-ion: aging mechanisms, Vol. 5. Elsevier B. V.
7. Leuthner S, Kern R, Fetzer J, Klausner M (2011) Influence of automotive requirements on test methods for lithium-ion batteries. Battery testing for electric mobility, Berlin, Germany

Materials and function

3

Kai Vuorilehto

Contents

3.1 Introduction

Lithium-ion batteries are hi-tech devices made of complex and highly pure chemicals and other raw materials. This Chapter aims to give a comprehensive picture of these materials and their functions. One might think that the lithium-ion battery is lightweight due to the small mass of its main component, lithium. However, this is not quite true: only 2 % of the battery mass is lithium, the rest being electrode materials, electrolyte, and inactive structural components.

3.2 Traditional electrode materials

The basic structure of the lithium-ion battery has changed little since 1991, when Sony brought the first version on the market. The main components of the lithium-ion battery are shown in Fig. 3.1.

K. Vuorilehto (✉)
Aalto University Helsinki, Kemistintie 1, 02150 Espoo, Finland
e-mail: kai.vuorilehto@helsinki.fi

R. Korthauer (ed.), *Lithium-Ion Batteries: Basics and Applications*,
https://doi.org/10.1007/978-3-662-53071-9_3

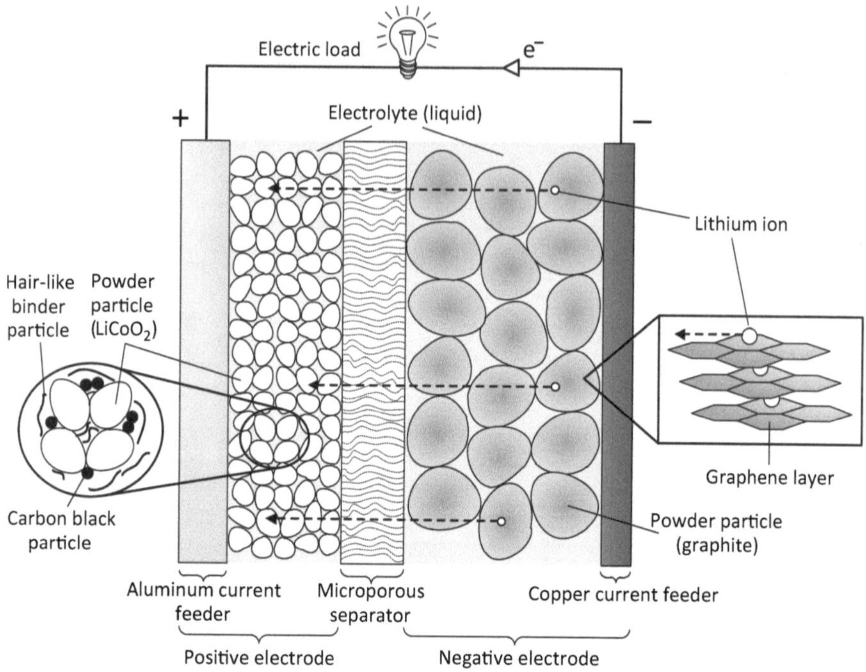

Fig. 3.1 Components of a traditional lithium-ion battery during discharging (Courtesy of Antti Rautiainen, Tampere University of Technology)

The positive electrode is often called "cathode" and the negative electrode "anode". These names represent reality only when discharging the battery. This is different during the charging process, where the positive electrode works as anode and the negative works as cathode. The misleading nomenclature stems from lithium primary batteries that are never charged.

The traditional positive electrode material is lithiated cobalt oxide, $LiCoO_2$.

It has a layered structure with alternating cobalt, oxygen, and lithium ion layers. During charging, lithium leaves the crystal (deintercalation); during discharging, it returns (intercalation). However, only 50 % of the lithium may be utilized. If more than half of the lithium leaves the crystal, the structure may collapse and liberate oxygen [1]. This can cause thermal runaway, as oxygen is able to burn the electrolyte.

For complete discharging, the reaction at the positive electrode is:

$$2Li_{0.5}CoO_2 + Li^+ + e^- \rightarrow 2LiCoO_2$$

Thus for one mole (7 g) of active lithium, two moles (189 g) of $Li_{0.5}CoO_2$ are needed as host for lithium during discharge.

By far the most common negative electrode material is graphitic carbon. It has carbon atoms in parallel graphene layers (Fig. 3.1, insert on the right). During charging, the lithium ions are intercalated into the graphite, between its layers. During discharging, lithium leaves the graphite.

Unlike cobalt oxide, graphite is stable even without lithium, so it can be almost completely discharged.

For complete discharging, the reaction at the negative electrode is:

$$LiC_6 \rightarrow Li^+ + e^- + 6\,C$$

Thus for one mole (7 g) of active lithium, there are six moles (72 g) of carbon that act as host for the lithium during charging.

3.3 Traditional inactive materials

The positive and negative electrode materials are powders that are applied as coatings on current collectors, resulting in composite electrodes. The positive current collector is aluminum foil, typically 15 to 20 µm thick. Aluminum has a high conductivity and it is rather stable even at the high potential of the positive electrode. The negative current collector is copper foil, typically 8 to 18 µm thick. Aluminum would be lighter and cheaper but it cannot be used at the low potential of the negative electrode due to parasitic formation of a lithium/aluminum alloy.

For the coating process, a mixture of electrode material, binder, conducting additives, and solvent is prepared. The binder is needed to ensure good cohesion of the electrode particles and sufficient adhesion to the current collector. Traditionally, polyvinylidene difluoride (PVDF) has been used as binder. PVDF forms hair-like structures that efficiently keep the coating together (Fig. 3.1, insert on the left). As PVDF is not soluble in water, N-methyl pyrrolidone (NMP) is used as the solvent. It is vaporized during the drying of the composite electrode, and thus the finished cell does not contain any NMP. Carbon black is used as conducting additive. The amount of additives is usually a trade secret. The order of magnitude is 1 to 5 % for carbon black, 2 to 8 % for PVDF. In energy optimized cells, the amount of additives is minimized as additives do not store energy. In power optimized cells, good contact and conductivity are more important, and the amount of additives can thus be higher. In research cells, the amount of additives may be up to 10 % as extra volume and mass are ignored.

The void volume between the positive and the negative electrode as well as the pores of the electrodes are filled with electrolyte. It is a solution of lithium salt in a mixture of organic solvents. In commercial cells, lithium hexafluorophosphate, $LiPF_6$, is used as the lithium salt.

Ethylene carbonate (EC), dimethyl carbonate (DMC), ethyl methyl carbonate (EMC), and diethyl carbonate (DEC) are the most commonly used organic solvents. Of these, EC is essential for cell stability as it protects the graphite surface [2]. However, at room temperature it is solid, preventing the use of pure EC. Typically a ternary mixture of EC and two of the other carbonates is preferred.

To avoid direct contact and a short circuit between the positive and negative electrodes, a microporous membrane is used as a separator. As organic electrolytes have a low conductivity, about 10 mS/cm, the inter-electrode distance must be short and therefore the common separators are only 12 to 25 μm thick. Very thin separators are used to minimize the resistance; thicker ones are used to maximize the safety. In commercial cells, polyethylene and polypropylene separators are preferred due to their chemical stability and reasonable price.

A lithium-ion cell needs hermetic casing. Especially the electrolyte and the lithiated graphite are damaged even by minute amounts of moisture. As water can diffuse through plastic materials, metal casing is used. Lightweight aluminum is preferred; heavier steel can be found in cheaper cells.

3.4 Alternatives for standard electrode materials

The main challenges of the traditional lithium-ion battery are safety, cost, and size. As the trend is to build larger batteries for electric vehicles and other large-scale applications, the importance of safety has grown. A burning mobile phone could be tolerated but a burning car may be fatal. The cost issue is similar. Small consumer batteries can easily be afforded, but the battery pack of a full-electric car is too expensive to be able to compete with the gasoline tank. A higher specific energy (Wh/kg) would be appreciated, too. However, the present value of up to 230 Wh/kg is satisfactory for most applications [3].

The use of cobalt oxide as positive electrode material is not safe. If it is kept "fully" charged as $Li_{0.5}CoO_2$, it reacts slowly with the electrolyte, thus losing performance. If it is slightly overcharged, there is a clear loss in capacity and service life. In case of severe overcharging, the cobalt oxide crystal collapses which can cause thermal runaway and fire. Overcharging easily happens, as there is no obvious voltage difference between normal charging and overcharging.

Cobalt oxide is expensive, as cobalt ore is scarce. This problem is getting worse as the demand grows. Economics of scale do not apply here. Last but not least, cobalt is toxic.

The main commercial alternatives for cobalt oxide are listed in Table 3.1. Each of the alternative materials solves some of the problems but they all are compromises. LMO is safer and very cheap but has a limited service life. NCM is safer and cheaper, but has a sloping discharge voltage. NCA is cheaper and lighter (more specific capacity, mAh/g) but it is hardly safer. LFP is very safe and slightly cheaper,

Table 3.1 Commercial alternatives for cobalt oxide

Compound	Abbreviation	Chemical structure
Manganese oxide	LMO	$LiMn_2O_4$
Nickel manganese cobalt oxide	NCM	$LiNi_{1/3}Mn_{1/3}Co_{1/3}O_2$
Nickel cobalt aluminum oxide	NCA	$LiNi_{0.8}Co_{0.15}Al_{0.05}O_2$
Iron phosphate	LFP	$LiFePO_4$

but it gives 0.5 V lower voltage than cobalt oxide. At the moment, NCM and LFP seem to be the most promising candidates for large-scale batteries. Positive electrode materials are discussed in detail in Chapter 4.

Graphite as negative electrode is not safe either. For graphite, the lithium intercalation potential is only about 80 mV more positive than the lithium metal plating potential. Even a small design failure or charging error causes deposition of metallic lithium on the electrode surface. Small amounts of metallic lithium increase the reactivity of the graphite surface, thus consuming electrolyte in secondary reactions. Large amounts of deposited lithium metal can grow as metallic peaks, "dendrites", that short-circuit the negative and positive electrodes. This might cause excess heating and ignite the electrolyte resulting in a fire.

The potential of lithiated graphite is far beyond the stability window of the common electrolytes (Fig. 3.2) [4]. During the first charging of the battery, graphite reacts with the electrolyte, building a protective layer on the graphite surface. This solid electrolyte interface (SEI) layer should prevent further secondary reactions. However, some secondary reactions take place throughout the lifetime of the battery, reducing its cyclic and calendar life.

Some commercial alternatives for graphite exist. Soft and hard carbons are used due to their slightly more positive intercalation potentials. This means less risk of lithium metal deposition and a possibility of faster charging.

However, the energy density is considerably lower when these materials are used. Lithium titanate is a very safe negative electrode material with an amazingly long service life, but the 1.4 V lower cell voltage limits the use of lithium titanate to very few applications. The newest commercial alternative, silicon, gives a formidable energy density, but low stability limits its service life.

Graphite is cheap and lightweight, especially when compared to cobalt oxide. Therefore, it can be expected that graphite retains its position as standard negative

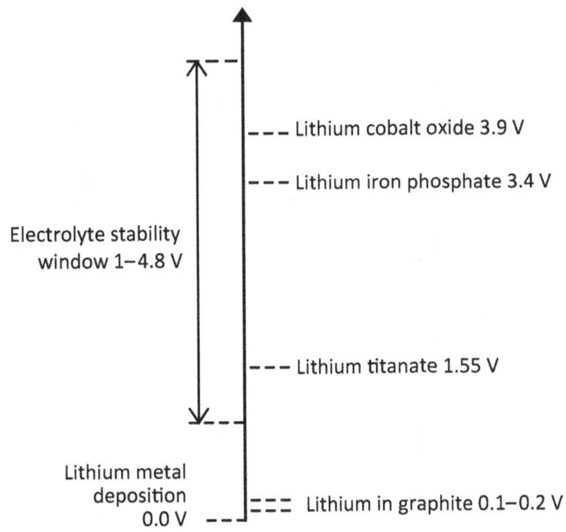

Fig. 3.2 The potential range of electrolyte stability, compared to the potentials of common electrode materials (Courtesy of Antti Rautiainen, Tampere University of Technology)

electrode material in the near future. Negative electrode materials are discussed in detail in Chapter 5.

3.5 Alternatives for standard inactive materials

Coating with PVDF binder requires the use of an organic solvent, usually NMP. Fresh solvent must be transported to the factory, and the vaporized solvent must be burned or recycled. Additionally, organic solvents cause safety and health hazards. A water-compatible binder system, carboxymethyl cellulose (CMC) combined with styrene-butadiene-rubber (SBR) solves these problems. CMC and SBR are "state of the art" for graphite electrodes. As shown in Fig. 3.3, the SBR binder forms contact points between graphite particles, instead of the hair-like structures of PVDF. For positive electrodes and lithium titanate, the solvent situation is more difficult as water tends to react with these materials during processing.

Some manufacturers coat iron phosphate positive electrodes using acrylic binders and water.

The conducting salt of the electrolyte, lithium hexafluorophosphate ($LiPF_6$), is decomposed by even minute amounts of moisture in a reaction that produces hydrofluoric acid (HF). This acid further deteriorates the cell. The organic solvents of the electrolyte are flammable causing safety problems, and they react with the lithiated graphite, reducing cell lifetime. In spite of active research, no superior alternative for the traditional liquid electrolyte has been found. Gel-polymer electrolytes of "lithium-polymer batteries" are one way to overcome the flammability problem, as the gelled electrolyte is less volatile. However, gelling increases the cost, and the salt-related problems still exist. To date, electrolyte problems are tackled by the use of electrolyte additives [5]. There are additives that scavenge HF or build a stronger SEI layer on graphite, and even flame-retarding additives are marketed.

Fig. 3.3 SEM micrograph of a negative graphite electrode. The large particles (20 μm) are graphite, the agglomerates (0.1-1 μm) on the surface are carbon black. Between the large particles there are contact points consisting of SBR and carbon black (Courtesy of Juha Karppinen, Aalto University, Helsinki)

Nevertheless, the need to replace $LiPF_6$ clearly remains. Electrolytes are discussed in detail in Chapter 6.

Polyethylene and polypropylene separators can melt if the cell temperature rises over ca. 150 °C. This could cause a "total short circuit" and a sudden release of all the energy stored in the battery. In small cells, this problem is solved by using tri-layer separators, in which the middle layer melts first, shutting down the current flow before the outer layers melt and lose their rigidity. In larger cells, the trend is to coat the separator with a non-melting ceramic layer, usually aluminum oxide [6]. On the other hand, the separators should have a highly controlled and even pore structure, and no defects can be allowed as they could cause short circuits. This makes the separator expensive. Tri-layer structures and ceramic coatings tend to further increase the cost. A cheaper and safer separator would be highly appreciated by battery manufacturers. Separators are discussed in detail in Chapter 7.

A rigid aluminum or steel casing for the battery is an excellent solution for cylindrical batteries, as the cylindrical form is dimensionally stable even under pressure. However, in large batteries the distance from the middle of the cylinder to the outer surface gets too long for efficient heat dissipation, so a flat battery design is preferred. A flat shape introduces the problem of stack pressure: If the inside pressure of the battery increases during cycling, the metal casing is deformed. This problem can be solved using thin aluminum laminate casing. During production, vacuum is produced inside the cell, so that the electrodes are pressed tightly together by the atmospheric pressure. The battery grade laminate consists of a thin aluminum foil (40 to 80 μm), coated on both sides with special high-stability plastic layers. A "pouch cell" with laminate casing is lightweight but needs an outer casing for rigidity, slightly increasing the final mass.

3.6 Outlook

The invention of lithium-ion batteries revolutionized the landscape of batteries, and the continued development of materials has brought us a battery performance that would have been difficult to believe in the era of lead-acid and nickel-cadmium batteries. For most small- and medium-scale applications, lithium-ion batteries are the right solution.

If lithium-ion material development is successful, most cars in the future will probably use lithium-ion batteries, and superfluous electric power from the grid can be stored in lithium-ion batteries. Those batteries will probably contain copper and aluminum foils as current collectors.

But, until then, the majority of the remaining materials must become safer, cheaper, or less heavy than those used today.

Bibliography

1. John B (2002) Goodenough: oxide cathodes. In: Walter A, van Schalkwijk, Scrosati B (eds.) Advances in lithium-ion batteries. Kluwer Academic

2. Tarascon J-M, Armand M (2001) Issues and challenges facing rechargeable lithium batteries. Nature 414:359–367
3. Dahn J, Ehrlich GM (2011): Lithium-ion batteries. In: Reddy TB (ed.) Linden's handbook of batteries, Mc Graw Hill
4. Scrosati B, Garche J (2010) Lithium batteries: status, prospects and future. J Power Sources 195:2419–2430
5. Ue M (2009) Role-assigned electrolytes: additives. In: Yoshio M, Brodd RJ, Kozawa A (eds.) Lithium-ion batteries, Springer
6. Zhang SS (2007) An overview of the development of Li-ion batteries. In: Zhang SS (ed.) Advanced materials and methods for lithium-ion batteries, Transworld research network

Cathode materials for lithium-ion batteries

4

Christian Graf

Contents

4.1 Introduction

Lithium transition metal compounds are employed as cathode materials. These composites can develop mixed crystals over an ample composition range and can deintercalate lithium ions from the structure during the charging process. The transition metal ions are oxidized because of the charge neutrality and therefore the oxidation state of the transition metal cation is elevated. Lithium is deintercalated while the battery is discharging, which in turn reduces the transition metal ions and decreases the oxidation number.

The following paragraphs describe the most important cathode materials (active materials), their structure and electrochemical performance as well as their advantages and disadvantages. The materials have been grouped in three classes based on their crystal structure: layered oxides, spinels, and phosphates.

C. Graf (✉)
Chemische Fabrik Budenheim KG, Rheinstrasse 27, 55257 Budenheim, Germany
e-mail: c_graf@gmx.net; batteries@budenheim.com

R. Korthauer (ed.), *Lithium-Ion Batteries: Basics and Applications*,
https://doi.org/10.1007/978-3-662-53071-9_4

4.2 Oxides with a layered structure (layered oxides, $LiMO_2$; M = Co, Ni, Mn, Al)

The most frequently examined system of cathode materials consists of layered oxides with the chemical formula $LiMO_2$ (M = Co and/or Ni and/or Mn and/or Al). The system's boundary phases, the important binary compounds, and the best-known ternary phase $Li_{1-x}(Ni_{0.33}Mn_{0.33}Co_{0.33})O_2$ (NCM) will be outlined.

Lithium cobalt oxide ($Li_{1-x}CoO_2$, LCO) has probably been the most widely used cathode material since the market launch of the first rechargeable lithium-ion battery by Sony in 1991. $Li_{1-x}CoO_2$ forms an α-$NaFeO_2$ structure (R-3m). In this structure, cobalt fills the 3a positons and lithium fills the 3b positions. Both are surrounded octahedrally by oxygen on the 6c layers. The two octahedral strata are stacked alternately along the c axis (in the hexagonal setting; along [111] in the rhombohedral cell). The lithium ions in this structure can move within one plane (2D), perpendicular to the stack direction between the cobalt octahedral layers. Therefore, they can intercalate and deintercalate the structure (Fig. 4.1) [1].

If lithium is deintercalated from the LCO structure, the redox pair Co^{4+}/Co^{3+} is formed, which in turn creates a potential of around 4.0 V vs. Li/Li^+. Almost all lithium ions are extracted from the structure electrochemically, resulting in a theoretical capacity of 274 mAh/g. Due to the instability of the low-lithium phase ($Li_{1-x}CoO_2$ $x < 0.7$), the charging voltage is limited to ≤ 4.2 V vs. Li/Li^+ in the usable voltage range. This means that only a little more than half of the available lithium is usable. Therefore, the maximum reversible capacity is 140 to 150 mAh/g [2, 3]. $Li_{1-x}CoO_2$ has a flat charging/discharging characteristic for a working level of 3.9 V and the lowest known working voltage range of 0.1 V for common lithium intercalation

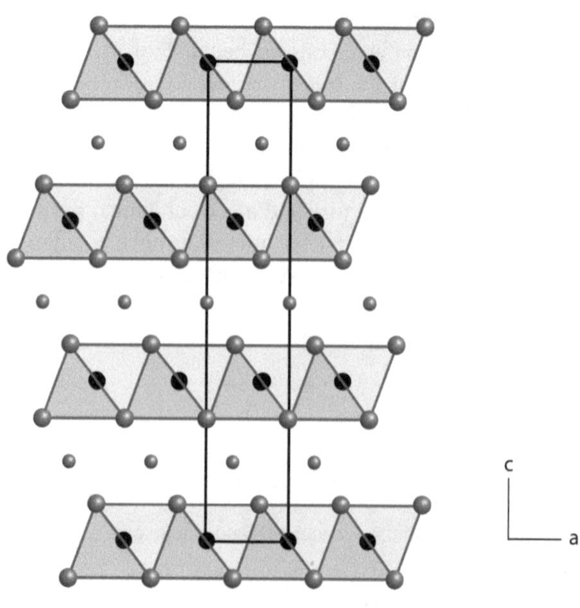

Fig. 4.1 Crystal structure of $Li_{(1-x)}CoO_2$ (LCO) with highlighted hexagonal unit cell (*black* Co; *gray* O; *light gray* Li)

bonds [4]. Due to its high working voltage, high relative density (5.1 g/cm^3), and bulk density (> 2.2 g/cm^3), the energy density of LCO could hardly be matched by other cathode materials [3, 5]. These aspects are less important for applications with unlimited space, e.g., stationary energy storage, where criteria such as stability and safety are essential.

Although LCO is successful, it is not the best cathode material, especially regarding criteria such as safety and stability. The band structure of $Li_{1-x}CoO_2$ and the decrease in the Fermi level caused by charging above 4.6 V vs. Li/Li^+ [6] enable the depopulation of 2p- oxygen states, releasing oxygen. Oxygen cannot escape from the cells and, in combination with the organic electrolyte, can react violently by causing flames or even an explosion [7]. Furthermore, the proven solubility of cobalt in commonly used electrolytes might cause the dissolution of cobalt ions from the LCO structure and result in a capacity loss and ultimately in failure of the cell [8]. Cobalt compounds are relatively expensive because cobalt is very rare in the Earth's crust (30 ppm) [9].

Due to the weaknesses of LCO in terms of safety and cost, it has been developed and modified to solve these problems and to improve it.

Substituting cheaper nickel in the structure for all cobalt results in lithium nickel oxide ($Li_{1-x}NiO_2$, LNO). Although LNO has the same structure, it displays a higher reversible capacity of around 200 mA/g [10]. During production of this compound there is a mixed population of around 12 % on the lithium positions. This is why the structure of LNO is better described as $Li_{1-x-y}Ni_{1+y}O_2$ [11]. This mixed occupancy is caused by the similar ionic radii of Li^+ and Ni^{2+} as well as an instability of Ni^{3+}, which effects a preference of Ni^{2+} in the lithiated material. The available amount of lithium in the structure is decreased during discharging due to the mixed population of Ni^{2+} on the Li^+ layers. This results in an irreversible loss of material capacity [12, 13]. Synthesis and controlling the synthesis parameters are paramount in order to minimize disorder.

Similarly to LCO, there is a risk with LNO that the oxide ions oxidize because of the position of the redox pair Ni^{4+}/Ni^{3+} in the band structure compared to the 2p- states of the O^{2-}. This may lead to the above-mentioned stability and safety problems caused by oxygen release.

Nickel has partially been replaced by cobalt up to a substitution degree of < 20 % to eliminate the negative effects of the nickel compound without losing its positive characteristics. The resulting compounds in their lithiated form always develop a structure similar to α-$NaFeO_2$. Cobalt substitution results in less mixed occupancy on the lithium layer and consequently in a higher reversible capacity [14]. In addition, the integration of cobalt increases lithium-ion conductivity as well as the concentration of charge carriers, which increases electrical conductivity.

Further modifications to improve stability have been employed to lower the 2p- oxygen bands in the band structure, thereby moving them out of the Fermi level range in order to minimize oxygen release. One means of achieving this is to incorporate aluminum into the structure. The compound $LiNi_{0.8}Co_{0.15}Al_{0.05}O_2$ (nickel cobalt aluminum; NCA) is available commercially and, in contrast to LNO, it exhibits better stability and cycling capability. Partial substitution of F^- and S^{2-} respectively for the O^{2-} layer could further improve cycling stability, because it prevents migration of Ni^{2+} to the Li^+ position [15, 16].

An $Li_{1-x}MnO_2$ compound would be extremely interesting from an economical and ecological point of view. Two possible phases are available for such a compound. They differ considerably with regard to their electrochemical activity. The thermodynamically stable orthorhombic form has very unfavorable electrochemical characteristics [17]. It is difficult to synthesize the electrochemically active form (type α-$NaFeO_2$), because it is not preferred thermodynamically [18]. During delithiation, $Li_{1-x}MnO_2$ exhibits a phase transformation when its composition is approximately $Li_{0.5}MnO_2$. In this range, the α-$NaFeO_2$ structure transforms into the more stable spinel phase $LiMn_2O_4$. This phase transformation is accompanied by a reduction of the voltage profile [19].

The concept of creating mixed crystals of layered metal oxides (which leads to the substitution of the positive characteristics of one ion for the negative characteristics of another) has resulted in the development of lithium nickel manganese oxides $Li_{1-x}(Ni_{0.5}Mn_{0.5})O_2$. Lithium nickel manganese oxide exhibits the highest capacity among LCO-analog materials (up to 200 mAh/g), albeit with low current densities [20]. The oxidation state of nickel in this compound is +2 and that of manganese is +4. Because the voltage can only be generated by the redox pairs Ni^{3+}/Ni^{2+} and Ni^{4+}/Ni^{3+}, the 2p- oxygen band is no longer in the range of the Fermi level. In comparison with NCA, this provides a more stable compound with less development of oxygen during charging. The nickel ions stabilize the α-$NaFeO_2$ structure. As a result, during discharging, neither a structural Jahn-Teller distortion nor a phase transformation into a spinel structure can be discerned for lithium nickel manganese oxide. However, similar to other nickel compounds, there is a mixed occupancy of nickel on the lithium position (3b position). It is between around 8 and 10 % for this compound. This mixed occupancy impedes the lithium diffusion and thus reduces the reversible capacity. Incorporating cobalt on the 3a layers can minimize the mixed occupancy of the lithium positions, but it cannot fully prevent it.

A commercially very successful cathode material with a layered structure is $Li_{1-x}(Ni_{0.33}Mn_{0.33}Co_{0.33})O_2$ (nickel manganese cobalt, NCM). The compound is similar to the one developed by all other layered oxides, namely type α-$NaFeO_2$. In lithiated state, the metals have the charges Ni^{2+}, Mn^{4+}, and Co^{3+} [21]. For reasons of structure stability, not all of the lithium can be deintercalated from the structure in NCM. This is also similar to other compounds in this class. NCM (theoretical capacity 274 mAh/g) only uses around 66 % of the lithium available in the structure and therefore has a gravimetric capacity of up to 160 mAh/g (Fig. 4.2).

In the range of $0 \leq x \leq 1/3$, the voltage curve is determined by the redox pair Ni^{3+}/Ni^{2+} in the range of $1/3 \leq x \leq 2/3$ by Ni^{4+}/Ni^{3+}, and in the range of $2/3 \leq x \leq 1$ by Co^{4+}/Co^{3+} [22]. The NCM has a lower lithium nickel disorder than $Li_{1-x}(Ni_yMn_{1-y})O_2$. This is because it comprises cobalt. Furthermore, cobalt contributes to the good electrical conductivity and consequently to a better electrochemical performance. Due to its oxidation state of + 4, manganese negatively influences the concentration of charge carriers and therefore conductivity. On the other hand, it stabilizes the structure and thus improves cycling stability. When compared to the boundary phases of the system (LCO, LMO, and LNO), the interaction of the chosen metal ions and the balancing of their advantages and disadvantages make NCM a material with higher reversible capacity and better cycling stability.

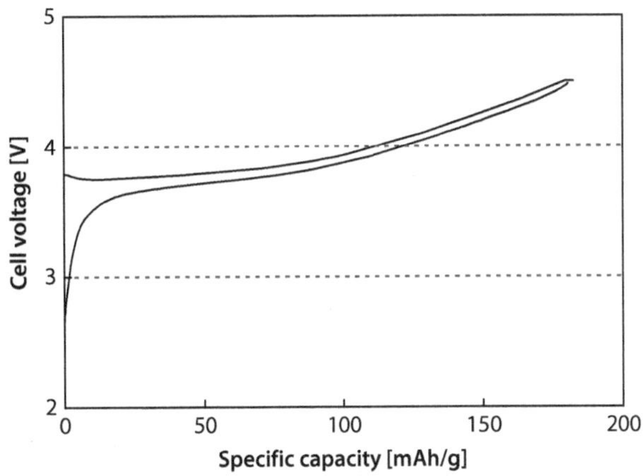

Fig. 4.2 Typical charging/discharging cycle of $Li_{1-x}(Ni_{0.33}Mn_{0.33}Co_{0.33})O_2$ (NCM) vs. Li/Li^+

Nevertheless, NCM has weaknesses that have not yet been remedied, leaving scope for small improvements. For example, in spite of decades of intensive research, NCM materials still exhibit an excessively high irreversible capacity due to the mixed population on the lithium position. In addition, phase transformations can occur in the delithiated material. NCM develops oxygen during charging. This has been remedied by incorporating aluminum into the structure for NCA.

Another problem is the discharge voltage (for NCM: 3.7 V vs. Li/Li^+). This is still low compared to that of LCO (3.9 V vs. Li/Li^+). These weaknesses are assessed differently depending on the specific application and weighting.

Current approaches are moving toward low-cobalt compounds with less than 25 % cobalt or less than 25 % manganese [23 and quoted therein]. Employing cheaper metals should decrease costs and increase capacity. Studies have shown that a high nickel content has positive effects on capacity, making values of up to 190 mAh/g possible (NCM: 160 mAh/g) [24].

4.3 Spinel (LiM_2O_4; M = Mn, Ni)

The $Li_{1-x}Mn_2O_4$ compounds (lithium manganese oxide, LMO spinel) crystallize in the spinel type (space group: Fd-3 m), where the oxygen atoms form a cubic close packing (ccp) on the 32e sites. The manganese atoms occupy the 16d sites and therefore are coordinated tetrahedrally by six oxygen ions. The lithium ions are located on the 8a positions and surrounded tetrahedrally by oxygen. The $MnO_{6/3}$ octahedra share edges and form a three-dimensional lattice (Fig. 4.3). The 16c positions, which are surrounded octahedrally by oxygen, remain unoccupied and, in combination with the lithium-occupied 8a positions, form a three-dimensional lattice, through which the lithium ions can diffuse. Lithium ions can be incorporated into high-lithium compounds such as $Li_{1+x}Mn_{2-y}O_4$ octahedra [25].

Manganese in its lithiated form takes on oxidation states + 3 and + 4. Trivalent manganese is oxidized during delithiation, and the redox pair Mn^{3+}/Mn^{4+} determines the working voltage of 4.1 V vs. Li/Li^+ (Fig. 4.4).

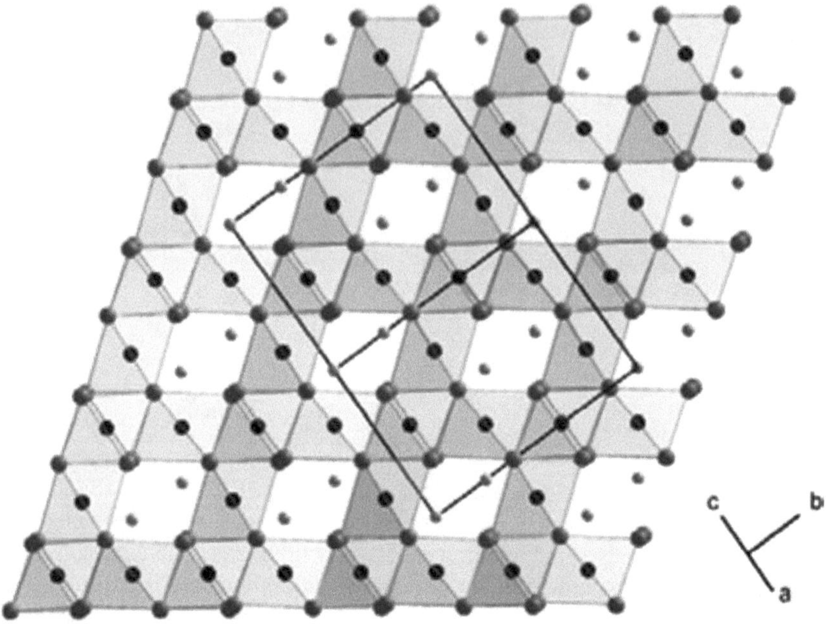

Fig. 4.3 Crystal structure of an LMO spinel with highlighted unit cell (*black* Mn; *gray* O; *light gray* Li)

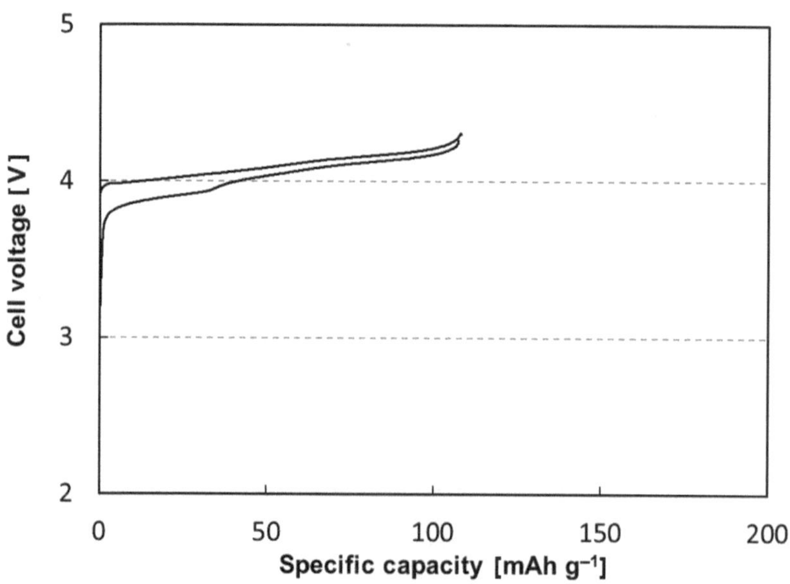

Fig. 4.4 Typical charging/discharging cycles of an LMO spinel ($Li_{1-x}Mn_2O_4$ vs. Li/Li^+)

$Li_{1-x}Mn_2O_4$ and its doping variants are, similar to LCO and NCM, established commercial cathode materials. Their implementation and establishment as first-generation cathode materials are the result of many years of research and development, which aimed to improve the electrochemical characteristics of the LMO spinel. Among other things, the researchers focused on the morphology control and surface chemistry of $Li_{1-x}Mn_2O_4$ crystallites for decades.

LMO exhibits poor cycling behavior as a cathode material, resulting in a continuous decrease in reversible capacity [26 and quoted therein]. This characteristic is even more pronounced at temperatures above 50 °C, which can easily occur in stationary and automotive applications. This behavior can be triggered by multiple factors. Probably the most important cause is the instability of $Li_{1-x}Mn_2O_4$ toward acids. For example, traces of water might react with the conducting salt in accordance with the mentioned reaction equations (4.2 to 4.5), resulting in hydrogen fluoride [27, 28].

$$LiPF_6 \leftrightharpoons LiF + PF_5 \tag{4.1}$$

$$PF_5 + H_2O \rightarrow POF_3 + 2HF \tag{4.2}$$

$$POF_3 + H_2O \rightarrow HPO_2\,F_2 + HF \tag{4.3}$$

$$HPO_2\,F_2 + H_2O \rightarrow H_2PO_3\,F + HF \tag{4.4}$$

$$H_2PO_3\,F + H_2O \rightarrow H_3\,PO_4 + HF \tag{4.5}$$

The resulting hydrogen fluoride reacts with the LMO spinel as shown in (4.6) [29].

$$2\,LiMn_2\,O_4 + 4\,HF \rightarrow 3\,MnO_2 + Mn_2 + 4\,F^- + 2\,Li^+ + 2\,H_2O \tag{4.6}$$

The Mn^{2+} ions dissolved during the electrolyte reaction can deposit on the cathode as MnO or MnF_2. At the same time, the spinel of the cathode material transforms into a deficient high-Mn^{4+} spinel (Li_2MnO_4 and $Li_4Mn_5O_{12}$, respectively). Depositions of MnO on the anode cause negative effects to the solid electrolyte interface (SEI) (Chapter 5) [30, 31].

A further reason for the loss of performance during the cycling of an LMO spinel is the Jahn-Teller instability of Mn^{3+} which leads to a distortion of the $MnO_{6/3}$ octahedra and to distortions of the structure [32]. The formation of two diverging cubic phases has been observed during different charging and discharging cycles. Their crystal structures vary considerably and cause micro-tension in the crystallites [33]. Oxygen can be released in the LMO spinel during charging as is the case with other oxidic cathode materials. This causes a loss in capacity and possibly a fire or an explosion of the cell [34].

Approaches to minimize the solubility of Mn^{3+} are lithium doping and the formation of phases with a composition of $Li_{1+x}Mn_{2-y}O_4$. Mn^{3+} is partially oxidized to the oxidation state +4, this leads to a decrease in capacity, as Mn^{3+} determines the electrochemical characteristics of the cell. Furthermore, other aluminum-doped LMO spinels and substitution variants (F^- instead of O^{2-}) that exhibit a decreased manganese solubility are known [35]. Coating the surface with transition metal oxides has been employed to try to incorporate acid scavengers in the cathode material. On a cellular level, researchers have tried to lower the manganese solubility by means of additives [35].

LMO spinels can be discharged with high currents (> 5 C). High-current charging of LMO is limited due to the small charging voltage range (0.3 V) predetermined by the compound [25].

Recently, researchers have become increasingly interested in LMO spinels with nanoparticles and/or nanostructures. These morphologies exhibit an even greater dischargeability at high currents [36]. Nanostructured materials have large surfaces, which makes Mn^{3+} solubility an important parameter. Therefore, nano-LMO spinels are mostly coated.

Other highly interesting materials for future applications are materials with higher working voltages (>4.0 V), including high-voltage spinels such as $Li_{1-x}(Ni_{0.5}Mn_{1.5})O_4$. These high-voltage spinels have a voltage of 4.7 V (vs. Li/Li^+), resulting in an energy density that is 12 % higher than that of commercially available materials. Other high-voltage spinels ($Li_{1-x}(M_{0.5}Mn_{1.5})O_4$ with M = Cr, Co, Fe, and Cu) are also possible alternatives. Compared to LMO, these compounds have a larger charging and discharging voltage range as well as maximum working voltages of up to 5.1 V vs. Li/Li^+ [37]. The stability of the electrolytes is currently the main obstacle faced by developers of high-voltage materials. The prevailing electrolytes are stable up to around 4.3 V vs. Li/Li^+, but they are electrolyzed themselves above this value, thus decreasing the cycling stability of the cell.

4.4 Phosphate ($LiMPO_4$; M = Fe, Mn, Co, Ni)

$LiFePO_4$ (lithium ferrous phosphate; LFP) was introduced as a cathode material in 1997 and crystallizes like the natural mineral lithiophilite in the olivine type in the space group Pnma (Fig. 4.4) [38]. The oxygen atoms form an almost ideal hexagonal close pack (hcp); the lithium atoms (4a layer) and iron atoms (4c layer) are located in a regular setting in the octahedron spaces of this pack. The tetrahedron spaces contain phosphorus atoms (4c layer), which, by means of covalent bonds, form a phosphorus ion with the four neighboring oxygen atoms and distort the structure (Fig. 4.5).

The distorted octahedral lithium coordination polyhedrons share edges [010] and form a one-dimensional diffusion path parallel to the b axis for the lithium ions [39].

The outer electrons of the oxygen ions are polarized in heavily covalent P-O bonds because of the inductive effect of the phosphorus atoms, weakening the covalence of the Fe-O bond. This leads to a decrease in redox energy of the redox pair Fe^{3+}/Fe^{2+} and a higher open circuit voltage vs. Li/Li^+ (Voc) [38]. This effect has also been observed in other poly-anionic lithium-iron compounds, but it is reinforced by the olivine structure for LFP.

Another characteristic of LFP is its extremely flat charging and discharging plateau (Fig. 4.6), which is caused by two structurally very similar boundary phases

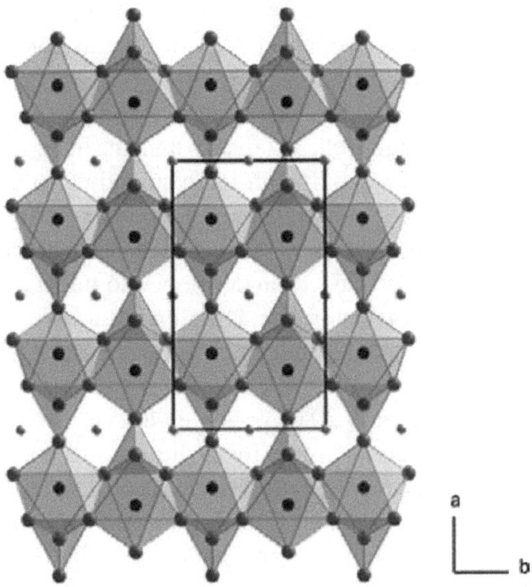

Fig. 4.5 Crystal structure of LFP with highlighted unit cell (*black* Fe; *gray* O; *gray* P; *dark gray and small* Li)

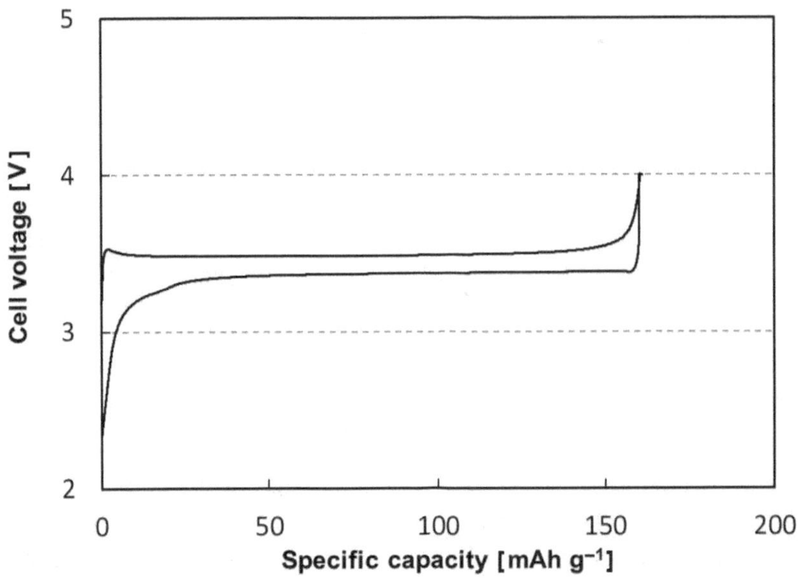

Fig. 4.6 Typical charging/discharging cycles of LFP ($Li_{1-x}FePO_4$ vs. Li/Li^+)

with their diffusion-driven proliferation. These boundary phases are characterized by minor differences in their crystal structures. The volume of the low-lithium phase is only 6.81 % smaller and its density is only 2.59 % lower than the high-lithium phase. Therefore, there are no structural defects or resulting tensions that might damage the material and decrease cycling stability. The stability of the delithiated phase ($FePO_4$) contributes to the extraordinary cycling stability of lithium iron phosphate. The diffusion-driven proliferation of the phases determines the characteristics only under extremely high currents (> 20 C) [38] but, at high temperatures (> 50 °C), LFP exhibits better charging and discharging characteristics than the materials discussed above.

However, the poor electrical and lithium-ion conductivity is the biggest problem of $Li_{1-x}FePO_4$. Smaller crystallites should reduce the distance for lithium-ion diffusion and also decrease the distance the electrons need to travel through the material. This is why many technologies have been developed to manufacture LFP particles in the submicron or even nanometer range. Usually, lithium iron phosphates mostly consist of primary particles with a diameter of ~200 nm that agglomerate into secondary particles with a diameter of 5 to 10 µm. These nanoparticles make capacities of > 100 mAh/g at high C rates (> 20 C) possible [40, 41].

Scaling down the particles compensates for the lithium-ion conductivity. However, it does not improve electrical conductivity. Poor electrical conductivity of a cathode material leads to poor cell performance. This is the reason why improving electrical conductivity has become one of the key goals of research into LFP. Carbon coating, for example, can significantly improve LFP conductivity. This technology enables capacities of up to 165 mAh/g (97 % of the theoretical capacity) and improves the charging and discharging properties at high currents [42]. The dependence of electrical conductivity on electrochemical characteristics makes the quality of carbon coating very important and a co-determinant for the cathode material's qualities.

During production, the coating also prevents the aggregation of particles and protects the cathode material from inadvertent oxidation and degradation [13].

There is no evidence that oxygen evolves in LFP compared to oxidic materials. This is due to the strong covalent bonds within the phosphate molecule [43]. Thus, LFP has an excellent thermal stability and is compatible with all common electrolytes [44].

In 1997, in the first publication [38], it was already shown that the olivine structure is also stable for other lithium transition metal phosphates. Incorporation of manganese, cobalt, or nickel enables higher working voltages and higher energy densities. Lithium manganese phosphate (LMP) $Li_{1-x}MnPO_4$ offers a working voltage of 4.1 V vs. Li/Li^+ [38]. This is equal to an energy density of 656 Wh/kg with the same theoretical capacity as LFP. LMP has a flat plateau, similar to LFP. In contrast to the spinels ($LiMn_2O_4$), the manganiferous olivines do not exhibit Jahn-Teller instability of the delithiated material and are extremely well cyclable [45]. But $Li_{1-x}MnPO_4$ possesses a weaker electrical conductivity than the iron analog, thus requiring considerably smaller particle sizes (< 80 nm) [46]. Current developments of phosphate-based cathode materials favor mixed iron-manganese phosphates with > 60 % manganese (LFMP) [47].

For olivines, the voltage plateaus of Co^{3+}/Co^{2+} and Ni^{3+}/Ni^{2+} are around 4.8 V vs. Li/Li^+ [48] and 5.1 V vs. Li/Li^+, respectively [49]. Most commercially available electrolytes disintegrate from around 4.3 V, as mentioned for high-voltage spinels. Therefore, new electrolytes need to be developed to allow the usage of 5-V olivines in cells with graphite anodes. But, in contrast to LFP, heavy metal ions pose a health hazard.

4.5 Comparison of cathode materials

Cathode materials are suitable for different applications with different requirements. This is due to their specific characteristics. But the decision as to which cathode material to use does not only depend on the electrochemical data (Table 4.1). It also depends on the costs, the service life, and the safety characteristics (Fig. 4.7).

Table 4.1 Capacity, working voltage, and energy density of the discussed cathode materials (typical data)

Material	Capacity/Ah kg^{-1}	Working voltage/V	Energy density/Wh kg^{-1}
NCA ($LiCo_{0.85}Al_{0.15})_2$	200	3.7	740
LCO ($LiCoO_2$)	160	3.9	624
NCM ($LiNi_{0.33}Mn_{0.33}Co_{0.33}O_2$)	160	3.7	592
LMO ($LiMn_2O_4$)	100	4.1	410
LFP ($LiFePO_4$)	160	3.4	544
LFMP ($LiFe_{0.15}Mn_{0.85}PO_4$)	150	4.0/3.4	590

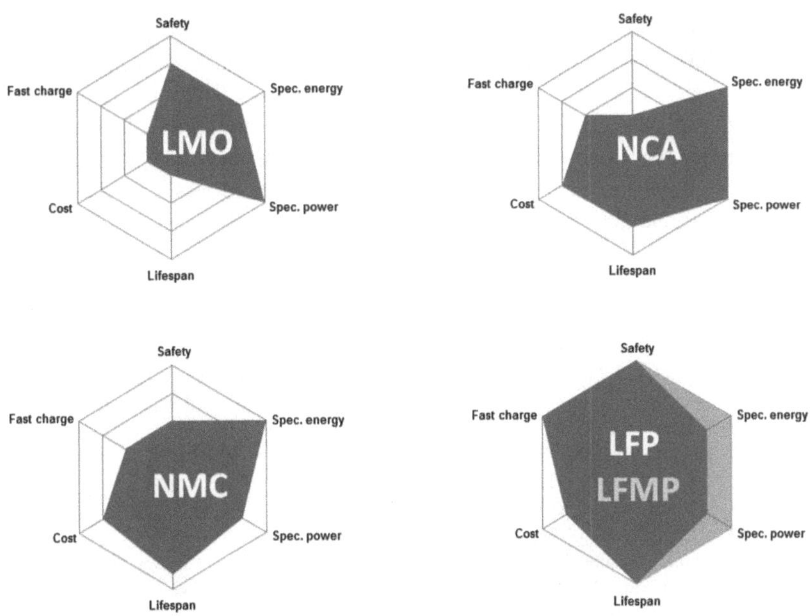

Fig. 4.7 Comparison of cathode materials pertaining to safety, specific energy, specific power, service life, costs, and fast chargeability (based on [50])

The layered oxides are advantageous in applications where a high energy density is of importance. Their high working voltages and resulting energy densities are explained by the redox properties of the nickel, cobalt, and manganese ions. In the lithosphere, these metals are very scarce, resulting in high prices on the global market (Co: USD 26 kg^{-1}, Ni: USD 10 kg^{-1}, Mn: USD 1.7 kg^{-1}, as of August 2016). Iron is the second most common element on Earth and, even when compared with manganese, extremely inexpensive (USD 0.057/kg^{-1}). (*Source*: www.InfoMine.com) Thus, LFP is the cheapest possibility for producing 1 kg of cathode material. Due to its lower energy density in comparison with NCM and NCA, this advantage is less pronounced.

Oxides pose a safety risk, as they can trigger the evolution of oxygen, thus causing a fire or even an explosion of the cell. There is virtually no risk of this occurring for LFP due to its intrinsic characteristics. LFP shows no thermal effects up to 300 °C as opposed to other cathode materials. Almost all oxidic cathode materials, however, exhibit strong exothermal effects. Furthermore, LFP and LFMP are non-toxic and non-hazardous to the environment, as is LMO. This is not the case for heavy metal-containing oxidic compounds.

Lithium diffusion is possible in one direction with LFP, while the other materials have two-dimensional, if not even three-dimensional, lithium diffusion paths. Nevertheless, LFP is the only cathode material that can be quickly charged and discharged. It has not yet reached the high energy (Wh/kg) and power densities (W/kg) of oxidic cathode materials. These values can be increased by 20 % in relation to LFP by using manganese (LFMP). Nevertheless, at this stage, LFP is a safe and resilient cathode material for usage in stationary and (plug-in) HEV applications.

Oxides have been a focal point of research and development since the Eighties. They have been modified many times, resulting in materials that can be produced on a large scale and are commercially successful. The development of the relatively "young" phosphate-based cathode materials is still in its infancy. It is possible to achieve higher working voltages with higher energy densities by substituting manganese, cobalt, or nickel for iron. This makes phosphate-based compounds very promising cathode materials for future high-voltage applications.

Bibliography

1. Akimoto J, Gotoh Y, Oosawa Y (1998) J Solid State Chem 141:298
2. Wang HF, Yang YI, Huang BY, Sadoway DR, Chiang YT (1999) J Electrochem Soc 146:473
3. Ohzuku T, Brodd RJ (2007) J Power Sources 174:449
4. Ohzuku T, Ueda A (1994) Solid State Ionics 69:201
5. Yuan LX et al (2011) Goodenough. Energy Environ Sci 4:269
6. Wang GX et al (2001) J Power Sources 97–98:298
7. Molenda J, Marzec J (2009) Funct Mater Lett 3:1
8. Amatucci GG, Tarascon JM, Klein LC (1996) Solid State Ionics 83:167
9. Breuer H (2000) dtv-Atlas Chemie, Vol. 1, 9th edition. dtv, München
10. Dahn JR, Vonsacken U, Michal CA (1990) Solid State Ionics 44:87
11. Molenda J, Marzec J (2003) Funct Mater Lett 115:115
12. Rougier A, Gravereau P, Delmas C (1996) J Electrochem Soc 143:1168

13. Pouilliere C, Croguennec L, Biensan P, Willmann P, Delmas C (2000) J Electrochem Soc 147:2061
14. Z. Lu, Macneil DD, Dahn JR (2004) Electrochem Solid-State Lett 14:A191
15. Naghash AR, Lee JY (2001) Electrochim Acta 45:2293
16. Park SH, Sun YK, Park KS, Nahm KS, Lee YS, Yoshio M (2002) Electrochim Acta 41:1721
17. Mishra SK, Ceder G (1999) Phys Rev B 59:6120
18. Armstrong AR, Bruce PG (1996) Nature 381:499
19. Ceder G, Van der Ven A (1999) Electrochim Acta 45:131
20. Makimura Y, Ohzuku T (2003) J Power Sources 119:156
21. Koyama Y, Tanaka I, Adahi H, Makimura Y, Ohzuku T (2003) J Power Sources 119:644
22. Hwang BJ, Tsai YW, Carlier D, Ceder G (2003) Chem Mater 15:3676
23. Wang L, Li J, He X, Pu W, Wan C, Jiang C (2009) J Solid State Electrochem 13:1157
24. Yoon WS, Paik Y, Yang XQ, Balasubramanian M, McBreen J, Grey CP (2002) Elektrochem Solid-State Lett 5:A263
25. Park OK, Cho Y, Lee S, Yoo H-C, Song H-K, Cho J (2011) Energy Environ Sci 4:1621
26. Ellis BL, Lee KT, Nazar LF (2010) Chem Mater 22:691
27. Gnanaraj JS, Pol VG, Gedanken A, Aurbach D (2003) Electrochem Commun 5:940
28. Thackeray MM, Dekock A, Rossouw MH, Liles D, Bittuhn R, Hoge D (1992) J Electrochem Soc 139:363
29. Benedek R, Thackeray MM (2006) Eectrochem Solid-State Lett 9:A265
30. Cho J, Thackeray MM (1999) J Electrochem Soc 146:3577
31. Cho J (2008) J Mater Chem 18:2257
32. Thackeray MM, Shao-Horn Y, Kahaian AJ, Kepler KD, Vaughey JT, Hackney SA (1998) Electrochem Solid-State Lett 1:7
33. Shin YJ, Manthiram A (2994) J Electrochem Soc 151: A208
34. Deng BH, Nakamura H, Yoshio M (2008) J Power Sources 180:864
35. Xia YG, Zhang Q, Wang HY, Nakamura H, Noguchi H, Yoshio M (2007) Electrochim Acta 52:4708
36. Kim DK, Muralidharan P, Lee HW, Ruffo R, Yang Y, Chan CK, Peng H, Huggins RA, Cui Y (2008) Nano Lett 8:3948
37. Liu GQ, Wen L, Liu YM (2010) J Solid-State Electrochem 14:2191
38. Padhi AK, Nanjundaswamy KS, Goddenough JB (1997) J Electrochem Soc 144:1188
39. Morgan D, Van der Ven A, Ceder G (2004) Electrochem Solid-State Lett 7:A30
40. Wang YG, Wang YR, Hosono EJ, Wang KX, Zhou HS (2008) Angew Chem Int Ed 47:7461
41. Kang B, Ceder G (2009) Nature 458:190
42. Ravet N, Chouinard Y, Magnan JF, Besner S, Gauthier M, Armand M (2001) J Power Sources 19:503
43. Koltypin M, Aurbach D, Nazar L, Ellis B (2007) Electrochem Solid-State Lett 10:A40
44. MacNeil DD, Lu ZH, Chen ZH, Dahn JR (2002) J Power Sources 108:8
45. Wang DY et al (2009) J Power Sources 189:624
46. Drezen T, Kwon NH, Miners JH, Poletto L, Graetzel M (2007) J Power Sources 174:949
47. Yamada A, Takei Y, Koizumi H, Sonoyama N, Kanno R (2006) Chem Mater 18:804
48. Amine K, Yasuda H, Yamachi M (2000) Electrochem Solid State Lett 3:178
49. Zhou F, Cococcioni M, Kang K, Ceder G (2004) 6:1144
50. Geoffroy D (2012) Phosphates

Anode materials for lithium-ion batteries

5

Călin Wurm, Oswin Oettinger, Stephan Wittkaemper, Robert Zauter, and Kai Vuorilehto

Contents

C. Wurm (✉)
Robert Bosch Battery Systems GmbH, Heilbronner Strasse 358-360,
70469 Stuttgart, Germany
e-mail: calin.wurm@de.bosch.com

O. Oettinger
SGL Carbon GmbH, Werner-von-Siemens-Strasse 18, 86405 Meitingen, Germany
e-mail: oswin.oettinger@sglgroup.com

S. Wittkaemper
GTS Flexible Materials GmbH, Hagener Strasse 113, 57072 Siegen, Germany
e-mail: stephanwittkaemper@gts-flexible.de

R. Zauter
Wieland-Werke AG, Graf-Arco-Strasse 36, 89079 Ulm, Germany
e-mail: robert.zauter@wieland.com

K. Vuorilehto
Aalto University Helsinki, Kemistintie 1, 02150 Espoo, Finland
e-mail: kai.vuorilehto@helsinki.fi

R. Korthauer (ed.), *Lithium-Ion Batteries: Basics and Applications*,
https://doi.org/10.1007/978-3-662-53071-9_5

5.1 Anode active materials – introduction

Secondary lithium cells initially had a metallic lithium foil as an anode (negative electrode) [1]. Pure lithium has a very high specific capacity (3,860 mAh/g) and a very negative potential, resulting in very high cell voltage. However, cycling efficiency decreases as lithium dissolves repeatedly while the cell is discharging and lithium is deposited as it is charging. This means that two or three times the normal amount of lithium must be used. In addition, lithium can be deposited as foam and as dendrites. The latter might grow through the separator [2, 3]. These dendrites can cause local short circuits, which might result in the cell completely self-discharging or, in the worst case, lead to an internal thermal chain reaction, fire, or explosion.

Today, only small cells (especially button cells) are mass-produced. They meet low requirements in regard to cycling stability and fast chargeability with lithium metal anodes. Recently, however, trials with high-capacity cells have been resumed. Special separators (cf. Bolloré [4]) are used in these trials. The most important issues for these LMP (lithium metal polymer) cells are still safety and cycling stability.

To produce safe cells with a good cycling efficiency, lithium metal is replaced by a so-called lithium intercalation material [5–8]. Generally, the intercalation process, e.g., in carbon, is a reversible and loss-free process, and lithium plating does not occur [9, 10]. The lithium-ion cell of the typical 3C market (portable consumer applications) is equipped with anodes made of graphite. There is an increasing focus on amorphous carbons (hard carbons and soft carbons) in the new applications with higher power and energy density as well as improved safety. In part, they

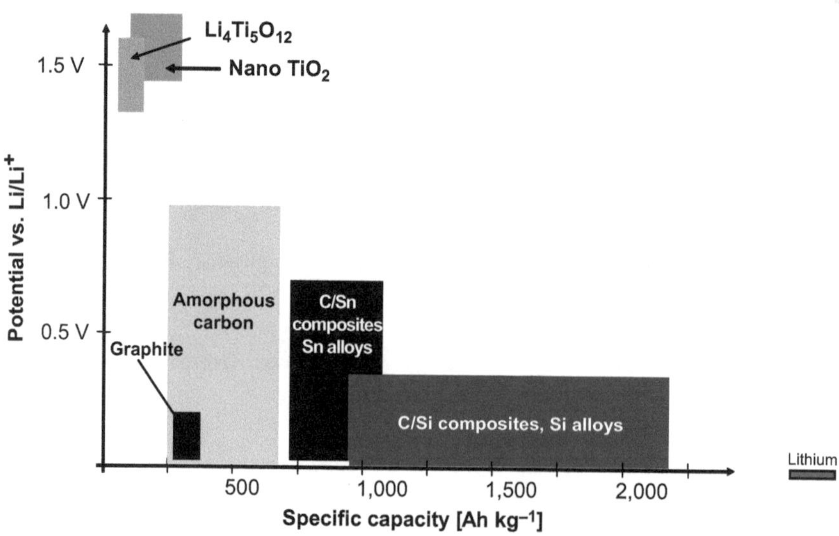

Fig. 5.1 Specific capacity and potential vs. Li/Li^+ of the most important anode materials [according to 11]

exhibit a better current capability and are safer and more stable in combination with novel electrolytes and cathode materials.

Their lithium storage capacity is much higher than that of graphite thanks to the usage of metals and alloys (intermetallic compounds) that can reversibly react with lithium. Metal-based systems are not yet being mass-produced in spite of intensive R&D efforts. So there is still a substantial need for research. Even if adding carbons (e.g., C/Si composite, C/Sn composite) enables improvement, cycling stability remains insufficient. Lithium titanate and titanium oxide are promising as anode active materials for improving cycling stability and meet extraordinarily high power and safety requirements. However, the specific capacity of these materials is very low and the potential vs. lithium is very high (Fig. 5.1).

The electrochemical characteristics of these materials play an important role. In addition, good processability (rheology characteristics, adhesion on metal foils, etc.) is a must during electrode production. Fig. 5.1 provides an overview of the specific capacity and the potential for the most important anode materials [11]. The following sections will provide information on the different anode materials.

5.2 Production and structure of amorphous carbons and graphite

Amorphous carbon (hard and soft carbons) and graphite both occur naturally, and synthetic synthesis of both is possible. A typical representative of naturally occurring amorphous carbon is anthracite. Naturally occurring graphite is called natural graphite. The largest natural graphite deposits are in Asia, mainly in China, but also in India and North Korea. In the Western hemisphere, natural graphite is mainly mined in Brazil and Canada. It is assumed that ca. 70 to 80 % of all natural graphite is deposited in China. For this reason, the European Union Raw Materials Initiative included natural graphite in its list of 14 critical raw materials in 2010. Natural graphite needs to be separated from the lode matter to make it utilizable for battery applications. It also needs to be purified chemically, thermally, and/or both.

The starting materials for synthesized amorphous carbon and graphite are usually byproducts of the coal and petroleum industry. In the latter, the most common materials are petroleum coke and resins with a high content of aromatic compounds, e.g., phenolic resins. In the coal industry, coal tar pitch and high-isotropic coke are the most important raw materials. The starting materials are carbonized or calcined at low temperatures (800 °C to 1,200 °C). The first product is amorphous carbon. If the carbonized starting material (e.g., furan resin or phenolic resin) remains amorphous during the subsequent heat treatment up to 3,000 °C, it is called hard or non-graphitizable carbon. If the carbonized starting material (e.g., coke or pitch) transforms into graphite above 2,500 °C, it is called soft or graphitizable carbon.

Fig. 5.2 provides an overview of the typical production process of synthetic graphite. The main raw materials are calcined coke as solid material and coal tar pitch as binder. The solid materials are preprocessed (ground, sieved, screened, and classified). The material is then mixed with molten, liquid binder pitch to produce

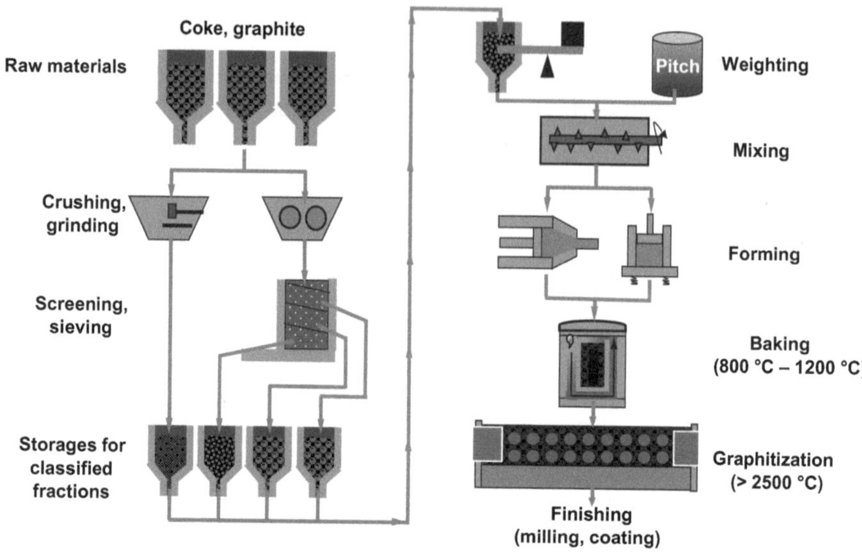

Fig. 5.2 Schematic diagram of synthetic graphite production

a plasticized "green" mass, which subsequently is shaped by extrusion or compression molding, for example.

The solidified green body is then fired at 800 °C to 1,200 °C in the absence of oxygen. The pitch carbonizes, creating amorphous carbon as binder phase. To transform amorphous carbon into synthetic graphite, the body is introduced into the so-called packing material and treated at > 2,500 °C. Contemporary graphitization technologies date back to Acheson or Castner [12, 13] and were invented at the end of the 19th century. The typical amount of electricity needed to produce 1 kg graphite in the Acheson furnace is 3 to 4 kWh and the required time in the furnace is around three weeks. The high temperatures cause the contamination to escape from the graphite, and the typical graphite layer structure develops (Figs. 5.3a and b).

Graphite is one of the best-known carbon allotropes. It is made of parallelly stacked graphene layers. Graphene is a hexagonal lattice of sp^2-hybridized carbon atoms. The widespread hexagonal form of graphite has the stack sequence ABABAB (Fig. 5.3b). The distance between the graphene layers is 0.3354 nm. The rhombohedral graphite modification with the stack sequence ABCABC is of lesser importance; its percentage can reach 20 % during forming processes such as graphite grinding. It is possible to reduce this rhombohedral ratio again with a high-temperature treatment [15]. The crystallographic density of both graphite forms is 2.26 g/cm^3.

Many times, additional finishing processes are employed depending on the application to use graphite optimally as an active material (Fig. 5.4). The final particle size and shape are modified after graphite grinding (Fig. 5.4a). This ensures that the specific surface area is as small and the surface morphology as smooth as possible

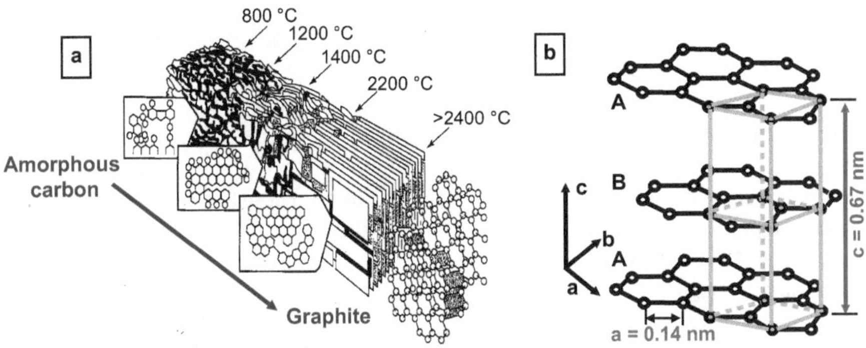

Fig. 5.3 (**a**) Development of the graphite layer structure from amorphous, graphitizable carbon according to Marsh [14], (**b**) hexagonal graphite structure

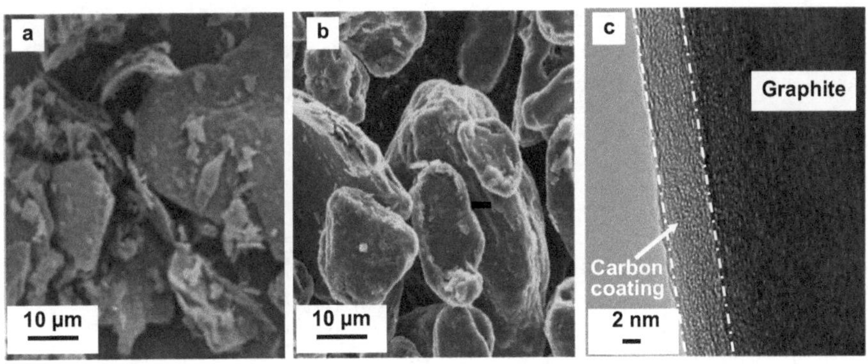

Fig. 5.4 Graphite particles (**a**) without rounding, (**b**) rounded, (**c**) with amorphous carbon coating

(potato-shaped particle design, Fig. 5.4b). Graphites or soft carbons based on mesophase pitch are an exception in this respect. They are almost round during their production, so they do not need to undergo a rounding process as powder after carbonization or graphitization. A further finishing process for graphite powder involves coating it with an amorphous or graphitic carbon layer (Fig. 5.4c). The reasons for the powder finishing processes (rounding and coating) are related to electrochemistry and will be set out in the next Chapter.

5.3 Lithium intercalation in graphite and amorphous carbons

The negative electrode of the first commercial lithium-ion cell, marketed by Sony Energytec Inc. in 1991 [1, 15, 16], was made of hard carbon, which was produced by carbonizing polyfurfuryl alcohol resin (PFA).

This success caused an increase in research and development activities in the fields of electrochemical characterization and in the intercalation processes of amorphous carbon and graphite.

The electrochemical intercalation of non-solvated lithium in graphite occurs within a potential range of 0 to 0.25 V vs. Li/Li^+. It follows different, well-defined two-phase plateaus with also well-defined chemical compounds at the beginning and end of the plateaus (Fig. 5.5a). Both phases simultaneously exist over the course of the plateaus. Experiments have shown that the intercalation steps are measurable (Figs. 5.5b and c) and easily discernible because of their color (Fig. 5.6). The hexagonal (ABABAB) and the rhombohedral (ABCABC) graphite structures, respectively, transform into an AAAAAA stack sequence with intercalated lithium during intercalation. The lithium's position is in the center of the C6 rings between two graphene layers. Therefore, the capacity of graphite depends on the number of available graphene layers. It is almost possible to achieve the theoretical reversible specific capacity of 372 mAh/g in practice, if a very well-structured graphite (e.g., natural graphite) and slow charging rates (low currents) are used.

Fig. 5.5b shows that there is a difference between the charged and the discharged capacity of the first cycle. This capacity loss results from the electrochemical reactions between the lithium ions from the cathode, the electrolyte components (organic carbonates, additives, etc.), and the anode surface.

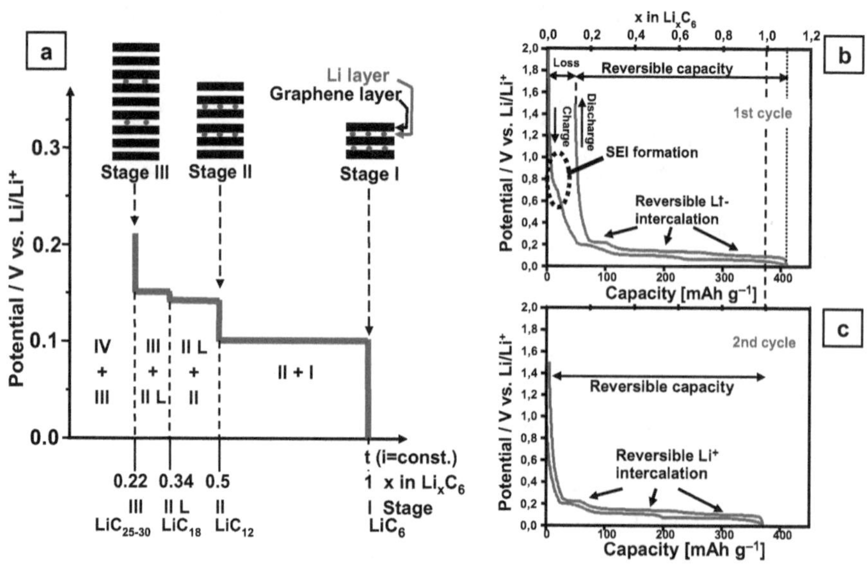

Fig. 5.5 (**a**) Thermodynamic lithium intercalation steps in graphite [17] with phases: I-LiC_6, II-LiC_{12}.II L-LiC_{18}, III-LiC_{25-30}, IV–LiC_{36-50} . Galvanostatic charging and discharging cycle of graphite with lithium metal as counter electrode and reference for the first (**b**) and second (**c**) cycle

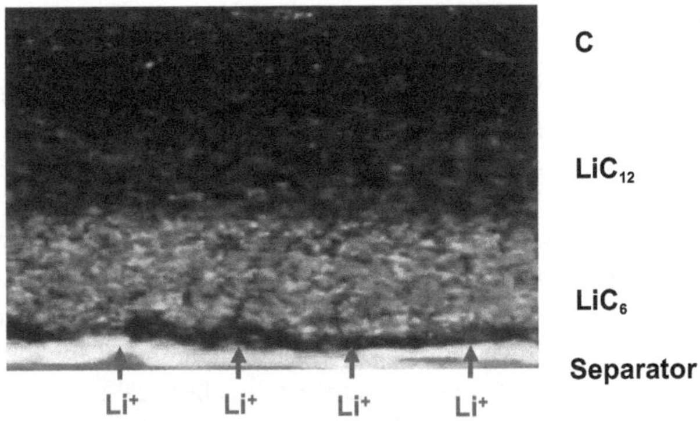

Fig. 5.6 Picture of a partially charged anode electrode [18]

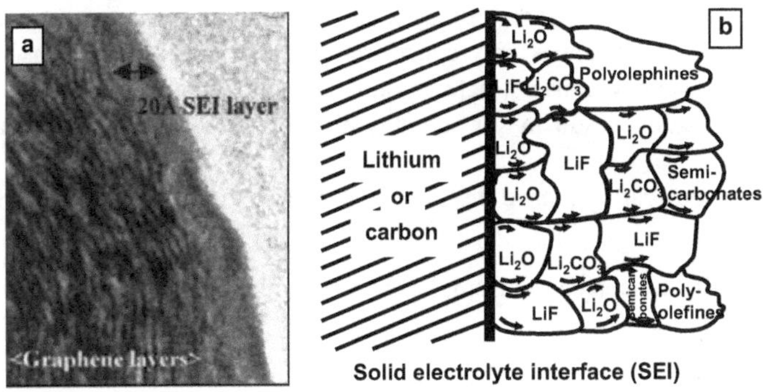

Fig. 5.7 **(a)** TEM picture [20] and **(b)** schematic diagram of the chemical composition of an SEI [21]

This electrochemical reaction results in a passivation layer between the electrolyte and the graphite particles. This layer (Figs. 5.7a, b) is called the SEI (solid electrolyte interphase) [19]. The quality of the SEI greatly influences the cycling stability, service life, power, and safety of lithium-ion cells. For example, the SEI must exhibit the best possible lithium-ion conductivity to achieve the lowest possible overvoltage.

At the same time, the SEI must work as a filter for the solvated lithium-ion solvent. This prevents the co-intercalation of solvents, which would otherwise destroy the graphite lattice. Furthermore, the SEI must adhere well to the anode particle and be expandable up to a certain degree (ensuring a volume increase without the anode particles cracking during intercalation) to prevent further losses during cycling.

The chemistry of the electrolyte and the physical and chemical anode surface greatly influence the thickness and chemical composition of the SEI. The most important characteristics of the anode surface are the specific surface (depending on particle shape, particle size, and porosity) and the surface chemistry (e.g., groups such as –COOH, –CO, –OH). The particle shape, particle size, and surface chemistry also affect the stability and quality of the slurry that is used as coating for the current collector (copper foil) during electrode production. Round particles flow better during the mixing and coating processes and also have a lower specific surface. This means that less binder is necessary in the electrode formulation, and fewer lithium ions and a lower amount of electrolyte are used for forming the SEI during the first charging cycle. Modification of the chemical surface during oxidation/reduction or by process steps such as the particle coating of the graphite anode material has an impact on the SEI quality as well as on the adhesion to the copper foil and the SEI itself.

Amorphous carbons (hard and soft carbons, Figs. 5.8 and 5.9) have no really consistent long-range order, contrary to graphite. The ordered areas are extremely small and the local layer distances vary considerably. There are distinct areas within the material with vacancy clusters, heteroatoms, and functional groups (e.g., –COOH, –OH). These factors result in considerably different electrochemical characteristics (Fig. 5.8).

All amorphous carbons have the same pattern and the following electrochemical characteristics:

- high specific capacity at extremely low charging rates in spite of a low degree of order

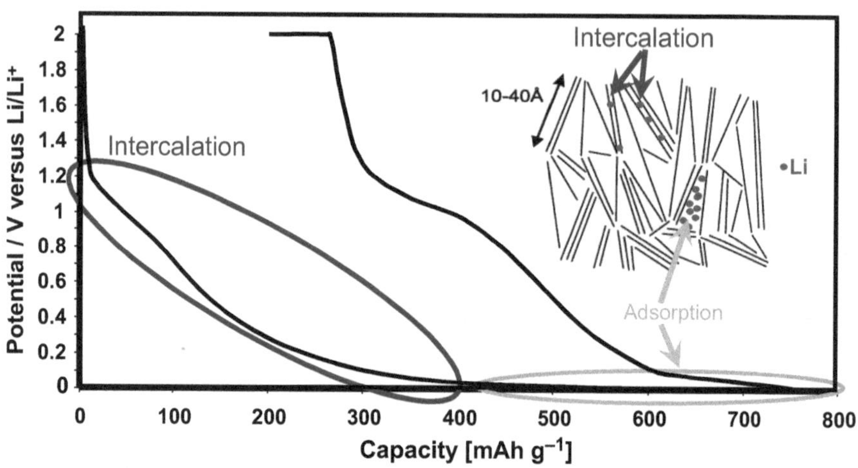

Fig. 5.8 Galvanostatic curve for the first charging and discharging cycle for amorphous carbons with lithium metal as counter electrode and reference

- high capacity loss during the first cycle in spite of a low specific BET surface
- large hysteresis of the galvanostatic curve for lithium intercalation and lithium deintercalation

Lithium intercalation in the areas of pronounced short-range order does not occur in defined stages but occurs more or less continually (Fig. 5.8), as opposed to graphite. An additional capacity is observed near the potential zero Volt vs. Li/Li^+. This is no longer caused by the intercalation of lithium between the graphene layers but by lithium adsorption in the inner nano porosity [22–24]. A separation of lithium metal clusters is also possible in the larger pores. This additional capacity is only available at very low charging rates (C rates), however, because the lithium has to diffuse all the way into the pores. Normally, only the intercalation part can be used to allow the high charging rates needed for typical industrial applications. In general, the specific BET surface of amorphous carbons is smaller or similar to that of graphite. Therefore, a similar capacity loss would be expected due to the formation of SEI during the first charging cycle. But, as shown in Fig. 5.8, the difference between the charging and discharging capacity is extremely high (here: ca. 200 mAh/g). The extremely high capacity loss during the first cycle is caused by a strong lithium reaction with surface defects (C-sp^3), with heteroatoms as well as functional groups on the inner and outer surfaces.

Because only the intercalation part is appropriate for industrial use and the required charging rates, the usable capacity of amorphous carbons is less than that of graphite. At very high charging rates, amorphous carbons integrate lithium much more quickly than graphite (this is shown in Fig. 5.9a). This characteristic is also observed when the structures of graphite and amorphous carbons are compared

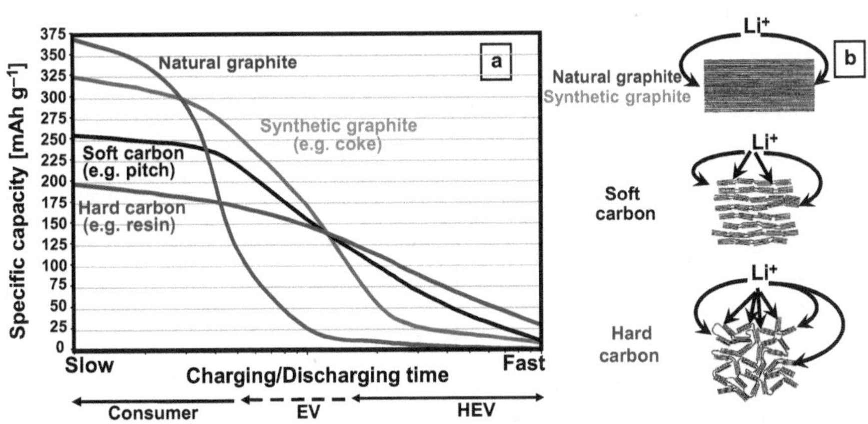

Fig. 5.9 **(a)** Schematic diagram of capacity vs. charging rate and **(b)** intercalation structure models of different carbon-based anode materials. The typical raw materials are in parentheses

(Fig. 5.9b). Graphite offers only a few "points of entrance" for lithium (at the front edges of the long-ranging layer areas) to enter the lattice, from where it has to diffuse into the center of the long-ranging crystalline domains. Amorphous carbons, on the other hand, offer more points of entrance. The distribution of lithium within the ordered strata is also very quick, since the diffusion distances are very short. Moreover, these domains with short-range order are also connected to each other, lowering the tendency to delaminate. Therefore, amorphous carbons exhibit a better cycling stability from an electrochemical point of view, especially at higher charging rates (C rates).

5.4 Production and electrochemical characteristics of C/Si or C/Sn components

As with graphite, lithium is inserted into silicon in different, well-defined two-phase plateaus with well-defined chemical compounds at the beginning and end of the plateaus (Fig. 5.10). The composition and the respective potential levels vs. lithium are as follows [25]: Si/$Li_{12}Si_7$ (332 mV); $Li_{12}Si_7$/Li_7Si_3 (288 mV); Li_7Si_3/$Li_{13}Si_4$ (158 mV); $Li_{13}Si_4$/$Li_{21}Si_5$ (44 mV). The theoretical specific capacity for the maximum amount of lithium intercalated into silicon is 4,212 mAh/g [26]. This currently is the highest known possible capacity of an alloy. The insertion of lithium in tin (Sn) follows several steps, similar to silicon. At room temperature, these are for the most part easily detectable in the first cycle, as shown in Fig. 5.10b. The theoretical specific capacity for $Li_{4.4}Sn$ is 993 mAh/g.

A large amount of lithium forms alloys with silicon and tin, thus resulting in a big change in volume during intercalation/deintercalation. This in turn causes the loss of the well-defined insertion steps during cycling and leads to amorphization of the material [27]. This destruction of the crystalline structure is similar to inner "grinding", which decreases the cycling stability considerably. Extremely

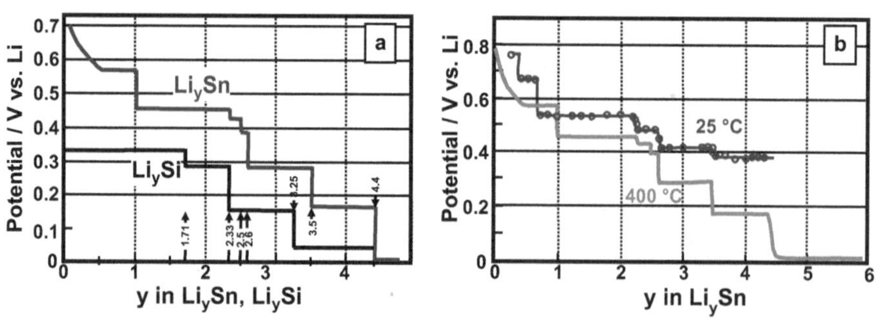

Fig. 5.10 (a) Lithium insertion in silicon and tin (Sn), (b) lithium insertion in Sn, measured at two different temperatures [28]

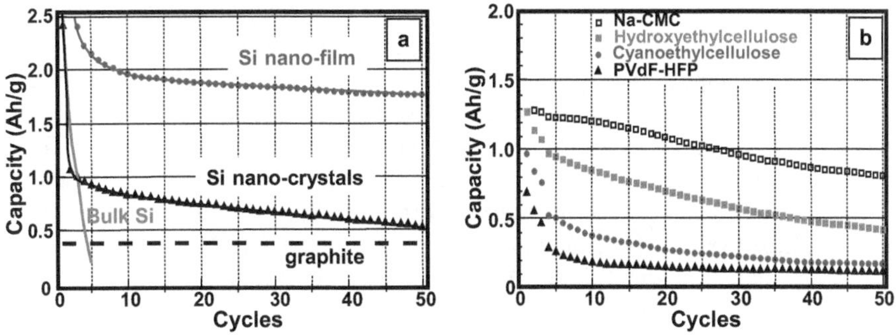

Fig. 5.11 (**a**) Influence of silicon particle size on cycling in comparison with the theoretical capacity for graphite [29], (**b**) influence of the binding agent on the cycling stability of anodes based on C/Si [30]

small particles (down to the nano range) are used to reduce this effect (Fig. 5.11a). A further improvement of the cycling stability is achieved by employing mainly carbon composites; i.e., nanoparticles are integrated into carbon or graphite matrices, for example. However, this reduces the specific capacity.

Reactive binding agents are used in another approach to improve cycling stability. They cause an additional stabilization of the composite anode (Fig. 5.11b).

5.5 Lithium titanate as anode material

The lithium titanate (LTO), which crystallizes as spinel ($Li_4Ti_5O_{12}$), reversibly intercalates lithium at 1.55 V (Fig. 5.12a) and reaches a reversible capacity of around 160 mAh/g ($Li_7Ti_5O_{12}$). Intercalation/deintercalation of lithium causes almost no change in particle volume in this process. The whole process consists of two phases with a distinct flat plateau within the stability window of the electrolyte. No SEI is

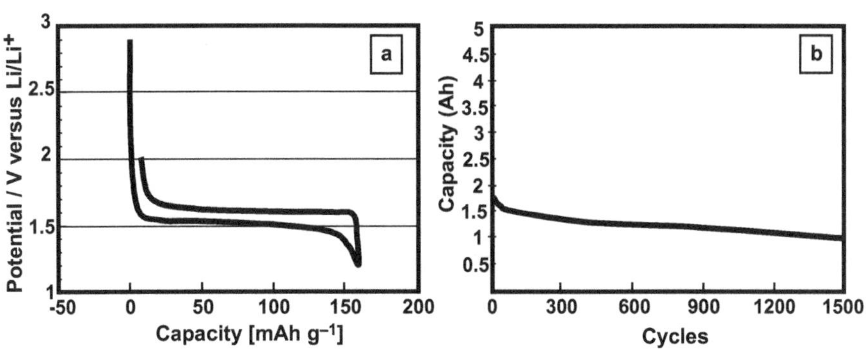

Fig. 5.12 (**a**) Galvanostatic curve of lithium titanate (LTO) with lithium metal as counter electrode and (**b**) cycling of lithium titanate at 60 °C [31]

formed, and cell impedance is very low. All these parameters are the reason for the cell's very good cycling stability (Fig. 5.12b) and safety characteristics.

The disadvantage of this material is its very low electrical conductivity and resulting relatively poor performance. Hence, the particles should be as small as possible, preferably in the nano range, and additionally coated with conductive carbon. These measures improve the power and service life of lithium titanate. Its energy density remains low, nevertheless, owing to the low specific capacity and the high potential level vs. Li/Li^+ (Fig. 5.1). Lithium titanate is therefore especially suitable for large cells in stationary applications or for cells in areas with very high power requirements (e.g., hybrid vehicles).

5.6 Anode active materials – outlook

Table 5.1 shows a qualitative comparison of the most important anode active materials in regard to the essential characteristics such as energy, power, service life, and safety. This assessment reflects the authors' state of knowledge at the time of writing and merely provides a basis for approximate orientation and classification.

Taking into consideration the above-mentioned property profiles, the following materials are currently the most significant anode active materials (depending on the application): synthetic graphite, natural graphite, amorphous carbon (hard and soft carbons), and lithium titanate. Graphite has the most balanced profile and, as a result, by far the largest market share. However, composites (in this context mainly C/Si) will most likely gain in importance in the long-term future, depending on the development. Examples of these are silicon alloys, non-silicon alloys such as

Table 5.1 Qualitative assessment of the property profiles of the most important anode active materials. This assessment is merely a snapshot that needs to be updated as active materials and the lithium-ion battery system evolve, since intensive research is currently being carried out globally

Characteristic	Energy	Power	Service life	Safety
Active material				
Synthetic graphite	++	+	+	+
Natural graphite	++	+	0	0
Amorphous carbon	0	++	++	++
Lithium titanate (LTO)	−−	+++	+++	++++
C/Si or C/Sn composite	+++	+	−	0
Silicon alloys	++++	+	−−	−
Lithium	++++	−	−	−−

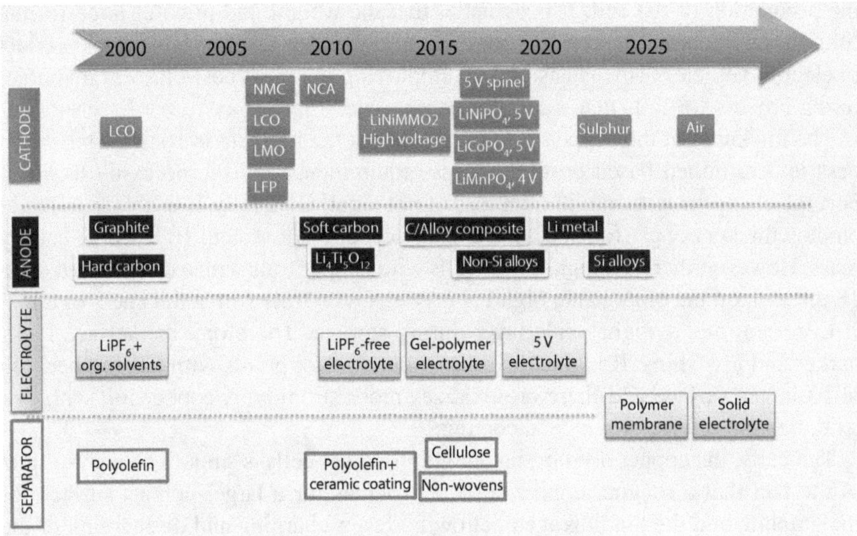

Fig. 5.13 Raw material road map for lithium-ion batteries [32] (*Source* Avicenne Compilation, Kai-Christian Moeller, Fraunhofer ISC)

tin-based alloys, and metallic lithium. An overview is presented in Fig. 5.13, the current raw material road map for lithium-ion batteries [32].

5.7 Copper as conductor at the negative electrode

In a lithium-ion cell there typically are two different metal foils that have the function of collecting the current. The negative electrode consists of copper and graphite; the positive electrode is made of aluminum which is coated with cathode material. The electrochemical characteristics of copper are an important reason for its use.

5.7.1 Requirements for the copper foil

The most important requirement for the copper collector is to collect electrons while remaining electrochemically stable. These requirements are fulfilled very well by pure copper. It has the second-highest electrical conductivity of all metallic materials (after silver) with 58 MS/m. Its positive electrochemical potential results in good corrosion resistance.

Furthermore, its surface must allow good adhesion of the chemically active anode material (slurry). Its mechanical strength must be high enough for the foil to survive the production of the cell. It is essential that the weight and price of large-format lithium-ion batteries are kept low for them to be used as an affordable mass product in electric vehicles. This is ensured by employing the most cost-efficient manufacturing process for foils that are as thin as possible.

The thickness of the copper foil mainly used at the moment is 10 μm. This thickness is determined based on the process requirements and the necessity to transport the electrons with a minimum loss of potential. Generally, it is not desirable to employ thicker copper foils because copper accounts for around 10 % of the battery mass. However, there are high-power cells with copper foils with a thickness of up to 18 μm and, on the other hand, high-energy cells with foils with a thickness of 8 μm.

Copper forms a slight oxide layer during storage. Therefore, the surface looks darker and less shiny. Rough copper has more contact points with oxygen because of its larger surface and therefore oxidizes more strongly. A copper foil with low oxidation is preferred for a good coating.

Typically, the copper used in standard high-power cells is smooth copper, in spite of the fact that a rougher copper surface provides for a larger contact surface for the graphite and the binding agent, allowing faster charging and discharging of the battery. The reason for this might be the size of the carbon particles, because adhesion is not influenced substantially with a typical particle size of 20 μm and foil roughness of 1 to 2 μm.

Impurities in the copper decrease the service life of lithium-ion batteries. Residues from production such as oils and fats (RA copper) are not acceptable.

Coating the foil with graphite is a critical process step. In this context it is important to be able to use long coils free of joints. Each stop and restart increases the reject rate. Furthermore, the copper should be mechanically stable and have no variations in weight per unit area.

5.7.2 Comparison of RA copper foil with electrolytically produced copper foil

Basically, there are two production processes for manufacturing copper foils: rolling and electrodeposition.

For the former, an industrial copper cast block is alternately rolled and heated until the required foil thickness is reached. The result is a well-rolled material with a fine granular structure that is elongated in the direction of rolling. Both surfaces are shiny and have the same typical rolling structure.

Electrodeposited copper is manufactured as follows: Copper ions are electrolytically deposited from a copper sulphate solution onto steel or titanium drums and rolled up into copper foil. This process is in accordance with Faraday's law. It stipulates that time and amperage determine the thickness of the deposited copper. The copper grows dendritically. It is possible to modify this typical structure by means of inhibitors in the solution. The resulting foil has differing surfaces. The surface facing the drum is smooth and shiny; the surface facing away from the drum is rougher. The grains are visible through the microscope.

Both RA copper and electrodeposited copper may be used to produce cells for lithium-ion batteries. Availability and costs, especially of the very thin foils (10 μm) that are currently demanded most, are the main reasons for the preference of electrodeposited copper.

5.7.3 Substitution of aluminum for copper?

Lithium ions are intercalated into the negative electrode (anode) while a lithium-ion cell is charging. The electrode material, typically graphite, expands by 10 % during this process. The graphite regains its original volume when the lithium ions deintercalate.

The lithium ions would not only be intercalated into the graphite if aluminum was used but also inserted into the conductor, thus forming an aluminum-lithium alloy. The reverse process would occur during discharging. The aluminum would be degraded after only a few cycles and would be useless as a current collector.

However, if the negative electrode consisted of lithium titanate instead of graphite, a different picture would emerge. The electrode potential of $Li_4Ti_5O_{12}$ is about 1.4 V higher than that of graphite (cell voltage is around 1.4 V lower). This would prevent the lithium ions from being intercalated into the aluminum. Therefore, aluminum is preferred over copper for cost-related and weight-related reasons. $Li_4Ti_5O_{12}$ is employed mainly in stationary applications because of its lower cell voltage.

Bibliography

1. Abraham KM (1993) Electrochimical Acta 38:1233
2. Matsuda Y (1989) Nihon Kagaku Kaishi 110:1
3. Peled E (1979) J Electrochem Soc India 126:2047
4. Bolloré (www.bluecar.fr/de/pages-innovation/batterie-lmp.aspx)
5. Armand M (1980) In: Broadhead J, Steele BCH (ed.) Materials for advanced batteries. Plenum Press, New York, pp. 145–150
6. Nishi Y (1998) In: Wakihara M, Yamamoto O (eds.) Lithium ion batteries. Wiley-VCH, New York, p. 181
7. Megahead S, Scrosati B (1994) J Power Sources 51:79
8. Dahn JR et al. (1994) In: Pistoia G (ed.) Lithium batteries–new materials, developments, and perspectives, industrial chemistry library. Elsevier, Amsterdam, pp. 1–47
9. Broussely M, Biensan P, Simon B (1999) Electrochim Acta 45:3
10. von Sacken U, Nodwell E, Sundher A, Dahn JR (1994) Solid State Ionics 69:284
11. Jossen A, Wohlfahrt-Mehrens M (2007) Overview on current status of lithium-ion batteries. In: 2nd International renewable energy storage conference, Bonn, 19–21 Nov 2007
12. Castner JH (1893) GB 19089
13. Acheson EG (1895) US 568 323, 1895
14. Marsh H, Griffiths JA (1982) A high, resolution electron microscopy study of graphitization of graphitizable carbon. International Symposium on Carbon, Toyohashi, p. 81
15. Omaru A, Azuma H, Nishi Y (1992) Sony Corp., Japan Patent Application: WO 92-JP238 9216026
16. Sekai K, Azuma H, Omaru A, Fujita S, Imoto H, Endo T, Yamaura K, Nishi Y, Mashiko S, Yokogawa M (1993) J Power Sources 43:241

17. Winter M, Besenhard JO, Spahr ME, Novák P (1998) Adv Mater 10:725
18. Harris SJ (LithiumBatteryResearch.com) (2009) on YouTube; S.J. Harris, A. Timmons, D.R. Baker, C. Monroe. Chem Phys Letters 2010 485:265
19. Peled E (1979) J Electrochem Soc., 126, p. 2047
20. Orsini F, Dupont L, Beaudoin B, Grugeon S, Tarascon J-M (2000) Int J Inorg Mater 2:701
21. Peled E, Golodnitsky D, Ardel G (1997) J Electrochem Soc 144:208
22. Dahn JR, Zheng T, Liu Y, Xue JS (1995) Science 270:590
23. Liu Y, Xue Js, Zheng T, Dahn JR (1996) Carbon, 34:193
24. Dahn JR (1997) Carbon 1997, 35:825
25. Wen CJ, Huggins RA (1981) J Solid State Chem 37:271
26. Weydanz WJ, Wohlfahrt-Mehrens M, Huggins RA (1999) J Power Sources 81–82:237
27. Limthongkul Pimpa, Jang Young-Il, Dudney Nancy J, Chiang Yet-Ming (2003) J Power Sources 119–121:604–609
28. Huggins RA (1999) J Power Sources 81–82:13–19
29. Graetz J, Ahn CC, Yazami R, Fultz B (2003) Electrochem Solid-State Lett 6(9):A194–A197
30. Hochgatterer NS et al (2008) Electrochemical and solid-state letters 11(5):A76–A80
31. Zaghib K, Simoneau M, Armand M, Gauthier M (1999) J Power Sources 81–82:90
32. Pillot C (2011) The rechargeable battery market past and future. Batteries 2011, 28–30 Sept 2011. Cannes

Electrolytes and conducting salts

6

Christoph Hartnig and Michael Schmidt

Contents

6.1 Introduction

"Chemistry gets it done." High-duty lithium-ion batteries of today and especially those of the future are not conceivable without advances in the development of materials. Chemistry will have to play a significant role in that. By employing innovative concepts for the materials, it will have to optimize the battery technology in terms of energy density power density as well as cycle and calendar life and thus open the door to electric mobility and stationary storage of renewable energy.

The development of new electrolyte systems for lithium-ion batteries is of paramount importance, alongside the development of new electrode materials and

C. Hartnig (✉)
Heraeus Deutschland GmbH & Co. KG, Heraeusstraße 12-14, 63450 Hanau, Germany
e-mail: christoph.hartnig@heraeus.com

M. Schmidt
BASF SE, GCN/EE – M311, 67056 Ludwigshafen, Germany
e-mail: michael.e.schmidt@basf.com

R. Korthauer (ed.), *Lithium-Ion Batteries: Basics and Applications*,
https://doi.org/10.1007/978-3-662-53071-9_6

separators. Lithium-ion battery electrolytes are more than colorless fluids that merely keep up the transport of ions between the electrodes. Today, they are high-purity multi-component systems with a multitude of requirements and tasks.

The requirement profile of the perfect electrolyte is manifold and encompasses, among other things:

- Function
  - high conductivity across a wide temperature range (−40 °C to + 80 °C)
  - cycling stability over several thousands of cycles
  - chemical and electrochemical compatibility with the electrode and inactive materials
- Safety
- Ecology
- Economy

Fulfilling all these requirements creates a set of challenges that can only be overcome by means of innovation. It is worth mentioning here that “the” perfect electrolyte does not yet exist.

It is the objective of this Chapter to provide the reader with an overview of modern, functional electrolytes, with an introduction to the different electrolyte components and to give some examples, including information on the role of additives, the so-called “master spit” of the electrolyte.

6.2 Electrolyte components

The tool box for each electrolyte for lithium-ion batteries consists of three classes of materials: conducting salt, organic aprotic solvents (or, partly, polymers), and additives. It is the combination of these components which largely determines the physico-chemical and electrochemical characteristics of the electrolyte and contributes to fulfilling the above-mentioned objectives.

6.2.1 Solvents

In keeping with the basic requirement profile for an electrolyte, it is possible to stipulate the minimum criteria for a suitable solvent [1]:

1. The solvent must be able to dissolve lithium salts in a sufficiently high concentration. This means, it should exhibit a high permittivity (ε) to ensure the correspondent solvation of ions.
2. At the same time, it has to have a low viscosity (η) to provide for an unimpeded transport of ions. This parameter plays an important role at low temperatures and for high-voltage applications, where (speaking from a microscopic point of view) a sufficient migration speed of the lithium ions is called for.

3. The solvent must be inert toward all other cell components in all operating conditions, especially toward the charged electrode materials and the current collector. Today's lithium-ion batteries generally have a charging potential of around 4 V, tomorrow's will probably have one of close to 5 V. This is why the electrochemical stability of the solvent plays a crucial role.
4. A wide liquid range is desirable. Therefore, a suitable solvent should exhibit a low melting point (T_m) and a high as possible boiling point (T_b).
5. Furthermore, the requirements concerning safety (toxicity, flash point [T_i]) and economy need to be taken into account.

In lithium-ion batteries highly reductive materials are used for the negative electrode (mainly lithiated carbons or graphite) and highly oxidizing components for the positive electrode (mainly lithium metal oxides or metal phosphates). This is the reason why solvents with an active acidic proton are unsuitable. This would immediately lead to a development of hydrogen. The same reason excludes water as a solvent.

Two classes of organic solvents that are simultaneously aprotic and highly polar have gained recognition as suitable materials for lithium-ion batteries: ethers and esters, including organic carbonates. While the technical literature discusses alternatives such as nitrile, functionalized silanes, sulfones, and sulfites, these are of purely academic concern. Fig. 6.1 shows the physico-chemical characteristics of selected battery solvents.

Ether-containing electrolytes for the most part exhibit a low viscosity and therefore a very high conductivity. But their electrochemical stability is restricted and they are already oxidized at potentials around 4 V vs. Li/Li^+. With the introduction of 4-V transition metal oxides as positive electrode materials, ethers have therefore disappeared as solvents for high-energy lithium-ion batteries.

Solvency	Structure	Melting point / °C	Boiling point / °C	Viscosity (25°C) / cP	Permittivity (25 °C)	Flash point / °C
Carbonate						
EC	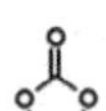	36	247-249	1.9 (40 °C)	90 (40 °C)	160
PC	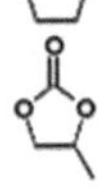	-48	242	2.53	65	135
DMC	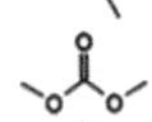	2-4	90	0.59	3.1	15
DEC	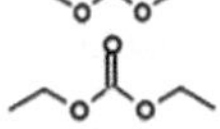	-43	125-129	0.75	2.8	33
EMC	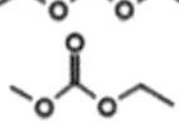	-55	108	0.65	3.0	23
Ester						
EA	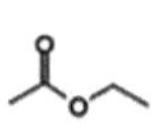	-83	77	0.45	6.0	-4
MP	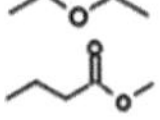	-84	102	0.60	5.6	11
Ether						
DME	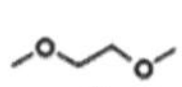	-58	84	0.46	7.2	0
THF	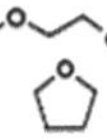	-108	65-66	0.46	7.4	-17

Fig. 6.1 Physico-chemical characteristics of selected battery solvents

Esters, especially organic diesters of carboxylic acid (so-called carbonates), are current state of the art. In general, the following are employed: blends of cyclic carbonates (e.g., ethylene carbonate [EC] and, to some extent, propylene carbonate [PC]) that exhibit a high dipole moment at moderate viscosity. Also, open-chain carbonates (dimethyl carbonate [DMC], diethyl carbonate [DEC], and ethyl methyl carbonate [EMC]) are used, which exhibit a moderate dipole moment at low viscosity.

Sometimes open-chain esters such as ethyl acetate (EA) or methyl butyrate (MB) are added as co-solvents to further improve the low-temperature characteristics of the electrolyte.

6.2.2 Conducting salts

The electrolyte provides for the lithium ion transport between the electrodes. Therefore, a suitable lithium salt must fulfill essential requirements:

- Maximum solubility and complete dissociation in aprotic solvents to ensure a high lithium-ion mobility.
- Very high electrochemical anion stability, especially in respect to oxidation, plus a high chemical stability in respect to the solvent.
- Good compatibility toward all cell components, especially the current collectors and the separator

From a chemical point of view, this requirement profile leads to mostly complex anions in which the negative charge is distributed to a high extent across the anions. This reduced charge density in turn causes a low attraction between the anion and the lithium cation, which determines the free movement of the cation and is therefore necessary for a high mobility.

Electron-withdrawing groups (simplest: flourine, complex: highly-fluorinated organic remnants, carboxyl or sulfonyl groups) lead to minimized interactions between the anion and the lithium cation. Although chemistry enables in principle a large number of different weakly-coordinating anions, in practice the number is much lower, due to reasons of stability, cost, and processing.

Lithium hexafluorophosphate In terms of the compounds possible in principle, lithium hexafluorophosphate ($LiPF_6$) plays a special role. Nowadays, commercial lithium-ion batteries are almost exclusively equipped with $LiPF_6$. This is on the one hand not by reason of a single outstanding characteristic, but of an unparalleled unique combination of characteristics. And on the other hand it is based on the willingness to accept individual disadvantages (Fig. 6.2).

Having a conductivity of 8 to 12 mS/cm (room temperature, 1 mol/l), $LiPF_6$ forms highly conductive electrolytes in blends of organic carbonates. These electrolytes are, in addition, electrochemically stable up to almost 5 V vs. Li/Li^+.

$LiPF_6$ is one of the few conducting salts that very effectively prevent the corrosion of the aluminum current collector of the positive electrode at potentials above 3 V vs. Li/Li^+.

 $LiPF_6$	Advantages: □ Highly conductive > 10 mS/cm @ RT in EC:DMC 1:1 □ Electrochemically stable > 4.8 V vs. Li/Li^+ □ Good compatibility toward other cell components Aluminum corrosion is effectively suppressed
	Disadvantages: □ Thermally unstable Degradation in the electrolyte from around 70 °C □ Very susceptible to hydrolysis Reacts with traces of water, forming hydrofluoric acid

Fig. 6.2 Property profile $LiPF_6$

It has already been used in lithium batteries as conducting salt since the late Sixties. Production process, quality, and purity, crucial for a battery's performance, have been improved over the course of the years. For example, high-purity $LiPF_6$ has been available on an industrial scale only since the end of the Eighties. Today's lithium-ion technology has been developed based on this conductive salt.

It was well known, from the very beginning of its use, that $LiPF_6$ only exhibits limited chemical and thermal stability: In its pure form it very slowly disintegrates into traces of lithium fluoride (LiF) and phosphorus pentafluoride (PF_5), creating an equilibrium. This already occurs at room temperature.

$$LiPF_5(g)(s) \leftrightharpoons LiF(s) + PF_5(g)$$

High temperatures promote this process.

$LiPF_6$ is somewhat more stable in organic solutions, e.g., in standard battery electrolytes based on organic carbonates, but still disintegrates above 70 °C. The strong Lewis acid (PF_5) initiates several other reactions during the subsequent process. A slow polymerization is observed especially in combination with cyclic organic solvents, noticeable by the slight yellow color of the electrolyte (Hazen > 50 APHA). The P-F compound furthermore is regarded as susceptible to hydrolysis and reacts with minute amounts of water, forming hydrofluoric acid (HF).

$$LiPF_6 + H_2O \rightarrow LiF + 2\,HF + POF_3$$

Lithium-bis(trifluoromethylsulfonyl)imide, -(fluorosulfonyl)imide, and their derivatives New formulas have increasingly been introduced into the market during the last years. These have been efforts to replace lithium hexafluorophosphate because it is highly susceptible to hydrolysis. The most promising approaches are those compounds that are based on sulfonylimides, for instance lithium-bis(trifluoromethylsulfonyl)imide (LiTFSI) and lithium-bis(fluorosulfonyl)imide (LiFSI). These new conducting salts have as an advantage their conductivity, which is similar to that of $LiPF_6$ and they therefore fulfill the high load current requirements. LiFSI demonstrates a conductivity of 12 mS/cm (0.85 M, 25 °C, EC/DMC). This is higher

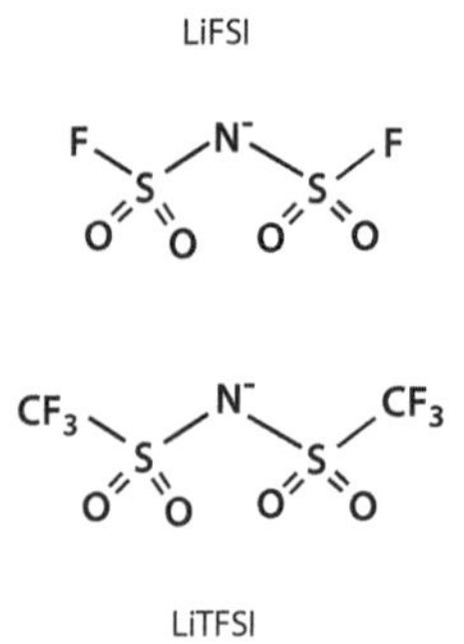

Advantages:

- Highly conductive
  > 10 mS/cm @ RT in EC:DMC 1:1
- Electrochemically stable
  > 4.8 V vs. Li/Li^+
- Hydrolysis-resistant
  No formation of hydrofluoric acid in the presence of water
- Thermally stable

Disadvantages:

- Insufficient passivation of the Al current collector
  Al corrosion
- Multi-stage production process – expensive

Fig. 6.3 Property profile lithium-bis(trifluoromethylsulfonyl)imide, lithium-bis(fluorosulfonyl) imide

than that of $LiPF_6$ and makes it especially suitable for high-current applications. The conductivity of LiTFSI is a little lower than that of $LiPF_6$ (Fig. 6.3). Dissociation in commonly used carbonates plays an important role at this point. High dissociation ensures a sufficiently high conductivity and battery performance even at low temperatures.

Both LiFSI and LiTFSI demonstrate high thermal stability. LiFSI is much more stable with its degradation temperature of more than 200 °C than the much-mentioned hexafluorophosphate. It therefore provides a significant improvement in terms of safety. Other applications of LiFSI and LiTFSI are ionic fluids, which mostly contain similar anions and cause a better solubility of the conducting salts. At the beginning of 2013, Nippon Shokubai of Japan put into operation the first large-scale plant for LiFSI, with an annual capacity of 200 to 300 t.

There are significant problems with these two compounds, however. On the cathode, increased corrosion of the aluminum conductors is observed. This local substrate damage is caused by an insufficient and not potential-stable formation of a passivation layer. It already starts at around 3.7 V for LiTFSI. In the literature, different potentials for the beginning of oxidation of the carrier material are mentioned for LiFSI: Degradation currents are measured at 3.3 V if the material is contaminated with chloride ions. This value might be above 3.8 V if the material is of higher purity (less chloride contamination).

New concepts for conducting salts

a) **Lithium fluorinealkylphosphate** A new class of conducting salts for electrolytes of lithium-ion batteries was presented at the end of the Nineties with the lithium fluorinealkylphosphates [2]. The structure betrays its relationship with $LiPF_6$. From a structural point of view, $Li[PF_3(C_2F_5)]$, LiFAP, derives from $LiPF_6$ by substituting three perfluoroethyl groups for the respective fluoride groups (generally speaking: perfluoroalkyl groups) (Fig. 6.4).

The considerable delocalization of the negative charge by the three perfluoroalkyl groups enables highly conductive electrolytes, this in spite of the size of the

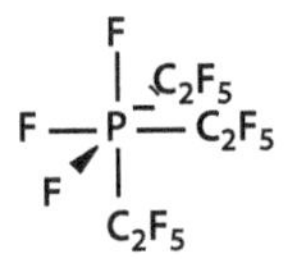	Advantages: ☐ Highly conductive 8–10 mS/cm @ RT in EC:DMC 1:1 ☐ Electrochemically stable > 5 V vs. Li/Li^+ ☐ Good compatibility toward other cell components Aluminum corrosion is effectively suppressed ☐ Hydrolysis-resistant No formation of hydrofluoric acid in the presence of water
LiFAP	Disadvantages: ☐ High molecular weight 450 g/mole (= threefold the molecular weight of $LiPF_6$) ☐ Elaborate production process – expensive

Fig. 6.4 Property profile lithium[tris(pentafluoroethyl)trifluorophosphate], LiFAP

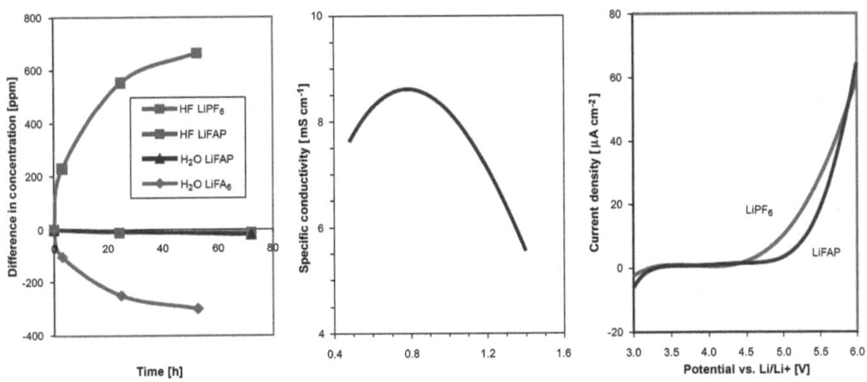

Fig. 6.5 Selection of physico-chemical characteristics of $Li[(C_2F_5)_3PF_3]$, LiFAP (*left* resistance against hydrolysis in the presence of 1,000 ppm water, *center* conductivity in EC:DMC (1:1) at 20 °C, *right* electrochemical stability in comparison to $LiPF_6$

FAP anion in typical battery solvents. The incorporation of electron-withdrawing groups, in addition, effects a slight increase in electrochemical stability. Ongoing research is trying to determine whether LiFAP is advantageous if used with the current 5-V electrode materials.

LiFAP does not react in the presence of water under formation of hydrofluoric acid. This is in contrast to $LiPF_6$-containing electrolytes (Fig. 6.5, left: no development of HF in the presence of 1,000 ppm water). This class of conducting salts has not yet found its way into an industrial-size production. One reason is the high molecular weight (for the production of 1 liter of electrolyte three times more LiFAP is needed than $LiPF_6$). Another is the elaborate and expensive production process. Currently, researchers are looking into alternative LiFAPs with a lower number of perfluoroalkyl groups with a lower molecular weight.

b) **Lithium-bis(oxalato)borate** Chelatoborates were developed to find environmentally friendly alternatives to existing conducting salts. They have boron as the central atom and the most successful variant is lithium-bis(oxalato)borate (LiBOB). It is

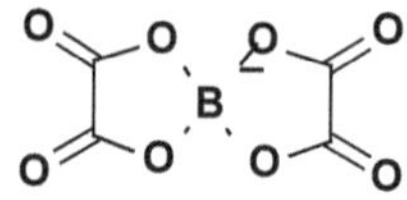 LiBOB	Advantages: □ Non-halogen, therefore environmentally friendly □ Electrochemically sufficiently stable 4.5-5 V vs. Li/Li^+ □ Good compatibility toward other cell components Aluminum corrosion is suppressed □ Thermally stable
	Disadvantages: □ Moderate conductivity 6-7 mS/cm @ RT in organic carbonates □ Can cause increased development of gas during the forming of the cell □ Susceptible to hydrolysis

Fig. 6.6 Property profile lithium-bis(oxalato)borate, LiBOB

possible to employ it as a functional additive with extraordinary film formation characteristics and as a non-toxic, non-corrosive conducting salt for specific applications.

Its optimized film formation characteristics may be used, for example, to reduce the share of ethylene carbonate (EC) in carbonate-based electrolyte compounds. EC is used specifically with graphite-based anodes to form a stable SEI boundary surface, which, in turn, prevents the graphite layers from fanning out due to intercalated solvent molecules (Fig. 6.6).

The film formation characteristics are not restricted to the active materials of the anode. They also are observed on the cathodic current collector foil. When LiBOB degrades, its typical degradation products induce a surface passivation via $AlBO_3$. This layer is stable enough to withstand potentials of up to 5 V. Conduction salts such as LiFSI and LiTFSI tend toward local corrosion of the aluminum surface, as described above. Additives such as LiBOB are instrumental in achieving a passivation and therefore a significant increase in service life.

There is another advantage in the usage of LiBOB as an additive in $LiPF_6$ electrolytes for manganese-based electrode materials: It attracts traces of water in the finished electrolyte. This means that the dissociation of $LiPF_6$ and the resulting formation of hydrogen fluoride (HF) are mostly inhibited. HF on the other hand causes heavy dissolution of manganese from, for instance, manganese spinels ($LiMn_2O_4$). The manganese might then diffuse onto the anode and act as a catalyst. This causes a reduction in service life.

Fig. 6.7 clearly shows the effect of LiBOB as an additive in $LiPF_6$-based electrolytes: The addition of 2 % LiBOB already results in a reduction of the manganese elutriation of around one magnitude for a deintercalation of lithium-manganese spinel in the following materials: pure $LiPF_6$ electrolyte, electrolyte with LiBOB as an additive, and pure LiBOB electrolyte. The degradation rate is extremely reduced in the pure LiBOB electrolyte.

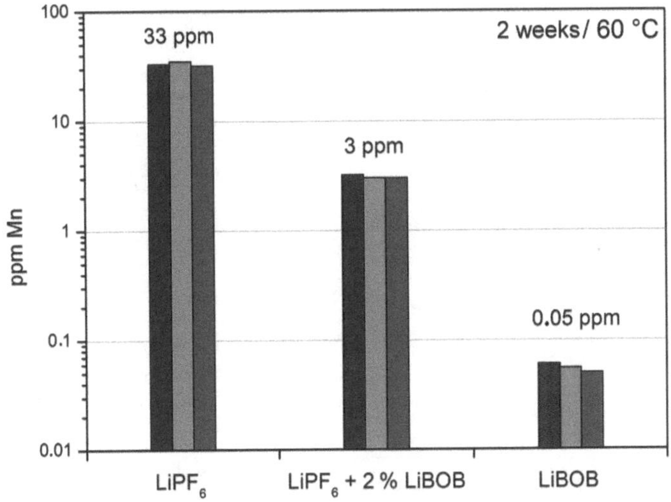

Fig. 6.7 Degradation rate of manganese from lithium manganese spinels after storage in respective electrolytes at increased temperature (EC/DEC 7:8)

6.3 Functional electrolytes

In addition to the transport of lithium ions, the electrolytes have to fulfill further tasks to achieve an increase in safety and performance.

The new applications for lithium-ion batteries such as electric mobility and stationary storage result in exacerbated requirements. Additives play the decisive role in fulfilling these requirements.

6.3.1 Calendar life and cycling stability

SEI film formation on the negative electrode Additives mainly are employed to optimize the so-called solid electrolyte interface (SEI), the boundary surface between the negative electrode and the electrolyte. The SEI significantly influences service life and performance of the lithium-ion cell.

Lithium ions do not exist in the form of "naked" cations in organic polar solvents, but rather as complex cation solvent adducts. This so-called solvate complex is many times larger than the naked lithium ion.

Solvated lithium ions penetrate into the outer structures of the graphite anode during charging of the lithium-ion cell (Fig. 6.8). The solvents (and in part also the anion of the lithium salt) degrade owing to the extremely reductive conditions and form hardly soluble precipitates. In general, hardly soluble lithium alkylcarbonates Li[OC(O)OR] and Li_2 [OC(O)O-(CHR)n-O(O)CO] form in organic carbonates [3]. They accumulate on the electrode and in the outer structures of the graphite and form the layer that is called SEI.

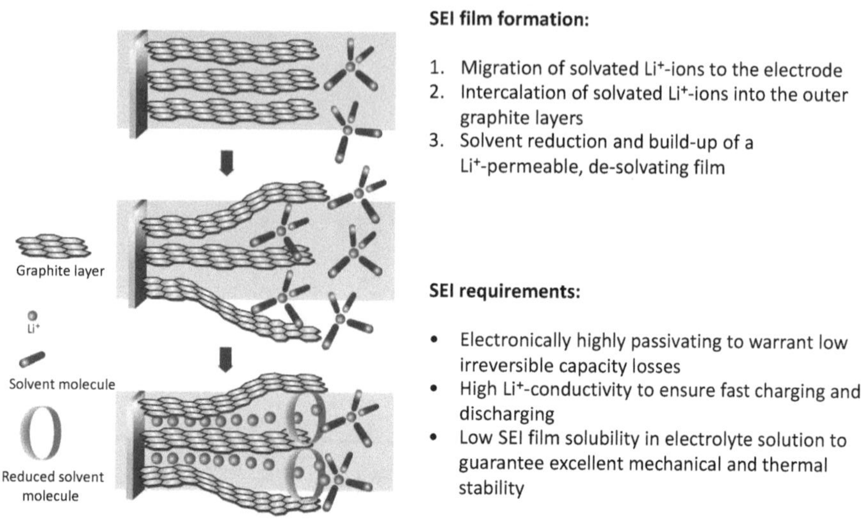

Fig. 6.8 SEI film formation

This layer is permeable for lithium ions. At the same time it is electrically isolating and prevents the direct contact of electrode and solvent, a characteristic that prevents further degradation of the solvent. Furthermore, SEI is desolvating, i.e., the lithium ion strips off the solvent molecules during its pass and diffuses into the electrode as naked cation.

During charging, there would be significant expansion of the graphite layers and a contraction during discharging of the cell without the SEI. This so-called "breathing" would lead to a "falling apart" of the electrode with an increasing number of charging and discharging cycles and therefore shorten the service life of the battery.

All carbonates employed in lithium-ion batteries nowadays form an SEI. The quality and composition of the layer, however, strongly depend on the chosen solvent combination. The SEI's layer quality decreases according to the reduction potentials of the solvents EC ≫ DMC > EMC > DEC ≫ PC.

EC holds a special position in this respect: It forms a high-quality, i.e., electronically passivating SEI that is still permeable for Li^+. This is due to the resulting bivalent lithium alkylcarbonates $Li_2[OC(O)O-(CH_2)n-O(O)CO]$. Accordingly, between 20 and 50 % of EC are added in almost all currently used battery electrolytes.

PC, on the other hand, does not form a suitable SEI. PC is less susceptible toward reduction because of the inductive effect of the methyl group. Caused by this stability (better by around 200 mV), PC, i.e., the Li^+-PC solvate, intercalates deeply into the graphite and only then disintegrates while forming CO_2. This causes the graphite structure to be expanded and destroyed.

The SEI's structure and characteristics might be influenced significantly by the additives. The basic theory for developing an SEI additive is simple. The SEI additives must be more reactive electrochemically than all other components of the electrolyte. Alternatively, the SEI must be reduced during the first charging cycle of the lithium-ion cell before the solvents. This way, an SEI is formed before the solvents react.

The best-known example of this class of additives is vinylene carbonate. It is used in almost every commercial lithium-ion cell and leads to a significant improvement of the cycling stability. It becomes effective during the first charging and discharging cycles of the lithium-ion cell. Vinylene carbonate is reduced slightly above the Li^+ intercalation in the carbon at potentials of around 1–1.1 V vs. Li/Li^+. A very thin, polymer-like (and therefore flexible) film is formed on the electrode.

Vinylene carbonate alternatives are manifold. Different functionalized organic carbonates, e.g., fluorine ethylene carbonate (FEC) and vinyl ethylene carbonate (VEC), are used. But also SEI-forming conducting salts such as the above-mentioned lithium-bis(oxalato)borate (LiBOB) are employed. All these efforts are still in their infancy (Fig. 6.9).

The usage of LiBOB as an additive is one example for a targeted change of the SEI structure: Different structures are observed in the presence of different solvents that exhibit different reactivities on the electrode surface. The boundary surface of EC-based electrolytes is dominated by EC degradation products. The reason is that these electrolytes have a higher reactivity and therefore a denser structure (Fig. 6.10, left). The boundary surface mainly consists of EC, even when LiBOB is used. This is different for EC-free solutions (Fig. 6.10, right): The boundary surface layer is made of both components. The film formation is controllable as a function of the forming protocol.

Film formation on the positive electrode Boundary surface chemistry on the negative electrode, especially on the carbon electrode, has been the focus in the development of new electrolytes for lithium-ion batteries during the last 20 years.

Now, however, the boundary surface between the electrolyte and the positive 5-V electrode (e.g., high-voltage spinels or $LiCoPO_4$) has moved to the foreground. The goal of this research is to design improved lithium-ion batteries with an even higher energy density and to develop 5-V electrode materials. It can be assumed that there is no electrolyte that is thermodynamically stable at 5 V vs. Li/Li^+. Approaches that exhibit a similar characteristic to the SEI film formation on the negative electrode,

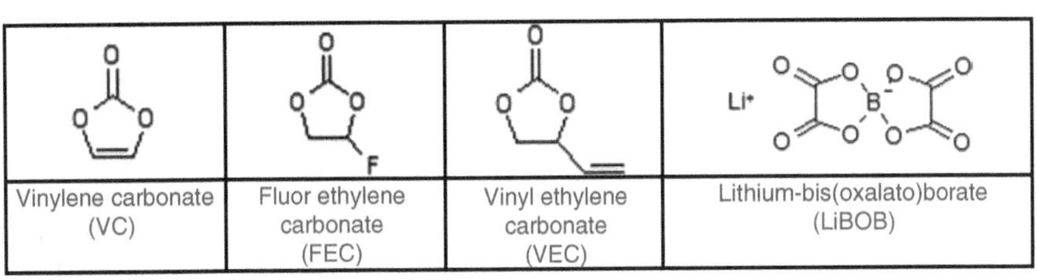

Fig. 6.9 Selection of additives that form SEI films

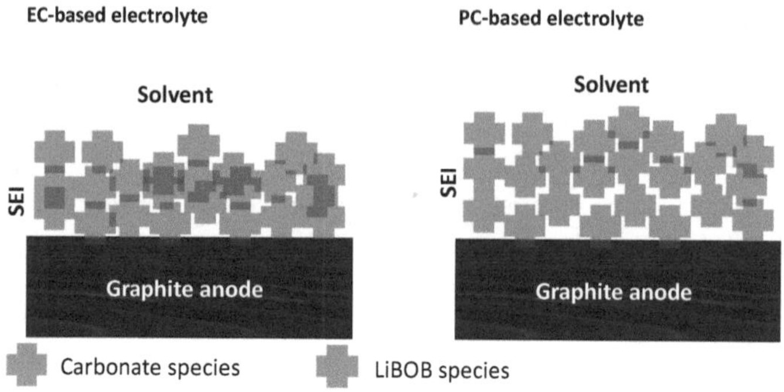

Fig. 6.10 LiBOB influences SEI film formation

i.e., passivation of the positive electrode by an oxidative degradation of special additives, currently seem to be the most promising.

This research is however still in its infancy. Potential film formers in the first publications were special borates and boron oxinates. Other authors focus on additives that are added to the electrolyte in conspicuously low concentrations (< 0.1 %) and that electropolymerize on the electrode at potentials around 4.5 V vs. Li/Li^+. Examples are biphenyl (BP) and its derivatives. These additives might also be employed as overcharging protection in higher concentrations (2 % and above).

6.3.2 Safety and overcharging protection

Not least because of accidents with e-bikes and burning laptops, questions about the future of the lithium-ion battery are intrinsically tied to the safety topic. A significant issue is the behavior of the battery during overcharging. Overcharging a lithium-ion battery might lead to a heating up of a cell that cannot be controlled. This so-called "thermal runaway" might, in the worst case, cause the cell to burst and to ignite.

It is of paramount importance to prevent this safety-relevant event. In order to achieve this, additives such as biphenyl and cyclohexylbenzene are employed in electrolytes as well as constructive approaches, e.g., safety valves and active cooling. Under regular operating conditions of lithium-ion cells, these additives are inert. They are electrochemically activated, however, when the cell voltage reaches a critical value.

Biphenyl for example electropolymerizes above 4.5 V. This polymerization forms an isolating layer on the positive electrode. At the same time, a larger amount of gas is released in a controlled event, which causes the safety valve to open (Fig. 6.11, left). This prevents further, uncontrolled overcharging of the cell. These so-called "shut-down" additives have a disadvantage: The battery is no longer usable after their employment.

Today's research is intensively focusing on finding concepts that do not exhibit this disadvantage of the "shut-down" additives. So-called "redox shuttles" have piqued the curiosity of many researchers (Fig. 6.11, right). Like the "shut-down"

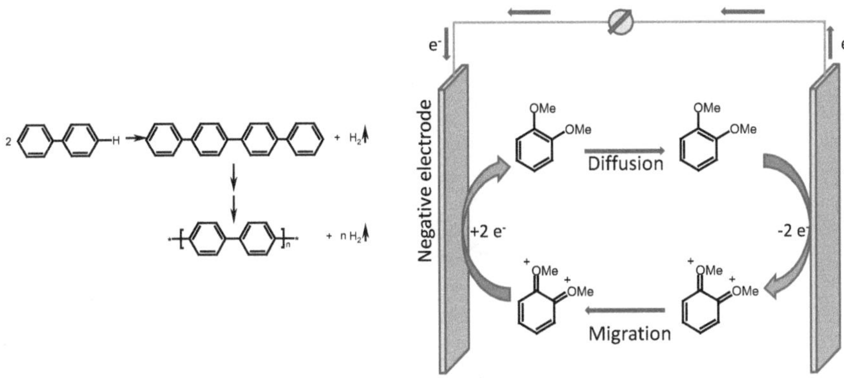

Fig. 6.11 "Shut-down" mechanism (*left*) vs. "redox shuttles" (*right*)

additives, this class of additives is activated at voltages above the charging voltage of the lithium-ion cell.

The name "redox shuttles" says it all: The additive is oxidized at the positive electrode and migrates in the electrical field to the negative electrode, where it is reduced. It then diffuses to the positive electrode and the process starts over. The additive "shuttles" charge through the lithium-ion cell by means of a redox reaction. The functionality of "redox shuttles" such as halogenated dimethoxybenzenes has already been documented, but still needs a large amount of research. Today's redox shuttle systems are not yet quick enough to reliably prevent thermal runaway. In addition to that, redox shuttle additives are limited to usage with 3-V battery systems because of their relatively low "activation voltage." Redox shuttles for 4-V systems play an important role in the current research.

6.4 Gel and polymer electrolytes

Standard battery systems consist of anode, separator, cathode, and liquid electrolyte. Advanced systems use polymer electrolytes that replace the separator and the liquid electrolyte. The most important requirements in this electrolyte class are a conductivity similar to that of liquid electrolytes, mechanical, and electrochemical stability. There are two different types of compounds; one of these is the standard polymer electrolytes that are employed in similar fashion in fuel cells. The other is hybrid materials, which contain additional solvents and gelling additives.

The most common materials are based on polyethyleneoxide (PEO) and are equipped with a correspondent lithium source. The charge carriers are added to the polymer as a salt in the standard, solvent-free systems. The binding strength between the anion and the cation muss be sufficiently low to ensure sufficient conductivity. Targeted anions with a suitable low charge density are, for example, the above-mentioned conducting salts LiTFSI and LiFSI [4].

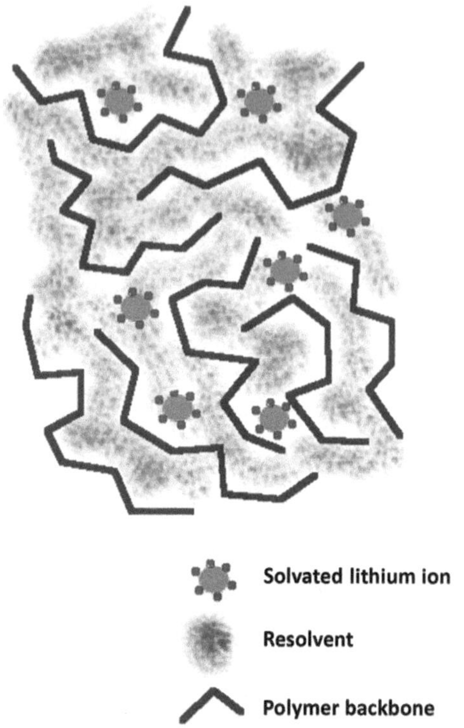

Fig. 6.12 Gel-polymer electrolyte with solvated cations and solvent domains

The ether bridges of the polymer lattice provide for the coordination of the lithium ions in these electrolytes. As a matter of course, the conductivity of these polymer electrolytes increases once the glass transition temperature T_g is reached.

Gel-polymer electrolytes differ from standard polymer electrolytes in their added additives. These enable a lower operating temperature by means of lowering T_g. The polymer matrix serves as an inert lattice for the electrolyte, which is immobilized in this lattice (Fig. 6.12). Thanks to the employment of suitable solvent additives, these electrolytes already reach a conductivity that is sufficient for high-energy applications at temperatures around 30 °C.

Systems with a lithium metal for the anode are the main area of application for polymer electrolytes. An interesting example in this respect is the "Bluecar" of the French Bolloré group: The lithium metal polymer battery (LMP) consists of an anode of lithium metal, a gel-polymer electrolyte, and a standard cathode.

This application shows which requirements the polymer electrolyte must fulfill: The safety of these cells is increased thanks to a great many factors.

First, a polymer layer softens impacts and vibrations better than a comparable hard separator. Second, volume changes during the charging and discharging cycles are better compensated and in general cause less stress on the components.

Third, this class of membranes has almost no permeable paths for the electrolyte solvent because of the polymer structure. This prevents the formation of dendrites.

So this approach is able to answer several safety-relevant questions and to effectively suppress dendrite formation, a typical problem of metal systems.

6.5 Electrolyte formulations – customized and distinct

In general, there is no such thing as "the" best electrolyte. Most of the time, improvements in one area lead to setbacks in another. Therefore, electrolyte solutions must always cater to the requirements of the customer and/or the requirements for the area of application. For example, the requirements for a lithium-ion battery in a cell phone are different in respect to service life than those for a traction battery in an electric vehicle.

The lithium-ion battery is generally chosen in respect to requirements for the lifetime number of cycles, aging, high-current stability, low-temperature and high-temperature characteristics.

Electrolytes for lithium-ion batteries nowadays consist of two to four different solvents. This "mixed solvent" approach for the most part enables the fulfillment of very different requirements, e.g., low viscosity with concurrent high permittivity, by blending solvents with greatly varying physico-chemical characteristics [5]. This is in contrast to conducting salts, where mixing is done only rarely. On the one hand, the number of suitable anions is limited. On the other hand, an advantage of a salt compound has yet to be proven [1].

Ethylene carbonate currently is a solvent component for almost every commercial electrolyte for lithium-ion batteries with a carbon-based negative electrode. The solvent demonstrates very good SEI film formation characteristics and ensures a very effective passivation of the graphite electrode. Furthermore, owing to its high permittivity, it enables the production of high-conductivity electrolytes. Then again, EC has a very high melting point at 36 °C. Electrolytes based solely on 1 M $LiPF_6$ and EC are solid at room temperature.

Typically, standard compounds used nowadays consist of 20 to 50 % EC and open-chain carbonates (DMC, DEC, or EMC) or esters (EA or MB). Depending on their mixing ratio, these compounds exhibit interesting conductivity values, electrochemical stability values, and film formation characteristics between −30 and +80 °C. There are no equivalent guidelines in respect to choice and concentration of additives. Their selection is always custom-made. The producers of electrolyte and cells need to collaborate very closely here.

Often, "trial and error" as well as the customer's requirements dictate the choice of suitable additives and especially influence the determination of the best additive concentration. This arises from the fact that the cell manufacturers have different requirement profiles. But not only that, even their cells are different (cylindrical, prismatic, pouch, hard case) in regard to the electrolyte/electrode layer ratio. "The more the better" does not do it here. Quite the opposite: too much additive is disadvantageous most of the time.

6.6 Outlook

Electrolytes are an essential component of a lithium-ion battery. The interplay of solvent, conducting salt, and the respective additives forms a complex system. This system must be diligently chosen, and its characteristics must be combined in the most efficient way. At the same time, the electrolyte is not an independent component of the cell. It needs to be chosen in dependence on the materials for the anode and the cathode side. That, in turn, calls for close collaboration between the electrolyte manufacturer and the cell and battery developers.

New challenges stem from the new materials of the next-generation batteries, e.g., high-voltage cathodes that demand suitably stable electrolytes not exhibiting a degradation tendency even at such potential levels. In addition to that, these electrolytes should ensure a passivation of the current collectors, which is a challenge on the cathode side at higher potentials.

Alongside all these technological requirements, to realize future applications in electric mobility and stationary energy storage, it must be assured that the necessary numbers can actually be produced.

Bibliography

1. Kang Xu (2004) Nonaqueous liquid electrolytes for lithium-based rechargeable batteries. Chem Rev 104:4303–4417
2. Schmidt M et al (2001) Lithium fluoralkylphosphates: a new class of conducting salts for electrolytes for high energy lithium-ion batteries. J Power Sources 97–98:557–560
3. Aurbach D et al (1995) The study of electrolyte solutions based on ethylene and diethyl carbonates for rechargeable Li batteries. J Electrochem Soc 142:2882–2890
4. Song JY, Wang YY, Wan CC (1999) Review of gel-type polymer electrolytes for lithium-ion batteries. J Power Sources 77:183–197
5. Ue M (2009) Electrolytes: nonaqueous in encyclopedia of electrochemical power sources. In: J.Garche (ed.) Elsevier B.V 71–84

Separators

7

Christoph J. Weber and Michael Roth

Contents

7.1 Introduction

Battery separators are flat materials situated between the positive and negative electrodes of a battery cell. Their function is to prevent physical contact and, therefore, short circuits. At the same time, they must enable ions to be transported as freely as possible within the electrolyte between the electrodes. This is essential for charge equalization and the electrochemical cell to work. To achieve this, separators are usually porous flat designs filled with an electrolyte. The following chapters first set out the basic characteristics of separators and the current status of conventional separator technology. Then, new separator concepts will be outlined and a currently available separator technology and its characteristics will be presented.

C. J. Weber (✉)
Freudenberg Vliesstoffe KG, Hoehnerweg 2-4, 69465 Weinheim, Germany
e-mail: christoph.weber@freudenberg-nw.com

M. Roth
Freudenberg Forschungsdienste KG, Hoehnerweg 2-4, 69465 Weinheim, Germany
e-mail: michael.roth@freudenberg.de

R. Korthauer (ed.), *Lithium-Ion Batteries: Basics and Applications*,
https://doi.org/10.1007/978-3-662-53071-9_7

7.2 Characteristics of separators

Separators must exhibit many mechanical, physical, and chemical characteristics to be able to fulfill their task [7, 22]:

Thickness: The separators of consumer application lithium-ion cells are relatively thin. They have a thickness of less than 25 µm. However, thicker separators (up to 40 µm) are used for the production of large-size lithium-ion cells. These have considerably higher mechanical stability and puncture resistance.

Porosity: Standard separators for lithium-ion cells have a porosity of around 40 %. Controlling the porosity is of great importance in separator manufacturing and immensely influences the porosity of the end product. In addition, high porosity enables a larger electrolyte reservoir. Non-uniform porosity, on the other hand, leads to non-uniform current densities which result in the electrodes aging more rapidly. The standard testing method is described in ASTM D-2873 of the American Society for Testing and Materials (ASTM).

Pore size and pore size distribution: Controlling pore size is important for lithium-ion separators. The pores must be small enough to prevent an electrical connection caused by loose electrode particles. Also, they need to prevent dendrite growth in the lithium-ion cell. Separators with a thickness of < 25 µm are assumed to have an average pore size in the submicron range. The pore size distribution in battery separators must be as homogeneous as possible, as is the case for requirements related to porosity. This enables a uniform current density and thus uniform aging of the cell. Pore size and pore size distribution can be measured by means of mercury porosimetry or with a capillary flow porosimeter [15].

Gurley (air permeability): For a given morphology, membrane air permeability is an indicator of the ionic resistance of the electrolyte in the separator. Once the relation between Gurley and resistance has been determined, it can be used instead of electrical resistance measurements. This is valid only within the narrow limits of the respective separator production process. The standard testing method is described in ASTM-726 (b). Air permeability in accordance with Gurley is defined as follows: the time it takes to press a specific amount of air through a specific surface at a constant pressure. The unit is [s/100 ml]. A low Gurley value stands for high air permeability, low tortuosity, and, as a rule, lower electrical resistance.

Dimensional stability/shrinkage: The shrinkage of a separator is an extremely critical value that immensely influences the safety of cells, especially of large-size lithium-ion cells. It can lead to direct contact between the electrodes and, as a result, to uncontrolled cell failure. A shrinkage test is conducted as follows: A separator's dimensions are measured before storing it at a defined temperature for a defined period of time. Shrinkage is then calculated based on the changes in dimensions with the following formula:

$$\text{Shrinkage [\%]} = \frac{L_v - L_n}{L_v} \times 100$$

L_v is the separator length before and L_n the separator length after storage at a fixed temperature.

Tensile strength and elasticity modulus Tensile strength can be determined by a number of standard testing methods. These tests are executed in machine direction (MD) and cross direction (CD). Testing is performed in accordance with ASTM D88-00. The unit for tensile strength is [N/cm^2] and/or [N/cm]. A separator has to be sufficiently strong and must not constrict in order to fulfill mechanical requirements during the winding process. This is why elasticity modulus at a given tensile stress is an important mechanical characteristic.

Chemical stability The separators must be chemically and electrochemically stable in both the battery and its electrolyte. In particular, the development of high-voltage materials has given rise to new requirements for separators. These characteristics can usually only be reliably determined by a post-mortem analysis because the battery materials, the battery production processes, and the service life conditions are decisive factors.

Wettability and electrolyte absorption Wettability, electrolyte absorption, and permanent storage capacity are important physical characteristics of a separator. Insufficient wetting can negatively influence the service life and the performance of a cell, because it increases the cell's internal resistance and the danger of dendrite formation. Good wettability, however, shortens production times required for filling the cell. It is determined by various factors such as the materials used (surface energy) and the pore structure of the separator. There is no standardized test for quantifying the wettability of a separator. A very good indicator is to observe the speed at which a drop of electrolyte is absorbed when placed on a separator.

Mix penetration strength A very important characteristic of battery separators is their susceptibility to particle punctures (usually electrode particles). This can occur during cell production, but mostly happens during the cycling of a cell. At worst, this can result in the cell discharging in an uncontrolled manner. This characteristic can be determined by the mix penetration strength test. This test procedure is described in [19]. A separator is placed between two commercially produced electrodes of a lithium-ion cell, and this compound is positioned on a polished and hardened stainless steel plate. Then a defined rounded stamp is pressed onto the compound. The pressure is raised continually and the electrical resistance of the compound is measured throughout the process. At a certain pressure, which is dependent on the material, a short circuit will ensue and the resistance suddenly drops. This value is called mix penetration strength.

Puncture resistance Puncture resistance is another physical parameter, similar to the mix penetration strength, for determining the mechanical strength of a separator.

This test measures the force that is needed to penetrate a "free-floating" membrane with a defined needle. The test results correlate more with the elasticity of the material than with the real puncture resistance to electrode particles, because it is executed with a free-floating membrane. According to [22], the mix penetration test is the more informative test for assessing the safety characteristics of a separator in a battery.

7.3 Separator technology

In lithium-ion cells, battery separators are mostly based on polyolefins into which submicron-sized holes are introduced by means of a physical process. These battery separators can be divided into two classes based on their production process: microporous polyolefin membranes and wet membranes.

Resulting from their chemical and physical compositions, these two separator classes exhibit in part extremely different characteristics. The production processes and characteristics of these membranes are presented in the following.

Microporous polyolefin membranes Fig. 7.1 shows the separate steps of the production of microporous polyolefin membranes in the process sequence [2, 22]: For the dry process, a polymer granulate, most of the time polypropylene (PP) or polyethylene (PE), is melted and then melt-extruded to produce a uniaxially oriented film. The morphology and orientation of the resulting film, the so-called precursor film, depend on the process conditions and the characteristics of the polymer melt. The precursor film requires a highly crystalline structure with lamellae arranged in rows. The longitudinal axis of the lamellae must be perpendicular to the machine direction (MD) of the extrusion. Such a structure is essential for producing micropores, because only stacked lamellae can be opened by a drawing process.

The films are subjected to an additional process step in which they are tempered just below the melting point of the used polymers. This further increases crystallinity. The next process step involves cold and then hot drawing. Thus, the required pore structure and porosity of the membranes is achieved. Next, these membranes are in part tempered again to at least partially reduce unwanted tensions within the material. Therefore, the porosity of the membranes depends on the morphology of the precursor film as well as on the tempering and drawing conditions.

Fig. 7.2 shows an SEM picture of a membrane produced in a dry process. The anisotropic slot-shaped pores are easily discernible. They are typical for the dry process.

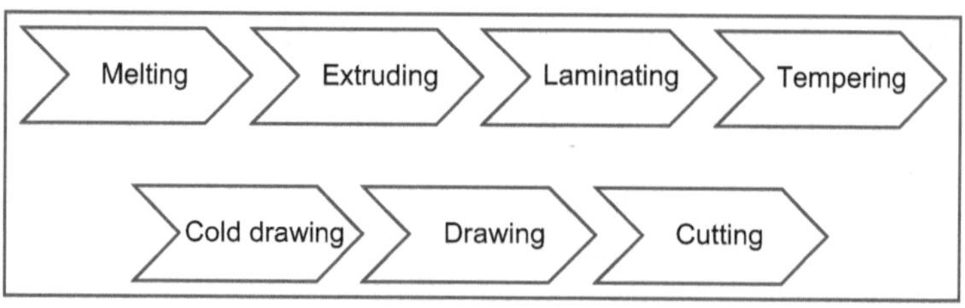

Fig. 7.1 Production processes of microporous polyolefin membranes

Fig. 7.2 Scanning electron microscopic picture of a microporous polyolefin membrane. The sample was metallized with gold to achieve a higher resolution

Microporous polyolefin membranes exhibit a thickness up to 12 to 40 µm, a maximum pore size of < 0.5 µm, and a high tensile strength. The low thickness positively influences the energy density of the battery. The pore distribution protects them against dendrite formation quite well. The tensile strength is beneficial for the production of cylindrical cells [19]. The disadvantages of these membranes are their low porosity (around 40 %), low melting point (around 160 °C for PP), and very high shrinkage (20 % at 120 °C/10 min) at higher temperatures. This is increasingly true for larger cells because they dissipate heat poorly.

If the separator loses its structure, there will be contact between the electrodes and thus a reaction of the electrode materials. This can cause the battery to explode. It is difficult to reduce the high shrinkage of these materials. The reason for this is that the production process, which involves drawing the membranes repeatedly, makes them anisotropic and highly oriented, thus inherently causing high shrinkage, because the materials relax when heated and lose their anisotropic characteristics. The maximum melting point – and therefore the separator meltdown – is limited by the melting temperature of highly isotactic polypropylene (around 160 °C).

Shutdown separator Several of these microporous polyolefin membranes can be laminated (mostly PP/PE/PP) to form a multi-layer structure for the production of shutdown separators [21].

The different melting temperatures of the PP and PE layers are utilized for the shutdown effect. PE has a lower melting point (130 to 135 °C) than PP (165 °C). The PE membrane seals the pores in the PP membrane after fusion (the so-called shutdown). The PP membrane maintains its mechanical integrity during this process. The temperature difference between the employed membranes is around 30 °C, so the shutdown can only function if the temperature increases very slowly in the lithium-ion cell, because it is essential that the mechanical integrity of the PP layer is maintained.

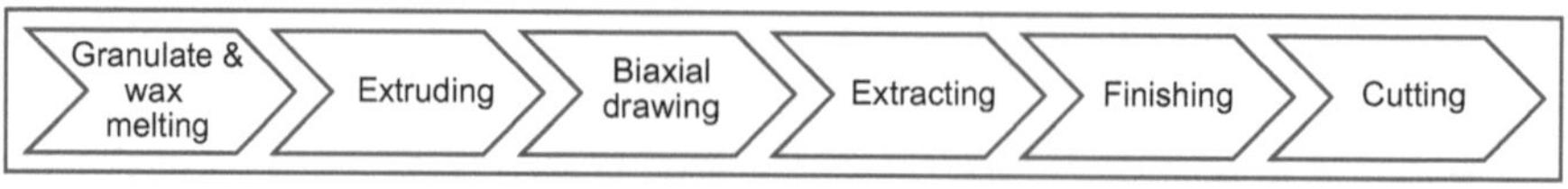

Fig. 7.3 Production processes of wet membranes

Fig. 7.4 Scanning electron microscopic picture of a wet membrane. The samples were metallized with gold to achieve a higher resolution

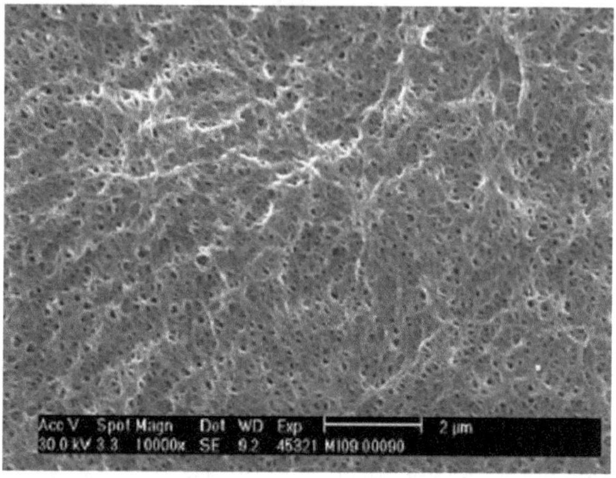

Wet membranes Another production technology for lithium-ion separators is the manufacturing of wet membranes. Fig. 7.3 shows the essential process steps of wet membrane technology. The wet process is as follows: A polymer compound is produced mostly from high-molecular and ultrahigh-molecular polyethylene (HDPE and UHMWPE), low-molecular waxes, and some process additives. This compound, like during the dry process, is then extruded to produce a film. It is advantageous to employ UHMWPE to achieve sufficient mechanical stability. In the next process step, the film is drawn bidirectionally and oriented. Then, the low-molecular waxes are extracted with volatile solvents (mostly dichloromethane) to obtain the required porous structure. This hydrophilization also increases the wettability of the separator in operating mode. Both crystalline and non-crystalline materials can be employed in the wet process. In comparison to microporous polyolefin membranes, the resulting wet membranes have a lower anisotropy of the pore structure and other mechanical characteristics. Fig. 7.4 shows an SEM picture of a wet membrane. The pore structure of the membranes is easily discernible.

Wet membranes have a thickness of up to 25 µm, uniform pore distribution with short distances (< 1 µm), and high tensile strength. The low thickness positively influences the energy density of the battery. The short pore distances effectively prevent dendrites from forming. As with the microporous polyolefin membranes, the tensile strength is beneficial for producing cylindrical cells. The disadvantages of these membranes are their low melting point (around 135 °C for PE) and very high shrinkage (7 to 30 % at 120 °C/10 min).

Separators for lithium-ion batteries are usually produced according to the procedures described above, and they dominate the current market for consumer applications. In 2012, this market had a material volume of 490 million m^2 [8]. Both technologies have about equal shares of this volume.

The most important manufacturers are located in Asia and the USA. No noteworthy amounts of separators are currently being produced with these technologies in Germany or Europe.

7.4 Electric mobility requirement profile of separators

Consumer applications require comparatively small batteries (low number of Ah), in small numbers (low number of units), and with a short service life (low number of years). By contrast, electric mobility requires cells with a capacity greater than 10 Ah, which are installed in a confined space. Cells for cell phone batteries contain a separator area of around 0.05 m^2 (single cell), while cells for electric vehicles require about 200 m^2. The automotive industry has very high quality requirements. In light of electric mobility, the requirement profile of batteries and separators needs to be reconsidered. Safety and reliability are the most important aspects for the development of lithium-ion batteries for HEVs/EVs. Service life, cost, and performance are also important.

A risk analysis of lithium-ion batteries for electric mobility [9] has shown that one of the main reasons for internal short circuits is the accidental introduction of metallic particles during the production process. Changes in the dimensions of electrodes and in their micro structure occur during the cycling of batteries. Changes in temperature also lead to additional mechanical stress of the cell. If there is a conductive particle in the cell wrap, it can thin out the polyolefin membrane over time and eventually penetrate the membrane and cause a short circuit. This, in turn, creates a hot spot in the cell and temperatures can rise to 200 to 300 °C within seconds [3].

Heat transmission becomes more and more difficult with increasingly large battery cells. This is important in respect to the new applications for large-size cells. Separators made of polyethylene (melting point ~135 °C) and/or polypropylene (melting point ~160 °C) melt over large areas under these conditions and enable the electrodes, whose surfaces then touch each other, to discharge in a totally uncontrolled manner.

This risk of internal short circuits has led to very expensive product recalls of laptop batteries which have attracted broad media coverage. Such a short circuit can happen any time during the service life of a battery (accidental overheating, fires, even explosions) [9]. Meticulous processes that employ clean-room technology, local extraction systems, and strict implementation of double-door systems reduce contamination. However, it is not possible to entirely prevent conductive particles from occurring in a cell production environment with conductive powders, cutting, and wrapping processes.

The shut-down function described above is closely connected to the risk of short circuits. The battery cell can slowly heat up internally if outer, quick discharging

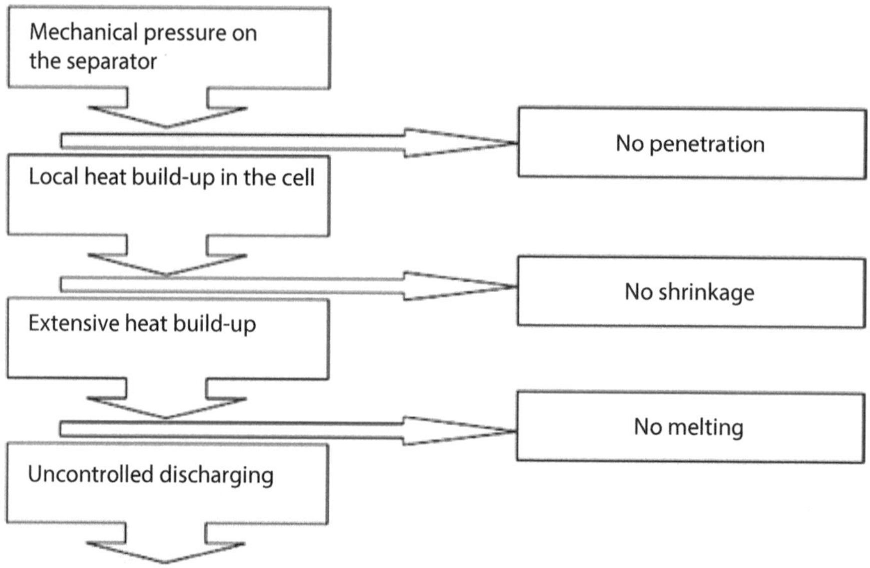

Fig. 7.5 Characteristics of the ideal separator to guarantee maximum protection against internal short circuits

or considerable heating occurs. This might also occur when dendrites form on the anode. If a dendrite grows, it is likely to cause a local increase in ionic currents which results in the release of Joule's heat. If the temperature increase is sufficiently slow, the polyethylene layer melts and the pore structure of the polyethylene collapses as desired. This prevents the two outer polypropylene layers from melting, and the shut-down mechanism has functioned as required. However, if the short circuit is caused by conductive particles, the separator shrinks first, then melts completely, and causes sudden discharging due to the immediate contact of the electrodes [17]. It is therefore at least questionable as to whether the shut-down mechanism described above actually provides additional safety, if the main cause for battery failure is taken into consideration.

Fig. 7.5 shows the typical scenario of a lithium-ion separator failure caused by particle penetration. In order to achieve maximum safety in the worst-case scenario of conductive foreign particles, the ideal separator must be penetration-proof, non-shrinking, and non-melting.

7.5 Alternative separator technologies

The particular requirements for lithium-ion batteries for electric mobility have led to intensive development initiatives by the current membrane manufacturers. The safety improvements, which are required immediately, must under no circumstances negatively influence today's separators' battery properties and battery service life.

Battery manufacturers are currently working on improving today's membranes and also on finding completely new approaches to fulfill the new application profile of electric mobility. The availability of new materials and independent assessment have been insufficient to this day. In the near future, new market actors are expected to emerge who aim at distinguishing themselves from the state-of-the-art technology providers. The essential design features of novel separator technologies will be set out below.

Coated membranes or electrodes One variant of the novel separators involves coating membranes with inorganic particles, which are fixed on the membrane surface by means of fluorinated polymers. This technology is known as the safety reinforced separator (SRS) [11].

Another procedure is quite similar: To prevent an internal short circuit caused by conductive particles, the anode is coated with a very thin layer of inorganic particles which are fixed with an organic binding agent. This additional layer is supposed to prevent the shrinkage of the PO separator which would cause a mechanical short circuit. The separator itself remains unchanged with this technology [6].

Integrated inorganic particles in membranes A promising wet process aims at integrating inorganic particles into the standard process.

This material is marketed under the designation "inorganic blended separator." The possible advantages are higher battery safety through decreased shrinkage and only partial melting. Further expected benefits are an increase in service life thanks to permanent wettability and an increase in power thanks to increased porosity [13].

Nanofiber non-woven fabric separators Another variant of novel separators is non-woven fabrics containing nanofibers. They contain high-melting polyimide fibers with a thickness of 200 to 1,000 nm. These fibers are meant to improve the power and energy density of batteries [5].

Non-woven fabric composite separators Another battery separator manufacturing technology is based on the combination of flexible polymer separator characteristics with the thermal and chemical stability of inorganic particles such as aluminum oxide (Al_2O_3). This technology employs a non-woven fabric that is impregnated with inorganic particles. In comparison to coated membranes, coated non-woven fabrics exhibit very high porosity and very low shrinkage [10, 20]. Fig. 7.6 displays the process chart for the manufacturing of coated non-woven fabrics. This procedure is effective for producing both purely inorganic and organic-inorganic impregnation.

As with the membrane production process, this process begins with the melting of a polymer granulate, from which the fibers are spun and cut. During a second step, an extremely thin and homogeneous non-woven fabric is produced through a wet lay process with subsequent thermal binding. The next process step is impregnation of the non-woven fabric with high-melting inorganic particles and a binding agent. Most of the time, the non-woven fabric consists of thermally stable PET

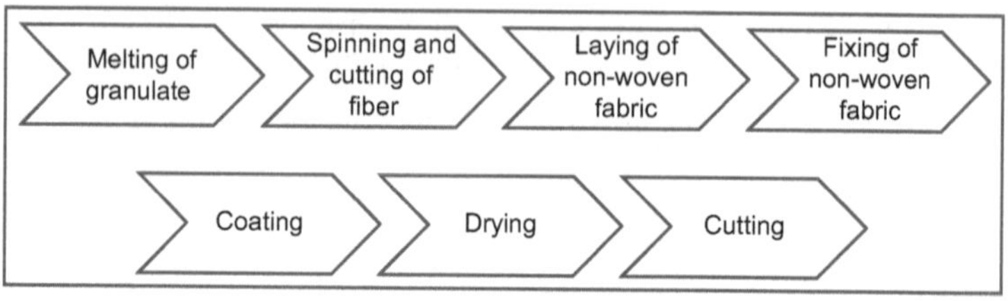

Fig. 7.6 Production process of inorganic composite membranes based on non-woven fabrics

fibers. The adhesive materials are either inorganic binding agents such as silane-based precursors or organic materials. Examples of inorganic particles are aluminum oxide, silicon dioxide, and zirconium dioxide. If silane-based precursors are used as binding agents, the result is a purely inorganic coating.

Research is also focusing on developing a technology for producing separators with organic binding agents. The object is to manufacture "organic-inorganic" composite separators, which are based on non-woven fabric and are very flexible and mechanically stable [18]. Fig. 7.7 shows an SEM picture of such a coated separator.

In response to the critical safety characteristics shown in Fig. 7.5, these separators were subjected to the relevant tests and their characteristics were compared to those of commercially available polyolefin membranes [16]. Both dry and wet membranes were tested.

For the mix penetration test, the separators were placed between commercially available electrodes (cathode: NCM/PVDF, anode: graphite/PVDF). The objective was to simulate the mechanical pressure on a separator (Fig. 7.5) in the experiment. The penetration probe had a Rockwell C hardness of 65, the supporting plate a Rockwell C hardness of 63. Fig. 7.8 shows that both membranes are no longer able to prevent a short circuit at a pressure of around 420 N. The Freudenberg non-woven fabric composite separators, on the other hand, dramatically increase penetration resistance. A short circuit in the compound only occurs from a pressure of 650 N and 730 N, respectively.

Fig. 7.7 Scanning electron microscope picture of coated non-woven fabrics

If the separator is not able to resist the mechanical pressure of a conductive particle, an internal short circuit is triggered, causing the cell to heat up locally. Form test specimens of the separators were punched out to assess their shrinkage characteristics under heat. These specimens were stored for 10 min at 120 °C and for 1 hour at 160 °C, respectively. Fig. 7.9 shows the test specimens after 10-minute storage at 120 °C. The shrinkage of the bidirectionally drawn membrane 1 and the unidirectionally drawn microporous polyolefin membranes 2 and 3 is easily discernible. The voltage, which is static at room temperature, was increasingly released in both cases.

This effect is particularly pronounced after storage at 160 °C. Understandably, the PE wet membrane melted fully. The PP-based microporous polyolefin membranes kept their original shape, but more than 40 % of the originally coated surface was exposed.

The Freudenberg separators do not shrink under these conditions. This is also valid for temperatures above 200 °C (Fig. 7.10).

To simulate the melting of a separator in a lithium-ion cell in accordance with Fig. 7.5, the separators were subjected to a very demanding test [4]: They were loaded with a 420 °C-hot soldering iron tip for 10 seconds on a surface of 0.126 cm^2. In this test, all membranes exhibited the expected characteristics because of their melting temperatures: Holes the size of the soldering iron tip melted into the loaded surfaces. Shrinkage also caused the material in the immediate vicinity of those holes to retract considerably.

The inorganic composite components of the Freudenberg separators do not melt under these conditions. This is why they did not exhibit any holes and the separators did not melt (Fig. 7.11).

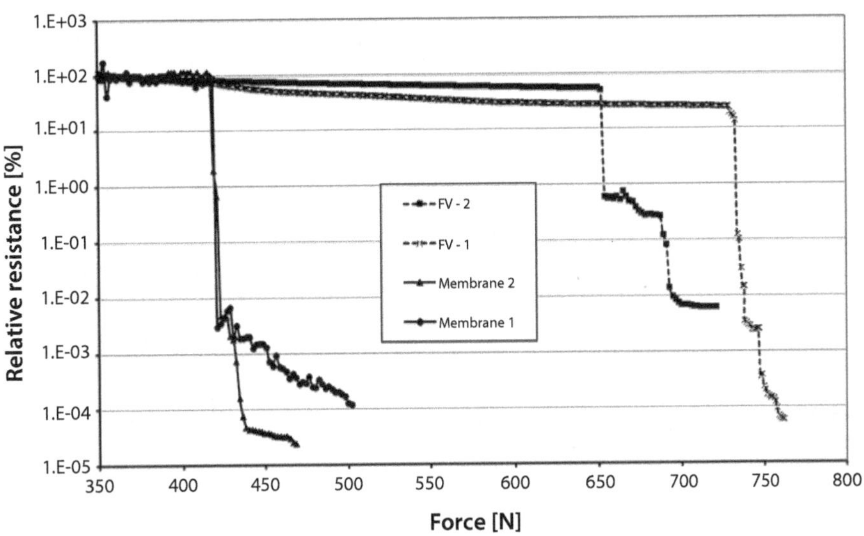

Fig. 7.8 Puncture resistance comparison of separators during the mix penetration test

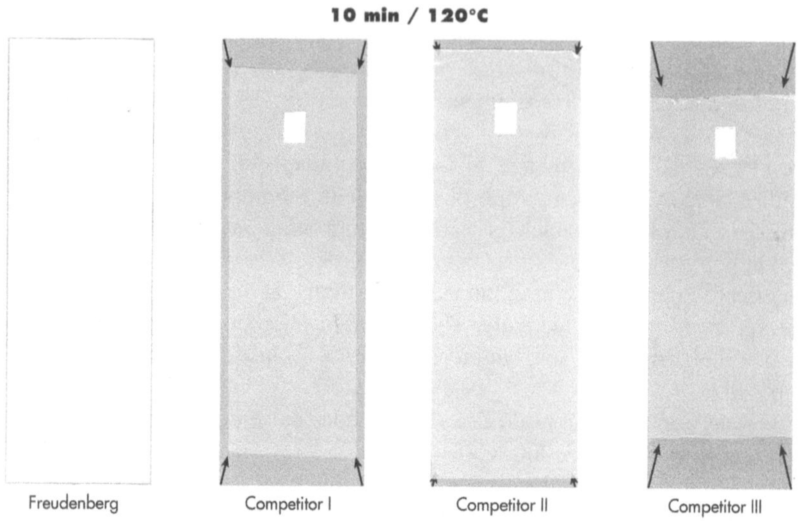

Fig. 7.9 Comparison of thermal shrinkage of lithium-ion separators after 10-minute storage at 120 °C. Membrane 1 is a PE wet membrane, membrane 2 a PP microporous polyolefin membrane, and membrane 3 is a multi-layer PE/PP membrane

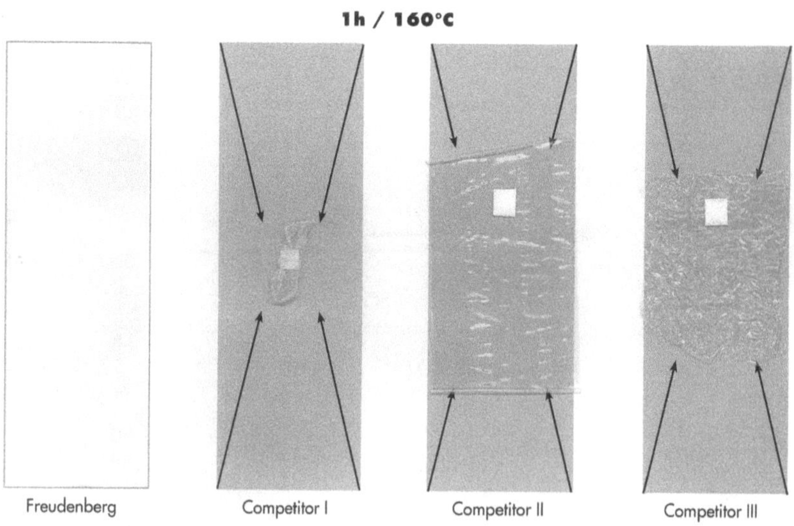

Fig. 7.10 Comparison of thermal shrinkage of lithium-ion separators after 60-minute storage at 160 °C. Membrane 1 is a PE wet membrane, membrane 2 a PP microporous polyolefin membrane, and membrane 3 is a multi-layer PE/PP membrane

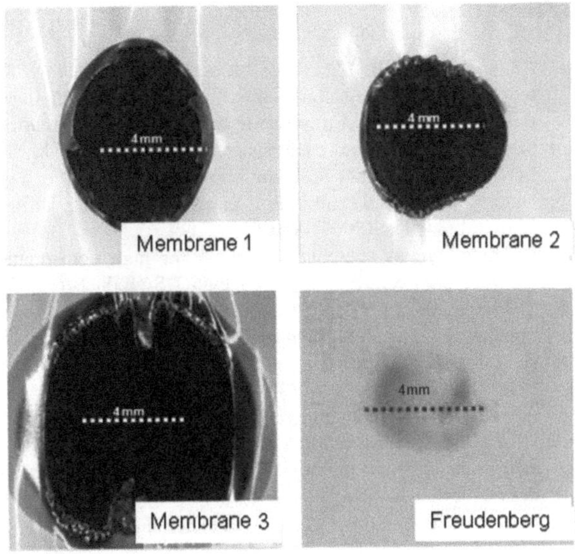

Fig. 7.11 Comparison of thermal characteristics of lithium-ion separators in the soldering iron test

7.6 Outlook

Typical separators for lithium-ion batteries are polyolefin membranes, which dominate the market for consumer applications. Two technologies are used for their production: the dry process and the wet process. The latter includes an extraction step with an organic solvent. Both procedures incorporate a drawing step which enables the manufacturer to specifically control porosity and pore structure.

Polyolefin membranes exhibit limited performance in terms of penetration resistance, shrinkage, and melting behavior due to the choice of material (polyethylene, polypropylene) and the production process (drawing). Employment of inorganic/organic coatings helps improve the shrinking and melting behavior of the membranes. This, in turn, improves battery safety.

A considerable number of companies all over the world are currently intensively developing new technologies to fulfill the safety requirements for large-size lithium-ion cell compounds without compromising on design and material costs. Freudenberg has developed non-woven fabric composite separators based on polyester non-woven fabrics. These have extraordinary thermal-mechanic characteristics that set them apart from the competition. These separators are expected to increase the intrinsic safety of lithium-ion batteries without requiring additional design work.

Bibliography

1. Anderman M (2011) Status of Li-ion battery technology for automotive applications. SAE international vehicle battery summit. Advanced Automotive Batteries, Shanghai
2. Baldwin RS (2009) A review of state-of-the-art separator materials for advanced lithium-based batteries for future aerospace missions. NASA/TM, S 215590
3. Barnett B, Sriramaulu S, Stringfellow R, Singh S, Ofer D, Oh B (2008) 25th International battery seminar and exhibit. Fort Lauderdale, Florida, USA
4. Choi S, LG Chem (2009) AABC
5. DuPont. Dupont energain separators for high performance lithium ion batteries. http://www2.dupont.com/Energy_Storage/en_US/products/products_energain.html. accessed on: 27 March 2012
6. Fujikawa M, Suzuki K, Inoue K, Shimada M (2006) Patent no. US 11396646
7. Ganesh Venugopal JM (1999) Characterization of microporous separators for lithium-ion batteries. J Power Sources 77:34–41
8. Avicenne (2013) The worldwide Battery market 2012–2025. BATTERIES 2013, Nice, France.
9. INERIS – L'Institut National de l'EnviRonnement industriel et des RisqueS (2011) Approche de la maîtrise des risques spécifiques de la filière véhicules électriques – Analyse préliminaire des risques
10. Kritzer P (2006) Nonwoven support material for improved separators in Li-polymer batteries. J Power Sources 161:1335–1340
11. Lee S-Y, Park P-K, Kim J-H (2009) Patent no. WO 2009/066916 A2
12. Opel. Maximaler Einsatz fur Ihre Sicherheit. http://www.opel-ampera.com/index.php/aut/ampera/how_use/safety. accessed on: 27 March 2012
13. Orendorff CJ (2012) The role of separators in Li-ion cell safety. In: The electrochemical society interface summer 2012, pp. 61–65
14. Penth B, Hying C, Hörpel G, Schmidt F G (1998) Patent no. EP 0946270B1
15. Porous Materials. Excerpt from Porous Materials. http://www.pmiapp.com/products/capillary_flow_po rometer.html
16. Roth M, Weber C, Berg M, Geiger S, Hirn K, Waschinski C, et al (2010) Patent no. 2012/019626 WO
17. Roth PE (2009) The 26th international battery seminar and exhibit. Abuse Response of HEV and PHEV materials and cells. Fort Lauderdale, Florida, USA
18. Roth M, Moertel R, Geiger S (2013) A new type of nonwoven separator. In: 5. Internationale Fachtagung Kraftwerkbatterie – Lösungen für automobil und energieversorgung
19. Schell W, Zhang Z (1999) Celgard separators for lithium batteries. In: IEEE (ed.) The 14th annual battery conference. Long Beach, California, pp. 161–169
20. Weber C, Roth M, Kritzer P, Wagner R, Scharfenberger G (2008) Patent no. EP 2 235 766 B1
21. Yu W-C, Geiger M W (1999) Patent no. EP 0715364B1
22. Zhang PA (2004) Battery separators. Chem Rev 104:4419–4462

Lithium-ion battery system design 8

Uwe Koehler

Contents

8.1 Introduction

The design of a battery system should ensure that an energy storage system operates efficiently, reliably, and safely during vehicle deployment for a very long period of time. Lithium-ion cells are the fundamental components of lithium-ion battery systems and they impose special requirements on battery design. Aside from electrochemical storage cells, the battery system comprises a multitude of mechanical, electrical, and electronic components with functions that need to be perfectly balanced. The electronic battery management system (BMS) not only monitors and controls the battery, it also provides data communication to the vehicle.

Battery system design and configuration take into account the specific technical characteristics of the lithium-ion cells in which the energy is stored. Suitable electrical and thermal management ensures that the storage cells permanently operate

U. Koehler (✉)
Conwitex GmbH, Am Leineufer 51, 30419 Hannover, Germany
e-mail: uwe.koehler@conwitex.com

R. Korthauer (ed.), *Lithium-Ion Batteries: Basics and Applications*,
https://doi.org/10.1007/978-3-662-53071-9_8

safely and reliably. Lithium-ion cells are sensitive to mishandling. This is the reason why the battery management system must be very reliable.

8.2 Battery system design

8.2.1 Block and modular design

Batteries can be designed in two different ways:with a block design or a modular design.

Block design: All storage components are combined to create a single block with an electrical collector structure, sensors, and other components. This block is then fitted with the necessary connections and peripheral components and is installed in the battery housing.

Modular design: A certain number of individual cells are combined to form a module, which constitutes a sub unit. These sub units are then used to build larger battery units. The advantage of the latter is that the components can be handled more easily during assembly and maintenance (replaceability).

Fig. 8.1 shows a typical block design example, the lithium-ion battery of the Daimler S-Class hybrid. Here, the cylindrical lithium-ion cells are connected to form a 35-cell block, which is the heart of the battery system. In Fig. 8.2, the battery of the Ford Transit Connect Electric is shown, an example of a modular design. This system consists of 18 individual modules, each containing 12 lithium-ion cells. Block designs are usually used for relatively small battery systems while larger batteries have modular designs, for the reasons mentioned above.

Fig. 8.1 Lithium-ion battery system with a block design, Daimler S400 hybrid (courtesy of Daimler AG)

Fig. 8.2 Lithium-ion battery system with a modular design (Ford E-Transit-Connect)

8.2.2 Serial and parallel circuits

The most simply designed battery systems consist of a certain number of cells connected in parallel. The battery voltage is the sum of the single cell voltages. The cell voltages depend on the specific electrochemical characteristics of the system they are based on and their specific electrode combination. For today's systems, they range from 2.2 V to 4.2 V per cell. It must be guaranteed that all lithium-ion cells comply with the upper and lower cell voltage limits. This can result in a high system outlay, especially for electrochemical systems that have lower cell voltages and, therefore, higher numbers of individual cells. Voltage and system characteristics also enable a parallel connection of lithium-ion cells. Series and parallel cell connections open up possibilities for different battery system designs.

Parallel connection of cells connected in series Here, two or more strings of cells connected in series are connected in parallel. The outlay of such a design with several strings is relatively high since the voltage of each individual cell must be monitored.

Additional outlay is caused by the connection of the individual strings to an entire system, requiring an additional higher-level battery management system to coordinate the individual strings function.

Serial connection of cells connected in parallel For this configuration, cells with the same design are first connected in parallel. This results in a higher capacity, depending on the cell quantity, for the same voltage. These cell packs are then connected in series. The circuitry permanently electrically connects the cells and therefore it suffices to monitor their overall voltage.

If voltage variations occur during operation or idling, an automatic charge exchange will take place, balancing the voltage and the state of charge (SOC).

This serial connection of cells connected in parallel is usually very simple and cost-efficient because the monitoring outlay is lower. However, it is usually used for cells with a relatively low capacity only.

8.3 Functional levels of battery systems

The basic requirements for a battery system and its management can be divided into four functional levels.

Mechanical integration This involves mechanically and purposefully integrating the individual components into a battery assembly. Designing the individual components and their connection ensures that the battery assembly fulfills the mechanical requirements over the entire service life of a vehicle without compromising on functionality and safety.

Electrical management Electrical management ensures the battery system's electrical functionality in all situations occurring during vehicle operation. This includes providing the electrical power required for the vehicle drive during operation and managing the charging procedure during external charging as well as regenerative reloading during vehicle operation, displaying safety-relevant statuses such as faulty electrical isolation, short-circuits, overheating, overcharging, deep discharge, and triggering the appropriate reactions to these statuses.

Thermal management Lithium-ion cells are based on electrochemical components. This makes their performance characteristics and service life extremely dependent on the ambient temperature. Their restricted current capability applies to both charging and discharging characteristics. Discharging performance is considerably reduced at lower temperatures. This is caused by the specific electrochemical cell kinetics, which increases internal cell resistance, but also reduces discharging capacity, especially for higher currents. Charging currents must be limited for lower temperatures because of increasing electrical internal resistance and restricted charge-accepting capability of the negative electrode. If the maximum permitted charging current is exceeded, metallic lithium can plate on the negative electrode's surface. This lithium plating directly impairs cell capacity and performance. Thus it is important that temperature and charging currents are reliably monitored.

In general, lithium-ion cell aging depends very strongly on the temperature. It is therefore necessary to always keep the battery in an appropriate temperature range. The thermal management system design must guarantee that the resulting heat losses are dissipated efficiently.

Communication with the vehicle The battery of an electric vehicle is one of the most important system components and is therefore directly integrated into the electrical vehicle environment. This requires permanent data exchange between the battery system and the vehicle. These data are of paramount importance for the vehicle's operation and the battery. Current data are exchanged on the state of charge, electrical performance, current-accepting capability, and internal resistance. Safety-relevant signals that guarantee that the overall system is safe in case of malfunctions are also essential.

8.4 System architecture

Fig. 8.3 displays the fundamental system architecture of a lithium-ion battery system. The most important system components and their functions are described in detail.

Cell block with electrochemical storage elements The cell block comprises the electrochemical storage cells; they are the battery system's core components. The battery system design's task is to ensure that they function optimally. The cells are connected by means of the electrical collector system. All thermal management components are usually integrated into the cell block. In air-cooling systems, these are the air ducts that are customized to meet the relevant requirements. Incoming

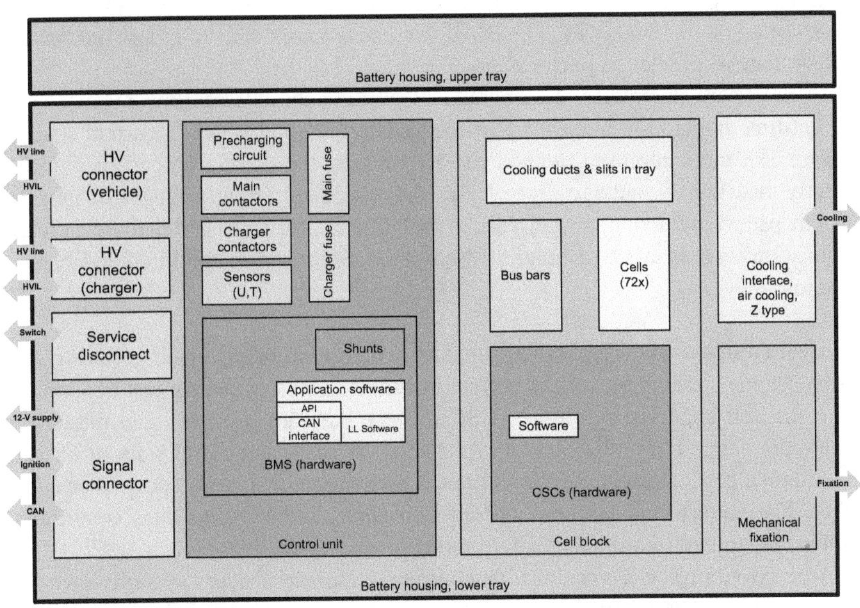

Fig. 8.3 Lithium-ion battery system architecture

air is guided across the cell surfaces, taking on heat that is then dissipated through air outlet ducts.

Liquid-cooled systems work differently: Chillers with a cooling medium flowing through them are installed in close mechanical contact with the cells. This cooling medium circuit takes on the cells' heat and dissipates it into the environment by means of an external chiller. Generally, the cooling medium is a mixture of water and Glycol®. More advanced chiller systems are also implemented; they use a technical refrigerant provided by the air conditioning system's compressor.

Monitoring components The battery monitoring system comprises the components that are necessary for monitoring the battery, its subcomponents, and components in the vehicle.

The monitoring components are voltage sensors which measure the cells' and modules' voltages. Other components are the temperature sensors, which measure the temperatures at characteristic points in the module and cooling system, and the current sensor, which registers the current flowing through the battery.

Cell voltages and temperatures are monitored by the cell supervisory circuit (CSC). This component's design, location, and functionality vary depending on the manufacturer. Figs. 8.4 and 8.5 display Johnson Controls lithium-ion modules, in which the cell supervisory circuit is directly integrated into the individual modules.

In lithium-ion systems, overvoltages and undervoltages pose safety hazards and can reduce the system's service life. This is why the voltage of each individual cell needs to be monitored. The cell supervisory circuit (CSC) transmits the measurement data to the battery management system or battery monitoring unit (BMU) by means of an internal data bus. The BMU uses these parameters to assess the current battery state to predict its performance.

An important parameter is amperage. Recording the charge flow by means of current-time integration makes it possible to determine the battery's current state of charge. The most common method for determining the current amperage is to indirectly measure the voltage drop at a precision resistor (shunt) integrated into the current path. Additional sensors can be installed in the battery, depending on the application. For example, a humidity sensor is often used in liquid-cooled battery systems.

Control components The battery management system or battery monitoring unit (BMU) is the core component of active control. It processes the signals transmitted from the cell supervisory circuit (CSC). These data are used to determine electrical parameters that are important for operation, such as current state of charge, maximum electrical discharge performance, and maximum charge acceptance capability. The battery management system also controls the main relays (main contactors) integrated into the battery's circuit.

Overcurrent protection is a passive control component. Battery systems have one or more fuses, which can disconnect the circuit in situations in which the battery

Fig. 8.4 High-performance lithium-ion module comprising cylindrical cells with integrated cell supervisory circuit and cooling system (Johnson Controls)

management system can no longer control a current that exceeds predetermined limits.

Interfaces Interfaces with the vehicle are those devices and components that are used to transmit electrical power, connect with the vehicle's coolant supply system, and exchange operational data. The electrical system comprises high-voltage connectors, by means of which the battery is charged and discharged, as well as the respective data transmission devices. An additional safety measure is the electrical service disconnect accessible from the outside of the vehicle. It is often used for service purposes and enables persons to disconnect the battery in a simple manner.

The interfaces play an important role in terms of data transmission. They enable the battery system to communicate with the vehicle's electrical control system. The data type and volume depend on the vehicle. Information on the state of charge, performance capability, and operating temperature is typically transmitted from the battery to the vehicle. Data needed for safety purposes are also included, such as data on electrical isolation resistance. Vice versa, the vehicle transmits information to the battery system that is important for its operation. It includes signals for

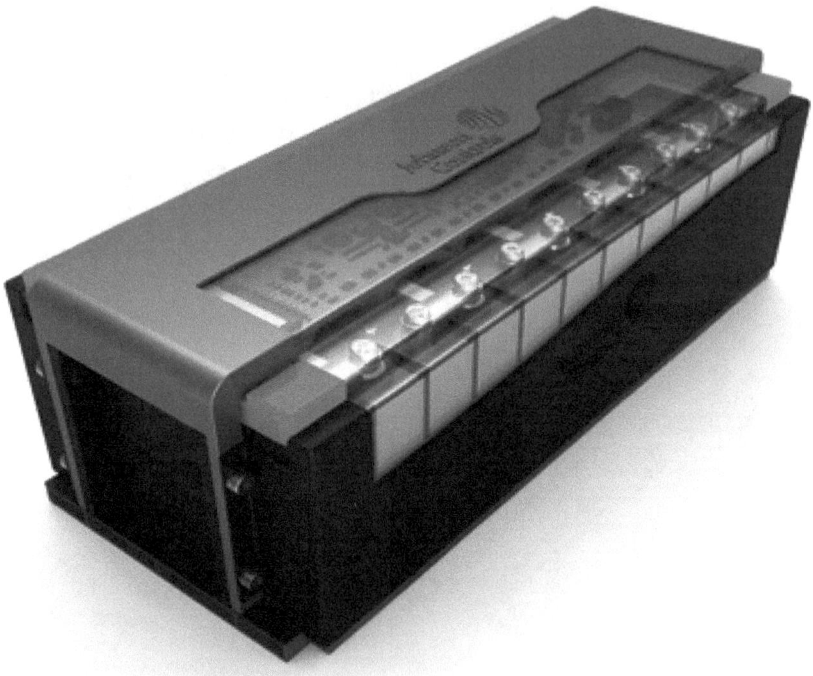

Fig. 8.5 Lithium-ion high-energy module comprising prismatic cells with integrated cell supervisory circuit and cooling system (Johnson Controls)

switching the battery on and off as well as signals that ensure immediate interruption of operation in an emergency. When such an emergency is signaled by the crash sensor, for example, the battery is switched off at once.

The cooling medium connection of the battery housing is the mechanical interface with the vehicle's cooling medium supply system.

Battery housing and fixation system The active and passive battery system components are installed in the battery housing, which therefore plays an important role in regard to functionality, safety, and the service life of the energy storage system. It protects the sensitive components from harmful environmental influences such as water, humidity, and dust, and is essential for the battery's long-term safe and reliable operation. Usually, traction batteries are installed outside of the passenger area, mainly in the vehicle's floor. Thus, the housing is exposed to extreme environmental influences such as temperature, humidity, spray, salt spray, dust, and stone-chipping. This means that considerable mechanical stability and corrosion resistance even under extreme circumstances are of great importance.

These environmental influences mainly concern the housing surface and its tightness. During driving, however, the internal and external fixation systems are subject to considerable mechanical and thermal loads. The overall system therefore has to be designed and configured in such a way that the battery system can operate without malfunctions throughout its entire service life. For this purpose, car manufacturer have drawn up mechanical load profiles (vibration, shock) in order to determine the system's mechanical stability, which undergoes extensive testing.

8.5 Electrical control architecture

Fig. 8.6 displays a modular lithium-ion system's basic electrical control architecture. As mentioned above, the cell block consists of the cells and their respective monitoring components. The modules are electrically connected with the power connector system by the main relays. When the main relays are closed, the connection with the vehicle's electrical drive system is established.

An internal data bus (private network) is used to exchange data between the modules and their measurement and control components on the one hand and the battery management system or battery monitoring unit (BMU) on the other hand. The vehicle's 12-V on-board electrical system provides power for the battery control system. It is independent of the lithium-ion traction battery.

The battery management system actively controls the following subsystems of the battery:

Main relays: The BMU operates the main relays that provide the connection to the high-voltage connectors. For safety reasons, a high-voltage battery has two main relays. One of them is arranged between the positive terminal

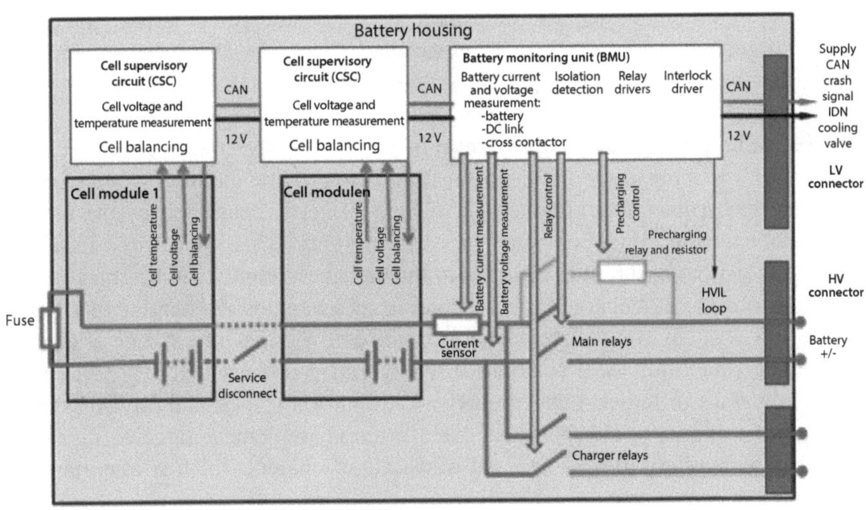

Fig. 8.6 Electrical architecture of a modular lithium-ion battery system

of the cell assembly and the positive external connector. The other one is located between the negative terminal of the battery pack and the negative battery terminal. When the battery voltage is switched onto the power connector, only one relay is switched on. The other one is initially precharged to operating voltage level. This technology prevents the main relays from closing under full system voltage because this would negatively influence their service life.

The actual switching routines (on and off) are customized to the technical parameters based on the vehicle's requirements and the battery system's possibilities. The battery system is only switched on when it is technically in perfect condition and when there is no malfunction. In the event of malfunction such as faulty electrical isolation, short-circuits, or defective cells, this switching process is blocked. Depending on the malfunction type, the switching-off processes can be immediate or gradual.

High-voltage interlock circuits (HVIL) are a very important safety measure for both the battery system and the entire vehicle. They immediately and automatically open the main relays and disconnect the battery terminals if the high-voltage system in the battery or in the vehicle is interrupted, for example because of a loose connection or a broken cable.

Thermal management system: The temperature data recorded by the battery management are used to control the cooling system. In a liquid cooling based system, these data are required to control the flow rate and the temperature of the cooland. In an air-cooled system, on the other hand, the power of the cooling fan and the airflow rate can be used to control the temperature. The heat is dissipated in a cooling liquid-based system by a chiller, which is installed outside of the battery or, if a technical refrigerant is used, by means of the air conditioning system compressor.

The liquid circuit can also be used to heat the battery, if it is configured and designed accordingly. This is especially advantageous for lower temperatures because the battery reaches its perfect operating temperature much more quickly.

Charge balancing electronic system: In Section 8.4, the cell supervisory circuit (CSC) was presented. Its main function is to provide monitoring technology, but it also has an additional active function. It is controlled by the battery management system or battery monitoring unit (BMU) and wired in a specific way to ensure that the cells in the battery assembly can be charged to a uniform state of charge. This is achieved by a targeted discharging of all cells of the system to that state of charge of the cell with the lowest state of charge. Over time, individual cells self-discharge differently, which is generally the reason for different states of charge. Lithium-ion cells usually self-discharge only minimally. Hence, most manufacturers implement targeted discharging technology to achieve a uniform state of charge, which is important for perfect operation.

8.6 Electric vehicle geometrical installation and operation

Designing and constructing a complete battery system is determined by the individual components' specific technical requirements but, most of all, by the vehicle itself. Space requirements play an important role in this respect. It is essential to fully utilize available space and to integrate the battery into the vehicle so that it requires as little space as possible and can still be fitted into the vehicles' safety architecture without any problems.

Fig. 8.7 shows how this can be achieved with the example of the Opel Ampera battery system. The battery was designed in such a way that all available space in the vehicle was utilized completely. Here, modular technology is used which involves connecting a certain number of prismatic lithium-ion cells (pouch cells) to modules. These subsystems are the basis for constructing the overall system described above, which is then integrated into the vehicle together with the other components.

Fig. 8.7 Opel Ampera lithium-ion battery system (courtesy of Adam Opel AG)

Bibliography

1. Westbrook MH (2001) The Electric Car, IEE Power and Energy Series No. 38. London
2. Naunin D (ed.) (2004) Hybrid-, Batterie- und Brennstoffzellen-Elektrofahrzeuge. Expert-Verlag, Renningen
3. Park J-K (ed.) (2012) Principles and applications of lithium secondary batteries. Wiley VCH, Weinheim
4. Aurbach D (ed.) (1999) Nonaqueous electrochemistry. Marcel Dekker Inc., New York
5. Wakihara M, Yamamoto O (ed.) (1998) Lithium-ion batteries – fundamentals and performance. Wiley-VCH, Weinheim
6. van Schalkwijk WA, Scrosati B (2002) Advances in lithium-ion batteries, Kluwer Academic/ Plenum Publisher, New York
7. Bergvald HJ, Kruijt WS, Notten PHL (2002) Battery management systems – design by modelling, Vol. 1. Kluwer Academic Publishers, Dordrecht
8. Jefferson CM Barnard RH (2002) WIT-Press, Southampton

Lithium-ion cell

9

Thomas Woehrle

Contents

9.1 Introduction

Lithium-ion technology has become indispensable in everyday life. A lot of devices are powered by lithium-ion cells nowadays. The following will discuss history, cell materials, cell electrodes, cell designs, market overview, applications, technology, requirements, trends, and further reading.

The development of the lithium-ion technology has been complex and interdisciplinary. Not only are electrochemistry and materials science of importance in terms of developing cells. Other areas in which know-how is essential are solid state chemistry (structure and operating principle of the active materials), organometallic chemistry (lithiated carbons and graphites), and chemical engineering (mixing processes, upscales, and rheology). Knowledge of all the above is necessary to develop

T. Woehrle (✉)
BMW AG, Petuelring 130, 80788 Munich, Germany
e-mail: thomas.woehrle@bmw.de

R. Korthauer (ed.), *Lithium-Ion Batteries: Basics and Applications*,
https://doi.org/10.1007/978-3-662-53071-9_9

a lithium-ion cell, to employ repeated production processes, and to introduce it on the market as a mass product.

9.2 History of battery systems

It was around 1800 when the Italian scientist Alessandro Volta described a battery for the first time. Around 1860, the lead-acid battery was invented and introduced on the market in 1880 with advanced technology. Fig. 9.1 shows the development of different batteries that are still or have been on the market for a long time. In comparison, lithium technology is still quite young. The first mass-produced primary lithium batteries entered the market in the mid 1960's. Michel Armand, Robert Huggins, and Stanley Whittingham were among the renowned scientists who researched intercalation electrodes based on lithium ions in the 1970's. It was Whittingham who suggested the combination of lithium with titanium sulfide (TiS_2) for the intercalation electrode. Intercalation is the inclusion of ions or smaller molecules in layer or channel structures. In 1990, the introduction of rechargeable lithium-metal batteries failed because of safety issues. One year later, Sony introduced the so-called lithium-ion cell with intercalation concept. The name lithium-ion had been chosen on purpose, since no metallic lithium is employed for the anode. Basis for this technology were the patent of John Goodenough (cathode) [1] and Sony's proprietary know-how about carbon electrodes. Lithium cobalt oxide ($LiCoO_2$) for the cathode and amorphous carbon for the anode were employed in Sony's lithium-ion cell of that time. The standard application of this cell was consumer electronics (video cameras etc.). The cell was a cylindrical cell with solid housing.

History of the battery Fig. 9.1 shows further milestones that were important for the development of the battery. Many of the battery systems mentioned stayed on the market many years or still are used today: lead-acid starter battery, primary zinc-carbon battery (e.g., cylindrical AA or AAA batteries in flashlights), alkaline manganese round cells (cameras, remote controls), and nickel metal hydride batteries (razors, telephones).

Lithium-ion technology is based on the intercalation principle, which is shown in Fig. 9.2. In this case, as opposed to a lead-acid battery, the electrolyte is not involved in a chemical reaction. The lithium ions intercalate and deintercalate reversibly into and out of the respective host lattices. This is known as the "rocking chair" principle. According to this, lithium-ion cells exhibit intercalation materials on both the cathode and the anode side.

Fig. 9.2 shows the lithium intercalation principle for charging and discharging the cell (example: graphite on the anode side, lithium cobalt oxide [$LiCoO_2$] on the cathode side). Lithium hexafluorophosphate is used as the so-called lithium conducting salt. It is dissolved in organic carbonates. When the cell is charged, the lithium

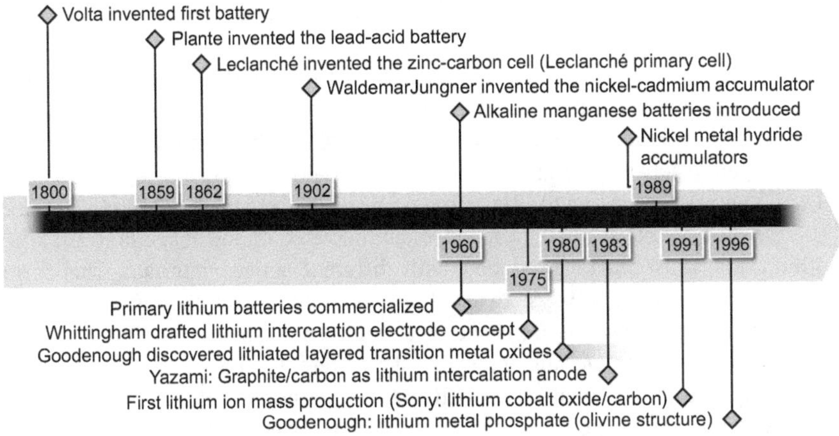

Fig. 9.1 Overview of battery development milestones

ions migrate out of the lithium cobalt oxide, through the separator (wetted with the electrolyte), into the graphite layer. The opposite happens during discharging.

There are several reasons for the success of the lithium-ion technology. Compared to sodium or potassium ions, the small lithium ion exhibits a significantly quicker kinetics in the different oxidic cathode materials. Another difference: As opposed to other alkaline metals, lithium ions can intercalate and deintercalate reversibly in graphite and silicon. Furthermore, a lithiated graphite electrode enables very high cell voltages.

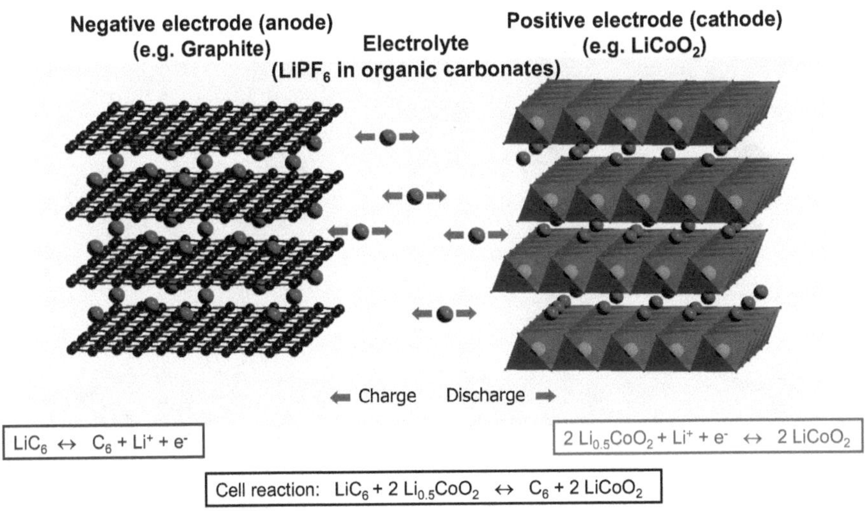

Fig. 9.2 Intercalation concept for a lithium-ion cell with reversible lithium ion migration

9.3 Active cell materials for lithium-ion cells

The outer cell voltage of a lithium-ion cell arises from the difference between the potentials of the positive and the negative electrode. Aside from the above-mentioned system with $LiCoO_2$ and carbon, there is a multitude of other combinations of positive and negative active materials.

Fig. 9.3 shows (according to [2]) an overview of the reference potentials (vs. Li/Li$^+$) on the ordinate. The specific capacities are depicted on the abscissa in Ah/kg.

Lithium-ion cells can be produced with different active materials. This is in contrast to conventional cell systems such as lead-acid, nickel-cadmium, or nickel metal hydride. These systems are based on a single chemistry.

The different active materials determine the different voltages of the respective cells. Example: Lithium cobalt oxide ($LiCoO_2$) vs. carbon enables a higher voltage than lithium iron phosphate ($LiFePO_4$) vs. lithium titanate ($Li_4Ti_5O_{12}$).

The energy of a cell is the product of its nominal capacity and the mean discharge voltage (nominal voltage). The unit is Wh. The specific energy and the energy density are important parameters for batteries. They are a measure of the useful electric power relating to mass (Wh/kg) or volume (Wh/l), respectively. Fig. 9.3 shows that it is possible to increase the energy density by employing materials that enable higher cell voltages or demonstrate higher specific capacities.

A lithium-ion cell consists of lithiated transition metal oxide (such as $LiCoO_2$) and graphite. This has the advantage that all active materials are processed into electrodes in a stable state. Right after the production the cell is uncharged

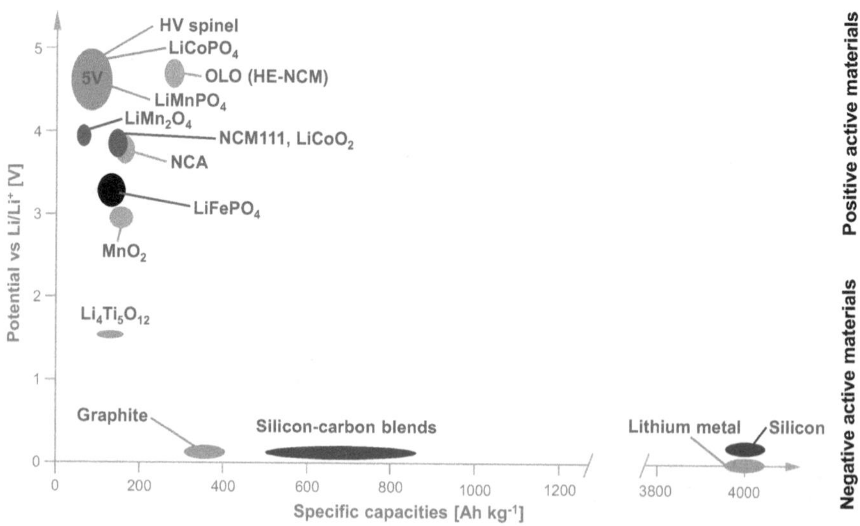

Fig. 9.3 Potentials and specific capacities of positive and negative active materials for lithium-ion cells

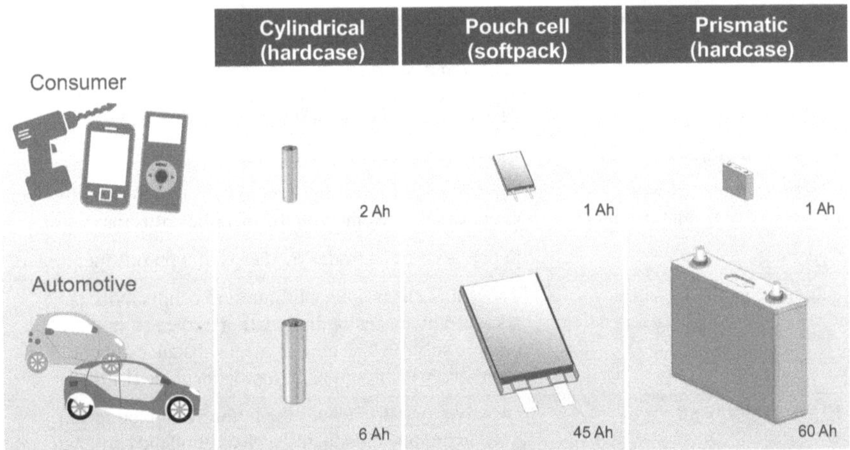

Fig. 9.4 State-of-the-art lithium-ion cell housing and packaging types

(0 V) and has to be converted into a ready-for-use state. This is achieved by the so-called formation, the initial charging of the lithium-ion cell. For graphites and carbons, this procedure is special. The first charging entails a protective layer (solid electrolyte interface, SEI) forming around the particle surface. Part of the lithium is "consumed" during this process and is no longer available for further cycling [3].

9.4 Passive cell materials for lithium-ion cells

The employment of so-called passive materials ("dead material") is essential to ensure the full function of a lithium-ion battery. Table 9.1 contains a list of these materials with a short function description.

9.5 Housing and types of packaging

All of today's lithium-ion cells exhibit metal-based housing and packaging material. One reason is that it prevents the entry of moisture into the cell, which would initiate a hydrolysis of the conducting salt $LiPF_6$ into hydrogen fluoride (HF). Another is the prevention of loss of solvent by means of diffusion from the cell. Only metal is able to fulfill these tasks. A housing made of pure plastics is not suitable because no plastic (not even polypropylene) is completely moisture leaktight. And none of them are diffusion resistant toward all organic solvents.

The current standard housing and packaging types for lithium-ion cells are depicted in Fig. 9.4.

Table 9.1 Examples of passive materials in a lithium-ion cell

Material	Function description
Carbon black	Electrical conducting additive in the electrodes
Conductive graphite	Electrical conducting additive in the electrodes
Electrode binder (dissolvable in organic solvents and in water)	Binds the active materials and the conducting additives as well as the electrode with the metallic collector
Separator	Separates the electrodes by means of a porous membrane
Lithium conducting salt	The actual electrolyte; a lithium-ion conductivity is achieved by means of dissociation of the salt in the organic solvent. The employment substance almost exclusively is lithium hexafluorophosphate ($LiPF_6$).
Electrolyte solvent	Dissolves and dissociates the lithium conducting salt. Mainly organic carbonates such as ethylene carbonate (EC) or diethyl carbonate (DEC) are used.
Aluminum collector	Dissipates the electrons from the cathode (positive electrode)
Copper collector	Dissipates the electrons from the anode (negative electrode)
Material of housing and packaging	Prevents entry of moisture into the cell and exit of solvent from the cell

The solid metallic housings (hardcase) are typically made of aluminum or stainless steel. The high-grade aluminum composite films of the pouch are made of several layers; an example for a standard film sequence for consumer cells is as follows: polyamide (25 μm), rolled aluminum (40 μm), polypropylene (50 μm). The single layers are laminated.

The introduction on the market of a cylindrical round cell in a stainless steel housing for consumer applications by Sony ensued in 1991. The round cell 18650 is a standard type and widely used cell on the market. This is why it can be produced more cost-efficient than other cell types. It is especially used in laptops, eBikes, and power tools. Fig. 9.5 shows its cylindrical shape and the outer dimensions.

9.6 Worldwide market shares of lithium-ion cell manufacturers

Until today, lithium-ion cells are manufactured almost exclusively in Asia. Fig. 9.6 shows the worldwide market shares of the most important lithium-ion cell manufacturers. These numbers contain the cylindrical and prismatic cells, hardcase, and pouch types.

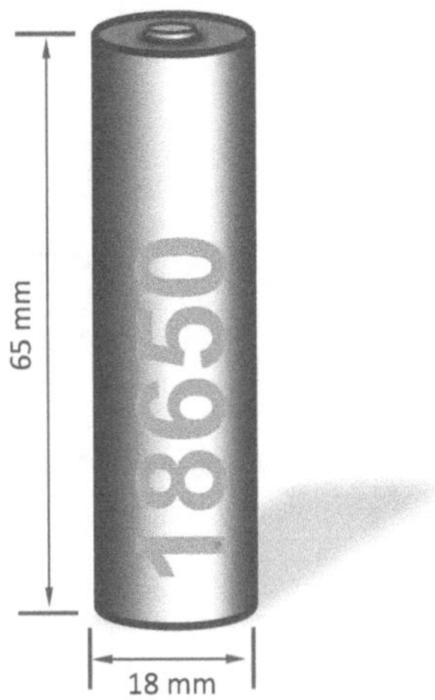

Fig. 9.5 Cylindrical round cell type 18650 (18 is equal to the diameter in mm; 650 equal to 0.1 mm units, thus the length is 65 mm)

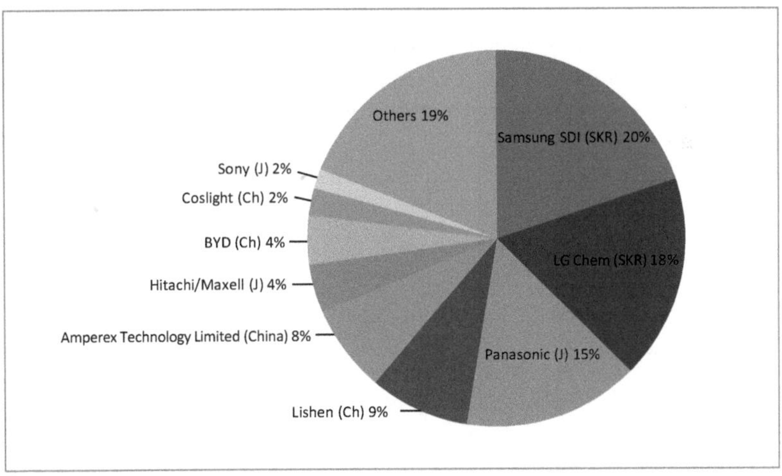

Fig. 9.6 Worldwide market shares of rechargeable lithium-ion cell manufacturers [4]

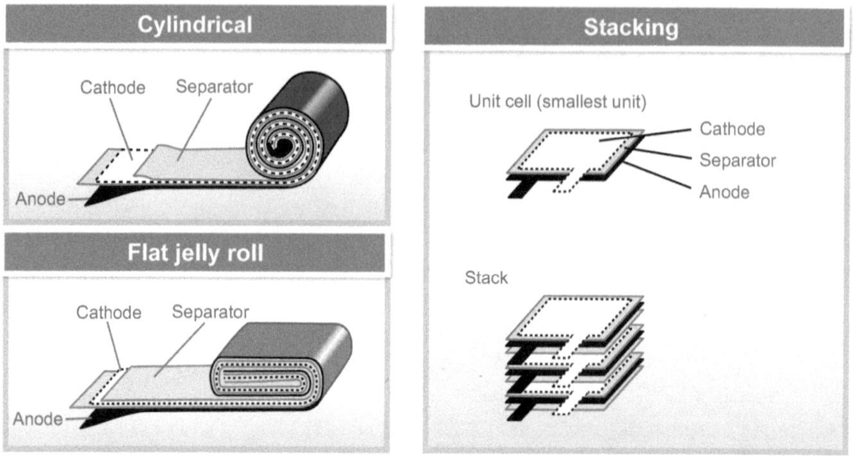

Fig. 9.7 Inner structure of lithium-ion cells

9.7 Inner structure of lithium-ion cells

The different inner structure of lithium-ion cells is shown in Fig. 9.7: jelly rolls, flat jelly rolls, and cell stacks.

In the mid 1990s, Sanyo manufactured prismatic housings based on deep-drawn aluminum and containing a flat jelly roll. Pouch cells (softpacks) for consumer applications were introduced on the market around the year 2000 as jelly roll and stacked type (Tables 9.2 and 9.3).

An innovative 4-V lithium-ion button cell has recently filled a niche. This cell is produced by VARTA Microbattery (CoinPower®). One of its applications is the Bluetooth technology.

Table 9.2 Examples of combinations of lithium-ion cell housings with inner cell structure for consumer applications

Cell housing	Design	Applications, examples	Manufacturers, examples
Round cell (hardcase)	Round jelly rolls	Laptops, eBike, power tools	Samsung SDI, Sony, Panasonic
Prismatic (hardcase)	Jelly roll	Mobile phones	Samsung SDI, Panasonic
Pouch (softpack)	Jelly roll	Smartphones, e-book readers, thin laptops, tablets	Sony, Samsung SDI
Pouch (softpack)	Stack	mp3 players, smartphones	Varta, LG Chem, ATL China

Table 9.3 Examples of combinations of lithium-ion cell housings with inner cell structure for automotive applications

Cell housing	Design	Applications, examples	Manufacturers, examples
Round cell (hardcase) type 18650	Round jelly rolls	Electric vehicle (Tesla)	Panasonic
Round cell (hardcase)	Round jelly rolls	Hybrid vehicles (HEVs) such as BMW ActiveHybrid 7	Johnson Controls
Prismatic (hardcase)	Jelly roll	Plug-in hybrid electric vehicles (PHEV) such as BMW i8	Samsung SDI, GS Yuasa Japan, Panasonic
Prismatic (hardcase)	Jelly roll	Electric vehicles (EV) such as BMW i3	Samsung SDI, GS Yuasa Japan, Panasonic, BYD
Pouch (softpack)	Stack	Electric vehicle (EV)	LG Chem
Pouch (softpack)	Stack	Plug-in hybrid electric vehicles (PHEV) such as Opel Ampera.	LG Chem

9.8 Lithium-ion cell production

The production process consists of the following steps: preparing the coating compound (slurry) for the electrode coating, slitting of the coated master rolls, equipping of the cell with positive and negative electrode and separator, filling with electrolyte, formation.

Fig. 9.8 displays a schematic of the production sequence of a prismatic cylindrical cell in a hardcase housing [5]. A stack made of electrodes and separators instead of the jelly rolls is produced for the variant with the cell stack. The other process steps are at least very similar, if not the same.

9.9 Requirements on lithium-ion cells

There are different requirements for cells. Among others: energy density, specific energy, cost, safety, dimensions, weight, maximum voltage, fast chargeability, power, and lifetime. Most of the time, the cell is developed in accordance with the requirements.

For example: If a lithium-ion cell for consumer applications is called for that needs to be fully charged via an USB interface within five hours, thicker electrodes (compared to those of a cell phone cell with a charging time of one hour) could be employed, while the cell chemistry could remain the unchanged.

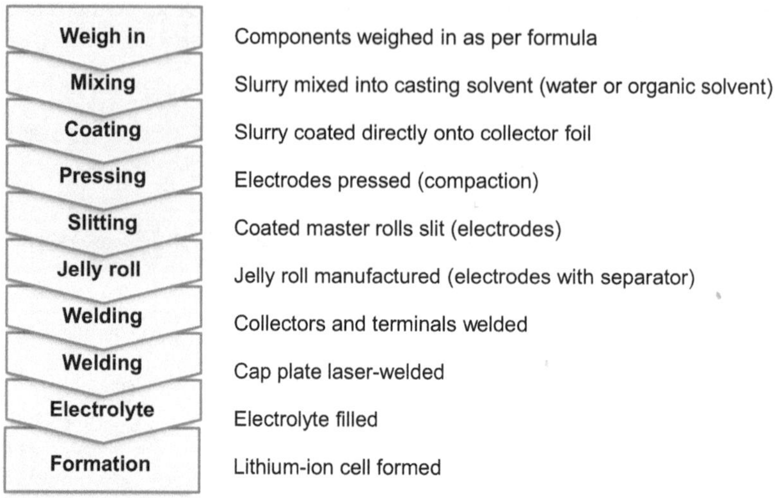

Fig. 9.8 Overview of the production process of a lithium-ion cell [5]

The product development for automotive applications is even more complex. Important requirements that need to be fulfilled simultaneously in this area are: cold start characteristics, maximum speed, necessary power, required range, costs, and passing safety tests. This means that a thorough selection of cell material, electrode, and cell design is of utmost importance. Most of the time, additional safety elements on cell level, such as bursting valves, overcharging protection components, and fuses, required. Furthermore, thin ceramic layers are often applied to the electrodes or the plastic separators. They provide an additional protection in case of separator failure.

It has often been forecast that fuel cells would become a competitor of the lithium-ion technology in the consumer market. This has not come to pass though. [6] compares fuel cell technology and lithium-ion technology.

9.10 Outlook

The requirements of the automotive industry on the lithium-ion technology will lead to considerable improvements in regard to material, design, and production technology. This will entail a cost reduction.

The energy densities on cell level have practically reached their maximum with today's established materials ($LiCoO_2$, NCM, graphite, separators, and metallic conductor foils) and housings (hard case, soft pack) for consumer cells. A further increase in energy density can only be achieved if active materials with higher specific capacities or higher cut-off voltages are employed. It is probable that consumer cells will be equipped with carbon silicon blends on the anode side in the near future. Carbon silicon blends already exhibit a silicon content of 10 % (in relation to weight) and a specific capacity of around 700 mAh/g. This is around twice as much as that of intercalation graphite.

On the cathode side, high-energy NCM, or OLO (overlithiated oxide), is currently being researched. This active material has a specific capacity of around 280 mAh/g at a mean discharging voltage of around 3.5 V vs. graphite. Simultaneously, high-voltage spinels are being developed (e.g., $LiNi_{0.5}Mn_{1.5}O_4$). They show a specific capacity of around 130 mAh/g at an approximate mean discharging voltage of 4.6 V vs. graphite. These are promising materials for financial reasons because of their high content of manganese. Other researchers are focusing on layered oxides with high nickel content, NCM. These materials exhibit a high specific capacity and can be employed at higher voltages up to around 4.4 V. However, electrochemically more stable electrolyte formulas remain to be developed for the above-mentioned cathode materials.

To improve safety even more, separators are currently being developed that are more stable mechanically, thermally, and electrochemically than the currently used separators. Innovations in this respect are non-woven separators (e.g., aramid-based) with a high chemical, electrochemical, thermal, and mechanical stability. Other concepts concentrate on conventional separators with ceramic coating, also to improve the electrochemical, thermal, and mechanical characteristics.

Cell systems based on lithium titanate ($Li_4Ti_5O_{12}$) might be of interest for special applications (e.g., low-voltage vehicles for 12-V or 48-V applications). Titanate is a material which enables positive cell characteristics such as a very high current carrying capacity, intrinsic safety, and a long service life. This is thanks to its very quick kinetics, extreme chemical and electrochemical stability, and lack of volume increase. A disadvantage of the employment of lithium titanate is the decrease in energy density and specific energy of the system.

Bibliography

1. J. B. Goodenough, K. Mizushina, J. P. Wiseman; EP0017400B1
2. Issues and challenges facing rechargeable lithium batteries J.-M. Tarascon, M. Armand, Nature, 414, 2001, pp. 359–367
3. Claus Daniel and J. O. Besenhard (Editors) Handbook of Battery Materials, Second Edition, Wiley-VCH, 2011 Chapter 15 Lithiated Carbons (Martin Winter and Juergen Otto Besenhard)
4. Source www.statista.com, 2016
5. Production of Automotive Li-Ion Batteries H. Fink, J. Fetzer, T. Wöhrle, WGP convention, May 06, 2010. Erfurt, Germany.
6. Fuel cells and batteries: Competition or separate paths? D. Ilic, K. Holl, P. Birke, T. Wöhrle, F. Birke-Salam, A. Perner, P. Haug Journal of Power Sources 155, 2006, pp. 72–76

Sealing and elastomer components for lithium battery systems

10

Peter Kritzer and Olaf Nahrwold

Contents

10.1 Introduction

Lithium batteries dominate today's consumer market. In the year 2014, around two billion lithium cells were produced for cell phones only. Off-the-shelf usage of lithium-based battery systems in vehicles began in the year 2009 with Daimler AG's S400 hybrid. In 2011, the first purely electric vehicles with lithium batteries were produced in series. As of today, all battery-driven and plug-in hybrid vehicles contain lithium-based energy storage systems. Table 10.1 compares consumer lithium batteries with automotive lithium batteries.

Automotive batteries need to be of much higher quality than those for consumer applications because of their considerably higher energy volumes and amperages, significantly more challenging environmental conditions, and substantially increased requirements in terms of service life and failure rates. Seals can, and must, substantially contribute toward fulfilling these tough requirements.

The following pages will discuss the main sealing components for cells and the entire battery system.

P. Kritzer (✉) · O. Nahrwold
Freudenberg Sealing Technologies GmbH & Co. KG, 69465 Weinheim, Germany
e-mail: peter.kritzer@fst.com; olaf.nahrwold@fst.com

R. Korthauer (ed.), *Lithium-Ion Batteries: Basics and Applications*,
https://doi.org/10.1007/978-3-662-53071-9_10

Table 10.1 Comparison of lithium batteries in consumer and automotive applications (typical values)

	Batteries in consumer applications	Batteries in automotive applications
Battery system weight	< 0.1 kg	> 100 kg
Battery system energy	0.002 to 0.02 kWh	HEV: 1 kWh; EV: > 20 kWh
System voltage	3.6 to 11 V	≫ 100 V
Maximum discharging currents	< 1 A	> 100 A
Maximum charging currents	< 1A	50 A (≫ 100 A)
Operating temperature	0 to 40 °C	– 40 to + 85 °C
Environmental conditions	Dust, spray	Dirt, oil, water, vibrations
Typically required service life	3 years	> 10 to 15 years
Typically tolerated failure rates (cells)	0.1 %	≪ 1 ppm

10.2 Cell sealing components

Cell sealing components must electrically isolate the two pole connectors from each other.

The sealing components used also have to be chemically stable toward organic electrolytes. In addition, during the battery's entire service life, the sealing material must not leach out contaminating substances into the battery electrolyte as this could have a long-term negative influence on the cells' electrochemistry.

Currently, thermoplastic materials such as polypropylene, polyamide (PA 12), or perfluoroalkoxy (PFA) polymers are generally used to seal solid housing cells. These non-elastic materials cannot fulfill long-term sealing functions, especially when taking into account service life and vibration resistance requirements because they can cause leakages due to the cell's long service life in combination with vehicle-specific vibrations. Current approaches use specially developed, polyolefin-based elastomers (ethylene-propylene-diene monomers [EPDM]) as cell sealing material. These materials reliably seal the pole feed-through even under the prevailing operating conditions.

10.3 Battery system sealing components

General remarks Generally, large-scale battery systems such as those used in electric vehicles consist of around 200 to more than 1,000 individual cells. These are mostly connected to form modules containing around 10 to 16 cells and are installed in a battery housing.

These systems' sealing components are housing gaskets, gaskets for electronic components such as plug seals and cable bushings, as well as seals for the coolant circuit such as connector seals or sealed tubes. Furthermore, there are several "sealing-like" components such as pressure-equalizing elements, system overpressure valves, and fixation elements for the individual cells.

All housing system gaskets must protect the battery interior over the entire service life against splash oil, splash water, and wading water. In general, automotive applications require at least protection rating IP67 (ISO 20653:2006–08) for battery housing gaskets. Thus, the battery housing must be dust-proof and also resistant to outside water pressure of 0.1 bar for at least 30 minutes. Also, like other automotive parts, the battery housing and its components are exposed to oil contaminations.

Components that can come into contact with leaking electrolyte must be resistant to such substances at least for short periods of time. Another relevant requirement is that production is carried out according to automotive standards such as ISO/TS 16949 of the International Automotive Task Force (IATF).

Housing gaskets Usually, it has to be possible to open and close the battery housing to easily repair minor defects such as loose electrical contacts or leaking coolant lines. Depending on the housing's position in the vehicle, stability, tightness, and temperature stability requirements vary. Due to the fact that in newly developed electric vehicles battery housings are designed to contribute to mechanical vehicle stability, deformation of the large housings (e.g. twisting) plays an important role. Other parameters that need to be taken into account for sealing design are the housing and cover material, and surface quality, respectively.

Mass production needs optimized and profiled gaskets that fulfill requirements concerning tolerance compensation over battery system lifetime. For example, it is important in this respect to keep the number of screws to a minimum without compromising on sealing effectiveness. New gasket requirements cover electro-magnetic shielding, which can directly influence gasket design. Seal assembly should also be quick and reliable.

Currently, there are two technologies for mass production: liquid gaskets that are applied by robots and large-scale profiled gaskets (Fig. 10.1).

Liquid gaskets are easily applied in full automation with existing equipment and are therefore frequently used. This technology, however, has several technical disadvantages: In general, the surface needs to be chemically or physically treated before gasket application (priming). Profile structures that can compensate local loads are not possible with such gaskets. Opening the housing usually destroys the gasket because it sticks to the lid or the housing. This causes battery maintenance problems because in order to seal the housing again, a new lid with sprayed-on gasket is required.

This is the reason why large-scale gaskets are used when tough technical requirements need to be met. Seal function redundancy is achieved through profile design. Also, customized gasket topography enables better tolerance compensation. This is especially important when the battery system has deformations that cause the gasket to "lift off" in certain areas. An electro-magnetic shielding can be implemented e.g. by metallic carriers and/or electrically conductive elastomer compounds. The disadvantages of large-scale sealing are that such gaskets are more complex to assemble and additional costs are necessary for manufacturing tools. Such large-scale gaskets are available for battery housing circumferences of more than six meters.

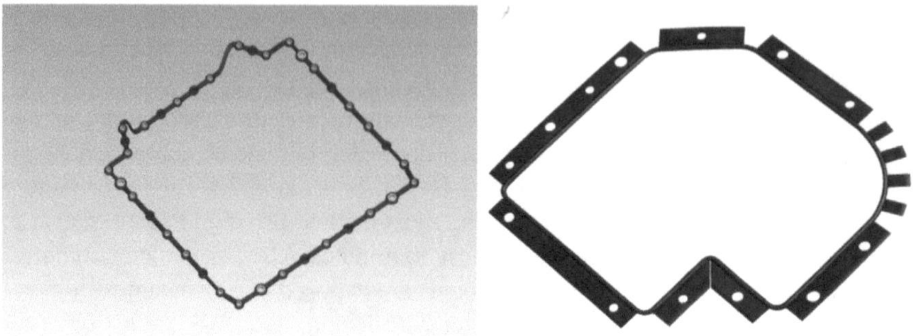

Fig. 10.1 Battery housing gasket solutions, *left* optimized flat gasket for mass production with locking pins and a circumference of around 2 m, *right* profile-based gasket for smaller lot sizes and/or very large housing dimensions

An alternative approach to battery housing gaskets for prototypes and small series is based on a gasket profile that can be manually customized for complicated housing designs (P2G - Profile to Gasket). The hollow profile used ensures suitable tolerance compensation and the system can meet protection rating IP 67 requirements.

Plug sealings/cable feed-throughs The sealings to connect power electronics are usually integrated directly into the plug. Silicon rubber-based components are used for this application in most cases. They have increased resistance toward high electrical voltages, and their surface does not carbonize, as opposed to carbon-based polymers. Higher voltages do not raise conductivity, hence eliminating short circuits. Silicon sealing components designed as multi-component parts are also used in plugs and in other high-voltage parts.

Simple cable feed-throughs can, however, be manufactured from carbon-based elastomers such as EPDM. Examples of these are cable bushings or bellows for sheathed cables and pipes into housings. It is important that components reliably seal the system even when the fed-through parts are vibrating.

Coolant system gaskets Large-scale battery systems require intelligent temperature management, which has two tasks: First, it dissipates heat from the cells and therefore protects them from overheating. Second, it enables cell heating at low temperatures. Quick-charge requirements with C-rates of two and higher pose new challenges, since they can result in high thermal loads for the battery system. There are three different technologies for cooling batteries: air cooling, direct liquid cooling with water-Glycol® mixtures, and direct temperature control with air conditioning systems or with cooling systems based on refrigerants such as partly fluorinated hydrocarbons like R1234yf and carbon dioxide. Main challenge is to keep all cells on a similar temperature level to prevent them from aging differently.

Electric vehicles mainly use liquid cooling for the battery systems. Air-cooled systems will be less and less able to meet future requirements for more compact

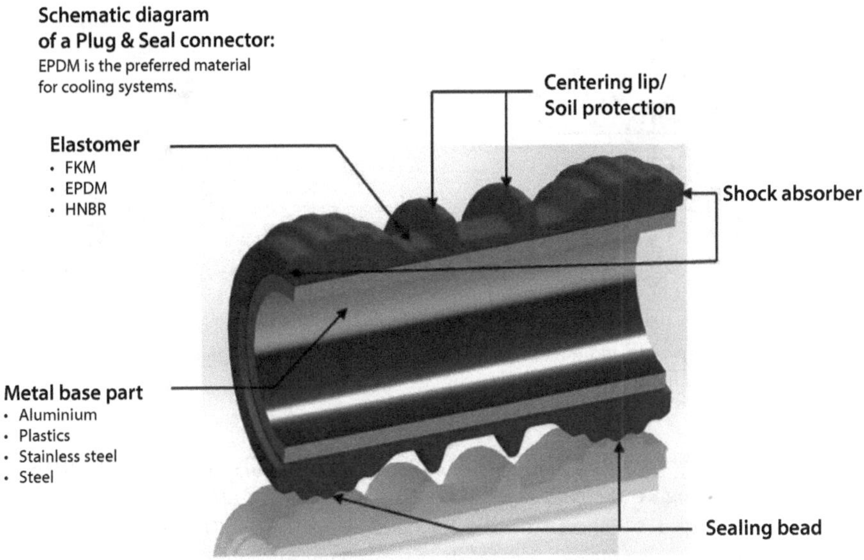

Fig. 10.2 Plug & Seal connector for water-based cooling modules [3]. Such connectors possess high tolerance compensation and excellent vibration strength

systems and quick-charge capability. A CO_2-based air conditioning system combined with the liquid battery cooling cycle can be especially efficient in this respect. It can provide effective heating for the battery and vehicle, and CO_2 is inert toward the battery's chemistry and is incombustible. It may even be possible to use the coolant to cool down cells that heat up uncontrollably in an emergency.

The modular design of large-scale batteries normally requires a modular cooling system with many connectors that need to be sealed. Welded piping does not enable the exchange of individual modules, and it is possible that its rigidity causes mechanical stress in the modules.

Components with metal or thermoplastic structures coated with elastomers such as the "Plug & Seal" connectors are a tried-and-tested solution for modular designs [1] (Fig. 10.2). They are made of a metal or plastic pipe with an elastomer seal. The seal is shaped in such a way that it compensates for production tolerances and thermal expansion. Once installed, Plug & Seal connectors are also resistant toward mechanical stress.

Furthermore, they ensure simple and reliable module assembly and enable the replacement of individual modules. They are available with inside diameters from 10 to 32 mm, a maximum operating pressure of around 15 bar, and an operating temperature ranging from −40 to +140 °C.

Plug & Seal components are already being used as standard in vehicle cooling systems and cooling modules of hybrid and electric vehicle batteries. Additional requirements for battery cooling systems can be met with sealed plastic pipe connectors and branched, flow-optimized components (Fig. 10.3). These products

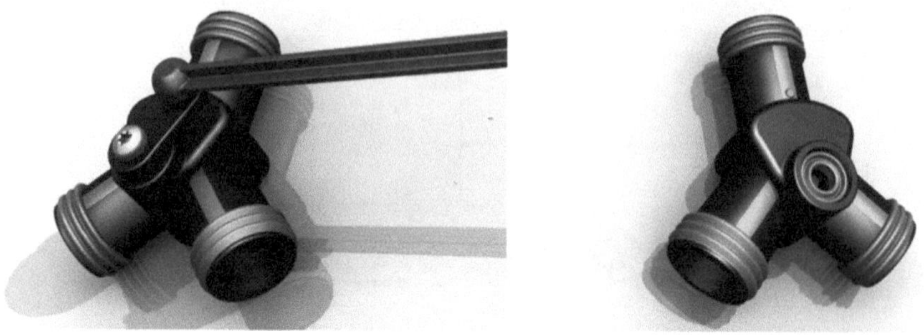

Fig. 10.3 Connectors with additional usage. The Y-shape enables a design that reduces flow losses; it can also be fitted with a valve

go beyond merely "cooling a system"; they also efficiently manage the released thermal energy, e.g., transfer it effectively to the passenger area, which is a future requirement. Fig. 10.4 shows plastic-based cooling tubes that allow free-form design, adaption to complex shapes, and integration of vibration reducing elements.

Pressure-equalizing elements with and without exchange of gases Automotive battery systems are subjected to pressure changes, which are inherent to such systems. They are mainly effected by atmospheric conditions, heating-up and cooling-down processes, uphill and downhill roads, entrance into and exit from tunnels, and air transport. Normally, "pressure-equalizing elements" balance out pressure fluctuations by enabling a gas exchange between inside the housing and outside. Nevertheless, short-term pressure peaks of up to 0.25 bar can occur. Components with microporous foils are mainly used to produce these pressure-equalizing elements. Their decisive advantage is that they remain watertight up to 1 bar overpressure, i.e., a water column of around 10,000 mm, while their main technical

Fig. 10.4 Cooling tube for efficient thermal management

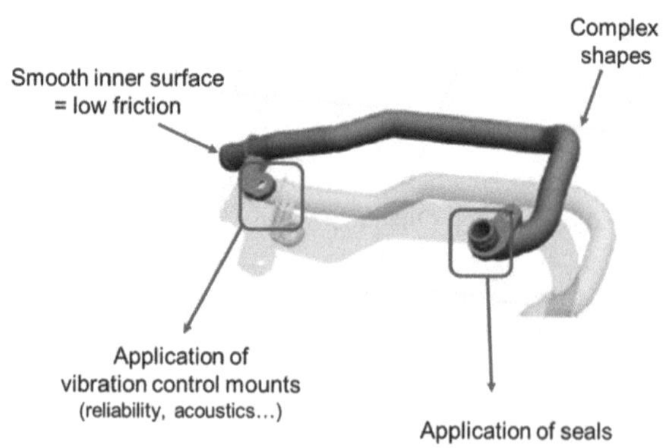

disadvantage is their low specific air permeability. This means that large elements are required for high gas exchange rates. Also, these elements are relatively susceptible to contamination.

Alternatives are robust pressure-compensation filters based on non-woven fabrics.

Both have a disadvantage that is inherent to the system: They are permeable to water vapor. The water vapor can precipitate inside the component and has to be bound by desiccants or led away from the housing by means of special condensate-removing elements.

New approaches use so-called "pressure-compensation elements" that promote the exchange of material with the battery housing [3]. This approach provides a variable volume so that pressure fluctuations between the housing's inside and outside can be compensated. One example is a spring-supported piston accumulator with an integrated overpressure valve. This technology is shown in Fig. 10.5. Installation inside the battery housing reduces the housing's dead volume and the additional space needed for the element itself. These systems are completely maintenance-free for their entire service life, which is a great advantage. They also prevent any build-up of contamination and condensate inside the housing. One technical disadvantage is the additional space requirement of such components.

System overpressure valves In rare cases, chemical or electrochemical processes can destroy a cell in the battery housing. In this scenario, a 40 Ah cell can release a gas volume of around 100 liters, which mostly stems from electrolyte degradation products.

The pressure is released very suddenly in solid housing cells because an overpressure of more than 10 bar is required to open the cell's bursting disk. Pouch cells with a flexible envelope already open at a relatively low overpressure of less than 1 bar; therefore, the pressure rises at a relatively low speed inside the battery housing.

The released toxic and combustible gases are emitted into the battery housing and can rapidly increase the pressure to, typically, more than 5 bar, depending on the dead volume of the battery system. This sudden overpressure cannot be compensated by the pressure-equalizing elements, and thus, the housing can burst. These gases cannot be released quickly enough by the above-mentioned pressure-equalizing elements. To prevent the housing from bursting, the gas volume must be released from the battery housing reliably, quickly, and safely. Overpressure valves with cross-sections of typically 5 to 10 cm^2 are necessary to achieve this. The components used here are bursting discs made of polymer or metallic foils, elastomer lids, and magnetic parts that open either autonomously or can be switched on and off manually.

Fig. 10.6 shows the new product DIAvent, which combines the functions of pressure equalizing and overpressure release in one product [4]. It comprises a non-woven-based pressure regulating section fulfilling IP67 requirements. An elastomeric umbrella valve lid, which opens at inner overpressure values of 50 mbar, secures the release of exhaust gases in case of cell malfunction.

Fig. 10.5 Example of a pressure-compensation element based on a piston accumulator. Gas is exchanged between the accumulator and the housing. If a certain pressure is exceeded, the gas can be released through an overpressure valve. Installation inside the battery housing reduces the housing's dead volume and the additional space needed for the element itself

Heat-shielding materials This new requirement for battery systems is owed to the increasing energy density both on cell and system level. The chemically stored energy is more and more "compressed". In case of a severe cell malfunction (the so-called "thermal runaway"), it becomes more and more difficult to protect the neighboring cells from the released heat. For this reason, heat shields placed between the cells become mandatory. Such heat shields can be consist of inorganic materials, e.g. mica foils. More recent products are elastomeric heat shields, which also take into account the surface expansion of the cell during charging and over lifetime. Such materials can shield heat coming from a 600 °C heat surface to values considerably below 200 °C after 30 seconds.

Fixation elements/frame-type gaskets for pouch cells Pouch cells have a flexible aluminum foil housing. The housing foils are coated on the inside and the outside with polymers to provide isolation. The advantages of these cells, namely high energy density, particularly easy scalability, and, compared to other cell types, lower production costs make them suitable for implementation in large-scale batteries and, in particular, in high-energy batteries such as those installed in electric vehicles.

There are some concerns regarding the flexible housing and sealing seam of the cell. The sealing seam must be completely tight during the battery's entire service life in order to ensure safe electric vehicle operation. It must compensate for cyclical loads such as fluctuations in the cell's thickness during cycling and pressure fluctuations from the outside. An especially critical situation is the blowing-off of combustible gas which can happen in the event of damage, such as when a short circuit occurs inside the cell. If this gas comes into contact with live parts, a fire cannot be ruled out.

In general, pouch cells are fixed in the battery housing in a metallic or plastic frame. On the one hand, this ensures efficient cell cooling, on the other hand, it subjects the sealing seam to a mechanical load, which can be problematic in the long term. This is why new approaches suggest "soft embedding" of the cells in a cell frame seal. In this case, cell and fixation frame are connected by means of an elastic sealing component (Fig. 10.6) [5].

Fig. 10.6 "DIAvent" element for pressure regulation and overpressure release. In normal operation, pressure regulation is performed through an IP 67 proven non-woven section (gas transfer typically 8 liters per minute). In case of overpressure release, an umbrella valve can be activated (activation pressure typically 50 mbar; gas throughput typically 40 liters/sec).

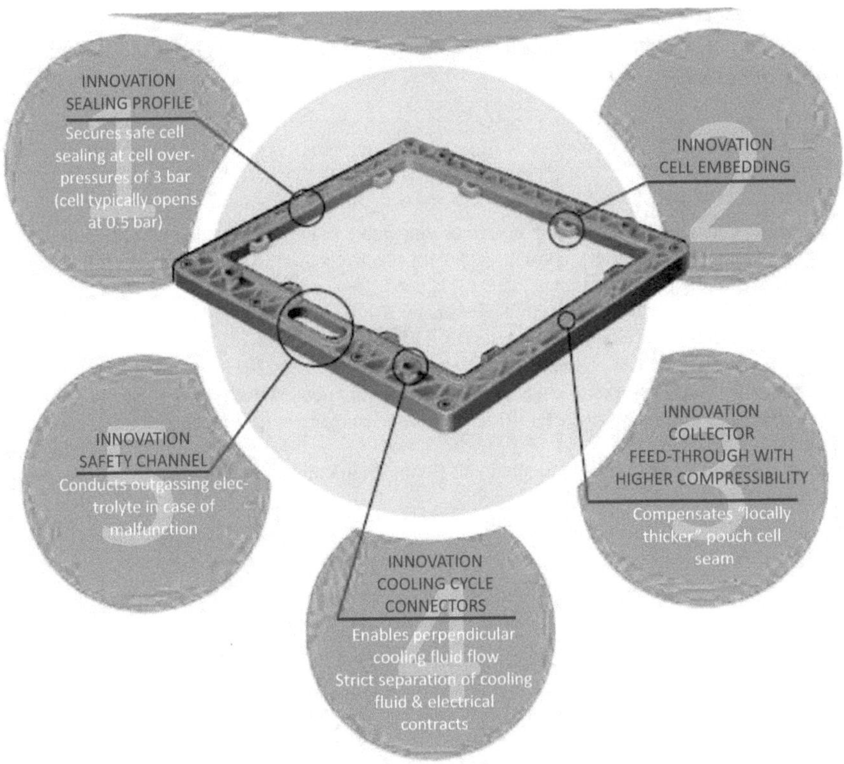

Fig. 10.7 Cell frame seal for embedding pouch cells with the additional functions of seam sealing and recess (predetermined breaking point). Modules produced with this technology also enable the integration of temperature management elements

This concept redundantly seals the cells and also flexibly and elastically fixes them. Its advantages are that the battery is more easily installed (tolerance compensation), its service life is extended (lower mechanical load, more reliable connections), and its safety increased (mechanical vibration compensation). Textile fabrics such as non-woven fabrics or the above mentioned heat shields inserted between cells can further improve this technology.

If a recess customized for the cell inner structure is integrated into an area of the gasket where there is no contact pressure on the sealing seam, a predetermined breaking point can be implemented, through which the cell can open in a controlled manner in the event of cell venting. For a stack design, an "outgassing conduct" can be integrated to enable the release of harmful gases to the outside.

Temperature management elements can also be integrated into the cell frame seal; this is an additional advantage. Temperature management channels can also be included in the frame. They are thermally connected with the cells via the sealing seam and enable cell cooling and heating. The cell surfaces can also be thermally connected if thermally conducting, compressible non-woven fabrics are inserted into the space between the cells. Also, it is possible to install flexible, sheet-type heating elements between the cells. These provide soft heating and slow down the cooling of cells (Fig. 10.7).

Bibliography

1. Kritzer P, Clemens M, Heldmann R (2011) Innovative seals: a robust and reliable seal design can provide efficient battery cooling cycles for electric vehicles and hybrid electric vehicles. Engine Technology International, June 2011, p. 64
2. Kritzer P, Gattringer R, Nahrwold O, Heldmann R (2015) Sealing and elastomer components in cooling circuits for electric vehicles. ATZ 117 (07-08): 54–58
3. Kritzer P, Rheinhardt H, Nahrwold O, Ewig T, Schreiner M (2011) Druckmanagement und Vermeidung von Flüssigkeiten in Lithium-Batteriesystemen. Mobility 2.0(3), pp. 42–47
4. Schäefer C, Kleinke C, Kritzer P (2017) Pressure Management in Battery Systems. Emobility-tec 4:48–51
5. Kritzer P, Nahrwold O (2011) Sealing and Fixing Elements for Flexible Cells in Large-Scale Lithium Batteries. ATZ 113 (06):26–28

Sensor and measuring technology 11

Jan Marien and Harald Staeb

Contents

11.1 Introduction

Lithium-ion batteries as an energy storage system represent one of the essential technological components in an electric vehicle and are the biggest expense factor. Therefore, measurement of state variables such as state of charge (SOC), state of health (SOH), and state of function (SOF) fulfills several requirements. Safety-relevant functions need to ensure protection against overcharging or deep discharging, for example. In addition, optimal utilization of the battery capacity is of great economic importance.

In contrast to lead-acid batteries, the total cell voltage of lithium-ion batteries is only marginally dependent on the state of charge. Current measurement is thus essential for diagnosing the battery state and for battery management. Total current measurement is used to balance the overall charge flow into the battery or from the

J. Marien (✉)
Isabellenhuette Heusler GmbH & Co. KG, Postfach 1453, 35664 Dillenburg, Germany
e-mail: jan.marien@isabellenhuette.de

H. Staeb
Seuffer GmbH & Co. KG, Baerental 26, 75365 Calw, Germany
e-mail: harald.staeb@seuffer.de

R. Korthauer (ed.), *Lithium-Ion Batteries: Basics and Applications*,
https://doi.org/10.1007/978-3-662-53071-9_11

battery. Current sensors employed must therefore fulfill a number of requirements. On the one hand, they must be able to measure a very large current range because peak currents can reach more than 2,000 A. On the other hand, they must be able to measure very low currents with high precision because in the passive state the currents are as low as a few mA. Furthermore, a current sensor should be offset-free. Every offset would considerably influence the charge balance over time because of the accumulation of the measured values.

Moreover, the current sensors must also fulfill the general requirements of the automobile industry. Examples of these requirements are electromagnetic interference resistance and measurement accuracy, because the steep switching edges of power semiconductors in power train electronics subject the battery to large voltage or current changes.

Two measuring principles are possible for these applications: shunt-based measurement and magnetic current measurement. The former is broadly used in Europe for measuring the current at the battery sensors for 12-V lead-acid batteries. Besides the commonly used solutions, novel approaches for measuring the current of lithium-ion batteries will be presented. The next two sections will outline shunt-based sensors specifically geared toward lithium-ion batteries. Furthermore, a magnetic sensor without magnetic materials that cause magnetic flux will be examined. In contrast to conventional magnetic current sensors, this sensor's accuracy is not negatively affected by hysteresis.

11.2 Galvanically isolated current sensor technology in battery management systems

Perfectly measuring the battery state constitutes the basis for implementing corresponding measures to optimize the energy storage system's usability. Charging strategies, predictions of load profiles, and optimization measures to increase the service life of a battery can be deduced from the measurement data. This requires a sound measurement data base. "Current sensing" thus is more than just a measurement. It is rather a quality criterion, like the quality of the battery itself. The following section will describe a current sensor technology based on Hall effect sensors that has been customized for automotive applications [1].

11.2.1 Hall effect battery sensors

Measuring the applicable data is the first step for ensuring the operational safety of the battery and the power supply in the vehicle. The objective is to generate these data as reliably and precisely as possible, even in extreme situations. Usually, three variables are necessary to dependably assess the state of a battery: current, voltage, and cell temperature.

An extremely low-impedance measurement is specifically required for measuring current. To minimally influence the system, a technology is preferable that does not interfere with the primary circuit. This is possible with Hall-sensor technology, which measures the magnetic field of the current and calculates the current therefrom [4].

Functional principle The following will examine the functional principle of Hall effect sensors based on an example of a commercially available integrated circuit. The approach consists of isolated current measurement by means of a magnetic field created by a live conductor. Thus, a low-impedance measurement is executed that is non-retroactive and fully decoupled from the primary circuit. The centerpiece of the sensor is a circuit specifically customized for this purpose, the so-called ASIC. Fig. 11.1 depicts a block diagram of the first-generation circuit. The ASIC can also simultaneously measure voltage and temperature [2]. It is therefore possible to generate the different variables of the data set at exactly the same time. This enables implementation of a full-blown battery management system in addition to mere current data logging.

Because no magnetic materials whatsoever are employed in this system, it is absolutely hysteresis-free and unsusceptible to overloads. The measurement data are transmitted via the bus without any problems even in extreme situations. Each of the three on-chip Hall effect sensors has a dedicated reference coil, which loads the Hall cell with a defined magnetic field (Fig. 11.1) [3]. This makes it possible to conduct a system function check at any time.

Fig. 11.2 displays the sensor unit which is integrated with the battery terminal. A Hall effect sensor has special characteristics. These characteristics now enable applications that were significantly harder to implement in the past or could only be implemented with significantly more time and effort. Even the standard task of measuring battery current is now considerably easier. Usually, measurement with a measuring resistor (shunt) is performed in the ground cable of the battery. The measurand is the voltage drop over this resistance. This analog value is very low for low and medium currents and therefore potentially susceptible to errors. For high currents, on the other hand, a voltage drop occurs that is inherent to the system. This considerably increases the temperature within the whole set-up.

Operating the sensor does not cause a voltage drop and therefore no temperature increase. This means that the electrical consumer load can use the maximum

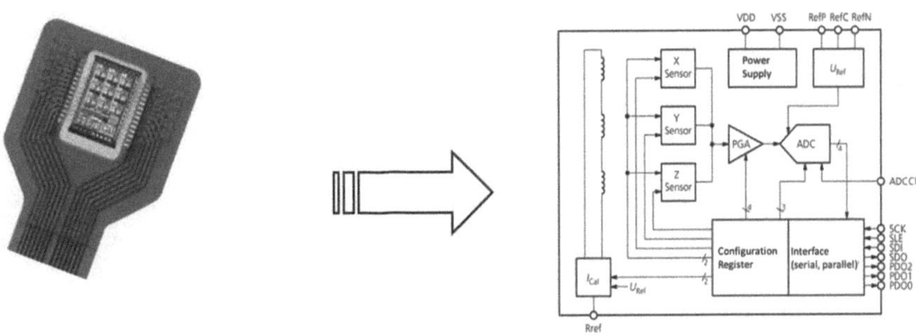

Fig. 11.1 Prototype and circuit diagram of an ASIC for current measurement

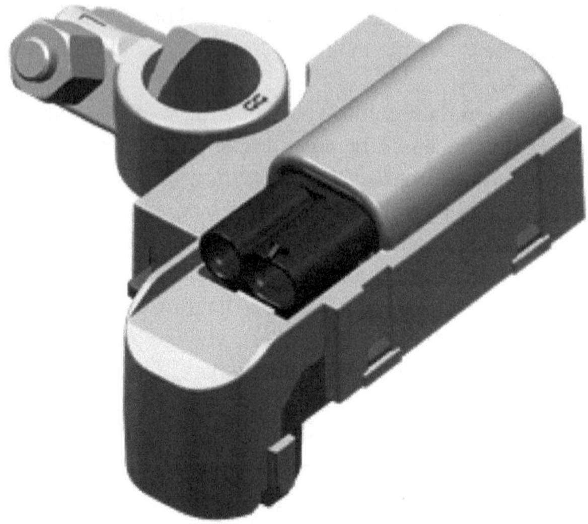

Fig. 11.2 Current measurement sensor unit, integrated in battery terminal

possible power of the battery. Implementation on the positive terminal's potential is possible without difficulties. The reliability of data measurements is further increased by the low rise in temperature of the device itself. These two factors allow implementation of a start-stop operation, for example, because battery status forecasts are considerably improved.

Installation in the positive line makes it possible to monitor individual consumer loads separately. Even measurement of current paths with floating potential is now very easy. The device can be employed between battery cells or between batteries in serial arrangement. The same applies to new on-board electrical systems with voltages of 48 V or to high-voltage applications.

The sensor measures current, temperature, and voltage. It provides reliable data for the current distributor and contactor, which are located downstream. One option is to use these data with a battery management system to predict the battery's starting capabilities.

Hall effect sensors do not influence battery performance; the internal resistance remains the same, and the battery can provide its maximum performance. Assessment can be performed on an ongoing basis in combination with a battery management system directly at the battery. All this results in an intelligent and reliable battery system that analyzes, safeguards, and displays its own status.

Galvanic isolation Galvanic isolation of current measurement is an essential characteristic of the sensor. It considerably facilitates employment in the assembly. It also improves accuracy in regard to environmental conditions such as temperature or overcurrents outside the measuring range.

Furthermore, internal resistance is not influenced, which is especially important during the starting procedure of the engine. Failure of the sensor does not affect the primary circuit.

While the engine is starting, the sensor is loaded with several 100 up to 1,000 A for a certain amount of time. The shunt heats up due to the electrical current flow and causes a voltage drop. Fig. 11.3 displays a thermography of a shunt system at 600 A (left). Here, temperatures above 200 °C are reached. In the same Fig. on the right, the temperature increase of a Hall effect sensor measurement system under the same conditions is shown. The temperature of this system with the same electrical current flow after 30 seconds is a mere 80 °C.

This increases accuracy in the event of repeated and increased loading, e.g., for start-stop operations. Thanks to the relatively low increase in temperature, the thermal load on the measuring system is considerably reduced. This in turn increases long-term stability.

Another advantage of galvanic isolation is demonstrated when high-voltage applications are used. Electric mobility requires specific levels of puncture resistance. The materials are commercially available. There is no need to isolate the voltage potentials with an isolating transformer, DC/DC transformer, optocoupler, etc. Because these devices are not necessary, technical performance and cost-effectiveness (data rate, losses) are improved.

Intelligently designing the cables in the primary circuit optimizes system precision. In combination with respective measures on the ASIC, external influences are restricted to the necessary minimum.

11.2.2 Shunt measurement battery sensor

A current sensor for measuring the total battery current is based on a low-impedance precision resistor and a high-accuracy measuring system. Typical continuous current loads for today's batteries range from 200 to 300 A; peak currents reach 2,000 A.

Measuring resistors in the 50 to 100 μΩ range are perfect here. The measuring system should be offset-free, highly linear, and should function with low noise. Furthermore, the sensor must ensure secure galvanic isolation of the high-voltage system from the on-board electrical system. Most of the time, further parameters are

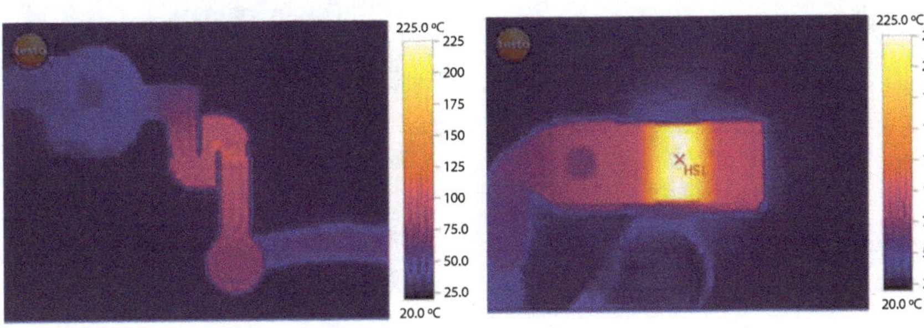

Fig. 11.3 Temperature increase of a shunt after 30 s at 600 A up to 225 °C (*left*), temperature increase of a current sensor after 30 s at 600 A up to 80 °C (*right*)

measured in addition to total current, for example, temperature or total cell voltage. Standardized digital interfaces are employed for communication with the superior battery management system.

Functional principle Fig. 11.4 shows an example of such a battery sensor on the left. This sensor is based on a precision resistor with 100 μΩ. It is made of the resistor material Manganin®, which exhibits a very low temperature coefficient of resistance (< 50 ppm) and is perfectly suited for copper at thermoelectric voltages of 0.3 μV/K. It is also extremely durable: Even under high temperature loads (> 125 °C) and very long operation times (> 2,000 h, loaded), the resistance varies by less than 0.1 %. The thermal resistance of such a resistor is around 2 K/W. This means that during regular operation at a load current of 300 A, the self-warming effect does not exceed 20 °C.

In addition, a high-accuracy measuring system is required. This system ideally is implemented by means of an ASIC; its block diagram is shown in Fig. 11.5. It is a multi-channel measurement system with a resolution of 16 bit. It operates with very high linearity (better than 0.01 %), very low noise (noise density of 35 nV/√Hz; thanks to the chopping operation there is no 1/f noise), and very low offset (better than 0.25 μV over the entire temperature range). The ASIC also has a programmable preamplifier, which enables precise measurement of input voltages from 7 to 800 mV.

The measured values are logged from the ASIC by means of a microcontroller, calibrated, and then outputted via a digital interface.

Although the system can be implemented with RS 485 and SPI interfaces, the trend is toward implementing a CAN bus. Analog signal feeds should be kept short because very low voltages are recorded. For the same reason, analog transmission of measured values is not recommended.

Digital transmission also makes it possible to prepare the data locally. Nowadays, raw data are no longer transmitted. Rather, the data are locally checked for

Fig. 11.4 Galvanically isolated IVT sensor for total current measurement with CAN interface, overcurrent detection, trigger input, and voltage measurement channel (*left*), IVT-F sensor, as used in the KER system in Formula 1 racing (*right*)

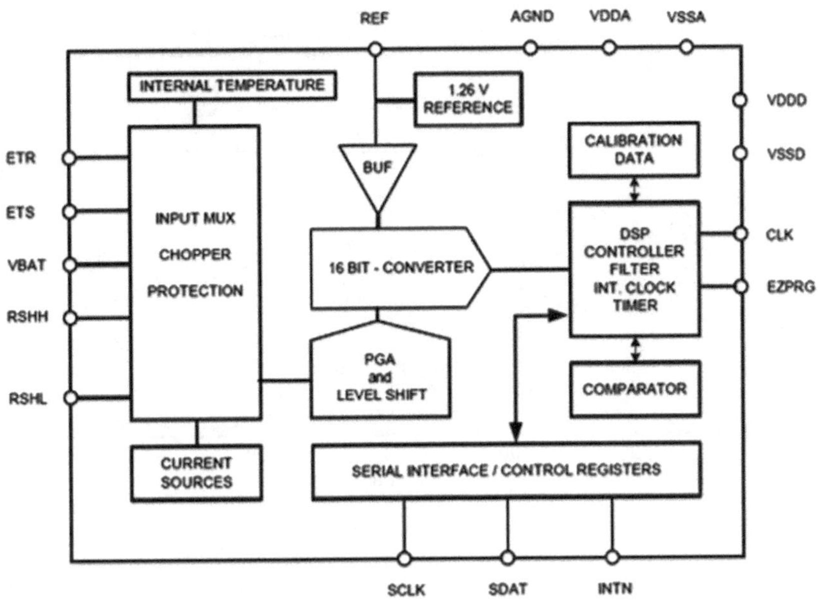

Fig. 11.5 Block diagram of the ISA ASIC

plausibility, averaged, summated, or preprocessed in other ways (depending on the application).

The sensors are galvanically isolated between the measurement system and the digital interface. This is achieved by today's optical, inductive, and capacitive transmitters. They are easily integrated into the sensor, do not require a lot of space, and are low-cost devices.

Measuring one or more voltages with the same sensor has also proved effective for many applications. For example, total cell voltage or voltage at the series relay can be measured with the same sensor and transmitted via the same interface.

The sensor also has an independent overcurrent monitoring system because battery safety needs to be guaranteed at all times. The user can use this functionality to parametrize two switching thresholds for positive and negative currents. A separate circuit monitors the current on the ASIC fully independently from the actual measurement and can trigger an overcurrent signal within a few microseconds.

Finally, it is often necessary to synchronize current measurement with other system measurements. This is why measurement can be triggered via a separate signal line independently of the digital interface, to ensure synchronicity of the measured values in the system.

Practical application Shunt-based sensors are already used nowadays in applications to measure lithium-ion batteries' total current. Their outstanding feature is the

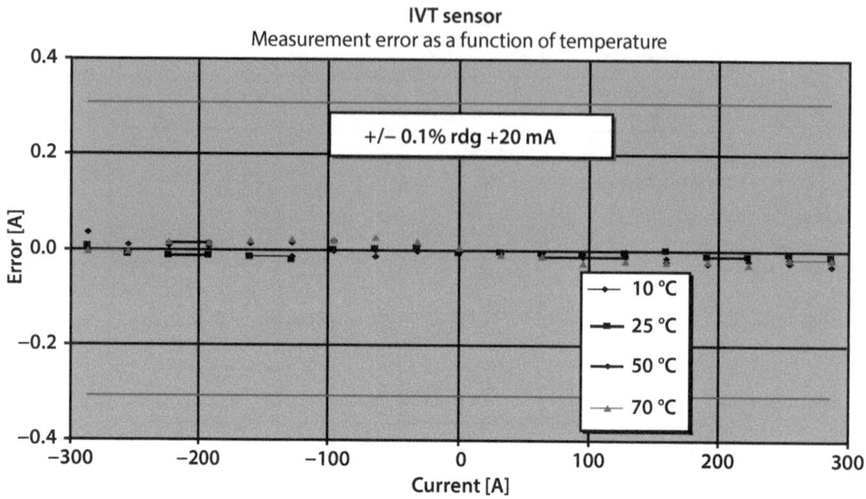

Fig. 11.6 Measurement of an IVT sensor's total accuracy

accuracy they achieve in practice. It has been shown that it is possible to guarantee a total accuracy of 0.4 % over the full specified temperature range and service life for a sensor with a 100-μΩ resistor and a continuous current load of 300 A. The offset of such a system is below 20 mA; RMS noise is around 5 mA. Fig. 11.6 shows an IVT sensor's total accuracy.

The Federation Internationale de l'Automobile (FIA) has decided to employ shunt-based sensors to monitor compliance with regulations of the electrical KER systems in Formula 1 race cars. When the race car is braking, the KER system partially transforms the vehicle's kinetic energy into electric power. This energy is stored in a lithium-ion battery. The driver is allowed to utilize a maximum of 400 kJ of energy for a maximum of 6.6 s per lap. He uses this as additional thrust, for instance when overtaking. The sensors permanently record total current and total voltage and calculate current power as well as accumulated power. During the race, these data are transmitted to the control units of the race car via the CAN bus and to the race officials via radio diagnostics. Fig. 11.4 displays an IVT-F sensor, as used in Formula 1 (right). The teams employ different voltages and currents. The mechanical adaptations also vary from team to team. This requires a very flexible sensor.

11.3 Outlook

Because the total current sensor is the main diagnostic instrument in the battery, there will be two primary trends in future. On the one hand, there will be efforts to increase sensor accuracy to the technical maximum. This includes implementing elaborate correction algorithms in increasingly powerful microcontrollers. On the other hand, the sensor is part of the battery's safety concept. Diagnostic abilities, self-monitoring abilities, or redundant current measurement will become increasingly important. In the future, magnetic and resistance-based measuring systems could be combined with a sensor in order to guarantee the required system stability and plausibility of measurement data.

Bibliography

1. Tille Th (2011) Sensoren im Automobil IV, expert Verlag Renningen, Germany. ISBN 978-3-8169-3066-2
2. Köhler U, Lorenz Th, Hohe H (2006) Neuartige dreidimensionale Magnetsensorik für Kraftfahrzeuganwendungen, Sensoren und Messsysteme 2006, Informationstechnische Gesellschaft im VDE
3. Hohe H (2005) Magnetfeldsensoren messen dreidimensional. Mechtronik 7
4. Popovic RS (1991) Hall effect devices. IOP Publishing, Bristol

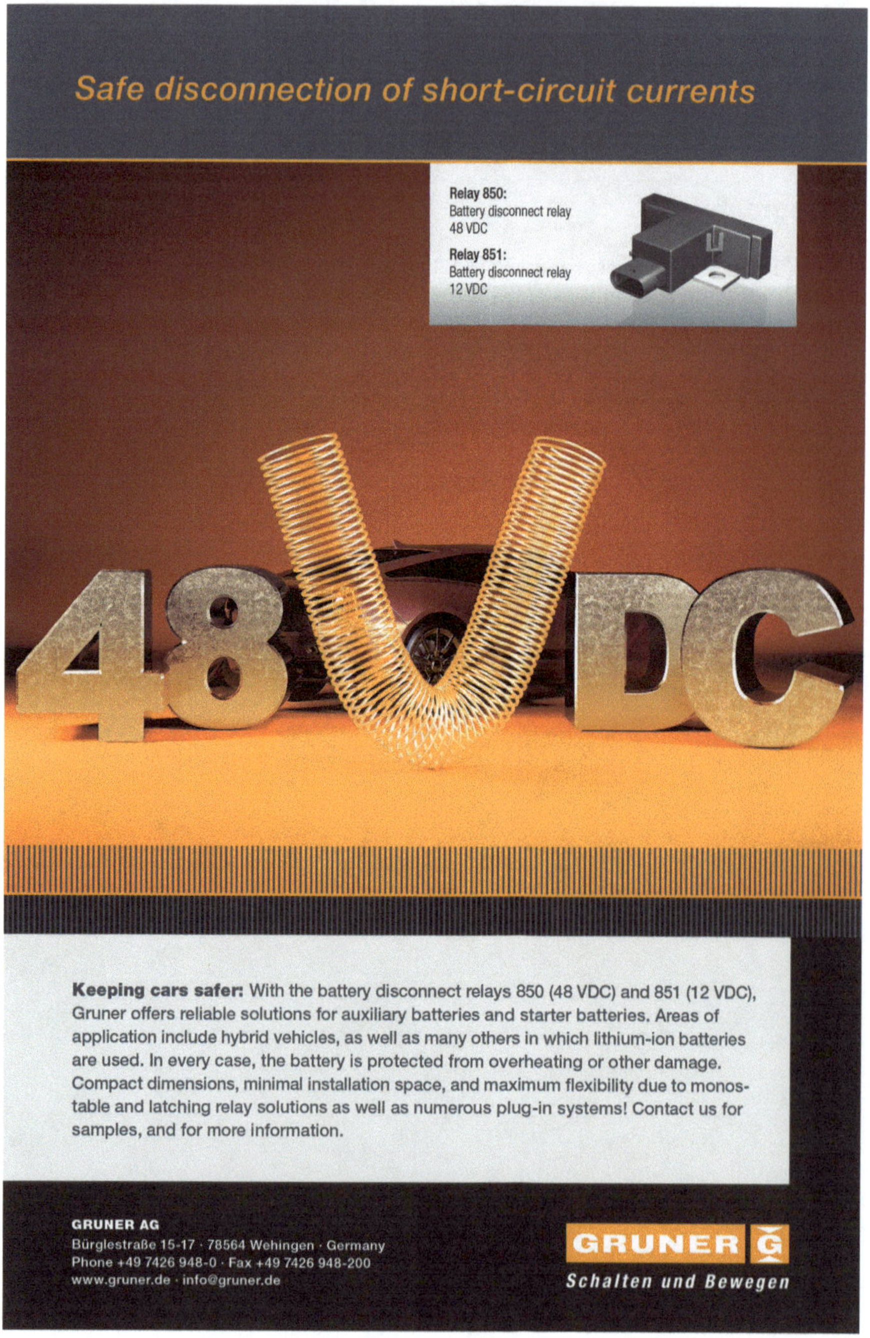
Safe disconnection of short-circuit currents
Relay 850:
Battery disconnect relay
48 VDC
Relay 851:
Battery disconnect relay
12 VDC
48VDC
Keeping cars safer: With the battery disconnect relays 850 (48 VDC) and 851 (12 VDC), Gruner offers reliable solutions for auxiliary batteries and starter batteries. Areas of application include hybrid vehicles, as well as many others in which lithium-ion batteries are used. In every case, the battery is protected from overheating or other damage. Compact dimensions, minimal installation space, and maximum flexibility due to monostable and latching relay solutions as well as numerous plug-in systems! Contact us for samples, and for more information.
GRUNER AG
Bürglestraße 15-17 · 78564 Wehingen · Germany
Phone +49 7426 948-0 · Fax +49 7426 948-200
www.gruner.de · info@gruner.de
GRUNER
Schalten und Bewegen

Relays, contactors, cables, and connectors 12

Hans-Joachim Faul, Simon Ramer, and Markus Eckel

Contents

H.-J. Faul (✉)
TE Connectivity Germany GmbH, Tempelhofer Weg 62, 12347 Berlin, Germany
e-mail: joachim.faul@te.com

S. Ramer
LEONI Silitherm S.r.l., S.S 10 – Via Breda, 29010 Monticelli d'Ongina (PC), Italy
e-mail: simon.ramer@leoni.com

M. Eckel
TE Connectivity Germany GmbH, Ampèrestrasse 12-14, 64625 Bensheim, Germany
e-mail: meckel@te.com

R. Korthauer (ed.), *Lithium-Ion Batteries: Basics and Applications*,
https://doi.org/10.1007/978-3-662-53071-9_12

12.1 Introduction

Electromechanical relays have always reliably switched electrical loads in automobiles. The special operating conditions of vehicles with a combustion engine already necessitate a special requirement profile for these relays. Their use in vehicles with electrical power trains requires these components to meet totally new demands. The voltage level in these vehicles is usually much higher, which plays an essential role in this respect. Combustion engine-driven cars have had a standard system voltage of 12 V or 24 V for many decades. The system voltage for hybrid and electric vehicles is generally several hundred volts. Commercial vehicles even have batteries up to 1,000 V.

At the same time, the main contactors in an electric vehicle must be able to carry and safely switch considerably higher currents than in non-electric vehicles. Another issue is the substantially greater requirements concerning electrical safety because the main contactors have the task of safely isolating the traction battery from the vehicle electrical system in case of malfunction.

Tried-and-tested technology has been available for switching high voltages for a long time. However, totally new questions have emerged in the automotive environment, which require customized solutions.

12.2 Main functions of relays and contactors in the electrical power train

There are different concepts for vehicles with electrical power trains: plug-in hybrid, pure electric, or fuel cell. What these concepts have in common is that the high-voltage battery must be connected with and disconnected from the power train by the main contactors. In general, two main contactors are used that switch both the positive and the negative load path (Fig. 12.1). Separately switching both load paths results in homogeneous redundancy which is important to comply with therefore such a system fulfills the ASIL safety requirements [1].

Vehicles with high-voltage batteries that have been planned and realized with only one main contactor, usually in the positive path, are exceptions. In Europe, however, two main contactors are standard, not least due to the requirements of the LV 123 standard, drawn up by the main German car manufacturers [2].

The precharge relay is another switching function that is usually found in all vehicles with an electrical power train. The power inverters for operating electric motors have large filter capacitors at the input stage, making it difficult for the contactors to close the main circuit. Initially, the filter capacitors are not charged, which could cause extremely high switch-on inrush currents. These would overburden the main contactor's switching contacts and could lead to contact welding. Precharge relays are used to prevent this. They employ a precharge resistor that precharges the filter capacitors to a voltage of around 80 to 98 % of the battery voltage (Fig. 12.1). Precharging takes only a few tenths of a second and ensures that the resulting switch-on inrush current for the main contactor amounts to only a few hundred

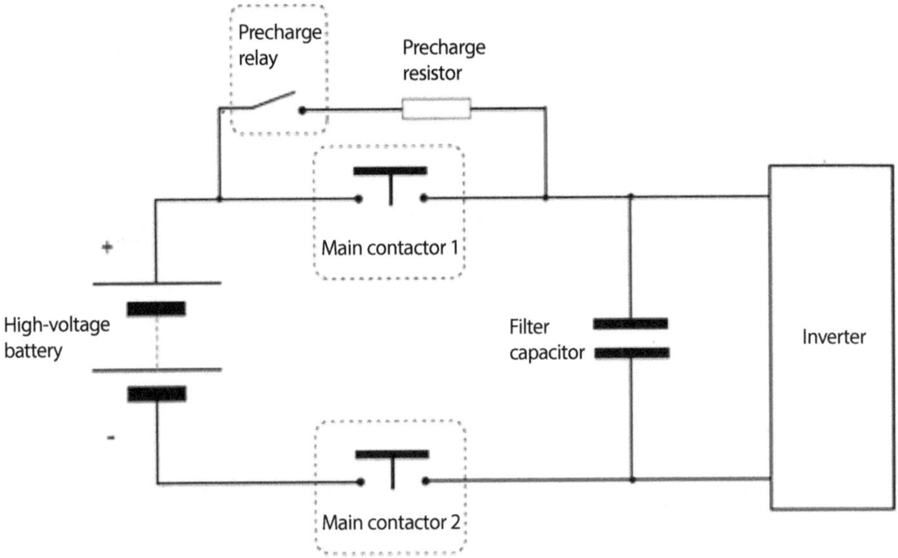

Fig. 12.1 Basic high-voltage circuit diagram

amperes (Fig. 12.2). The precharge relay usually switches a current peak, which is limited by the precharge resistor to around 10 to 20 A, but when the precharge relay is opened, the current already has decreased to 0 A thanks to the main contactors that are then closed. The switching sequence and the electrical current flow are depicted in Fig. 12.3.

High-voltage relays and contactors are used in a variety of applications in hybrid and electric vehicles. The charging connection of electric vehicles or plug-in hybrid vehicles is usually powered on and off by means of high-voltage contactors. In certain cases, high-voltage relays are also used in discharging stages, which ensure controlled discharging of the filter capacitors when the on-board HV system is

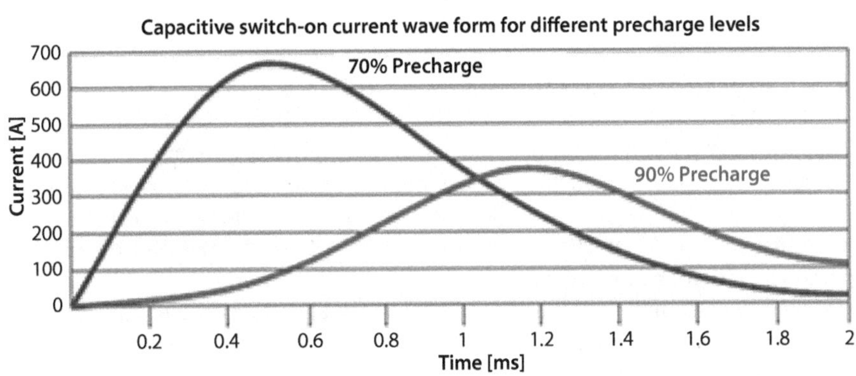

Fig. 12.2 Switch-on inrush current after precharging

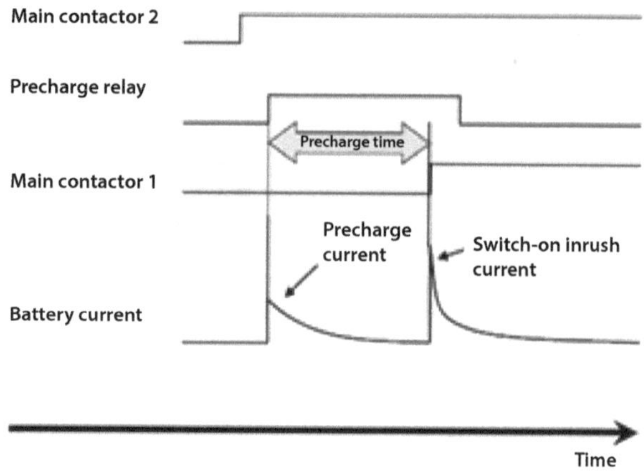

Fig. 12.3 Precharge time sequence

powered off. This procedure is similar to that of the precharge circuit described above.

Here, it is important to reduce the residual voltage to below 60 V within a maximum of 5 s. Fuel cell vehicles require a similar circuit to ensure safe discharging in the fuel cell circuit.

To monitor the traction battery, the voltage at the battery's terminals is usually transmitted to the battery monitoring unit (BMU) by means of measuring lines.

These measuring lines must be switched by high-voltage relays, however, because they constitute a connection between the high-voltage and the low-voltage system of the vehicle. The relays must be capable of switching high voltages, but not of switching zero or very low currents.

Last but not least, a large number of secondary devices, e.g., electrical heaters, which are fed from the high-voltage network, are also switched by means of high-voltage relays. The switching capability required in these applications is lower than that of the main contactors, but can still be in the kilowatt range. Fig. 12.4 shows an overview of the described applications.

12.3 Practical applications

There are many different requirements in regard to high-voltage contactors installed in electric vehicles, of which the most important are related to electrical conductivity. These will be described here in more detail. Additional aspects on other requirements such as mechanical characteristics, environmental and operational conditions will not be examined.

At all times, the main contactor must be able to safely carry the traction motor's operational current in standard operational conditions without being overloaded. The resulting peak currents depend on the system voltage and maximum power of the traction motor. Of course, today's vehicle concepts cover a broad range of nominal power values: For small and compact vehicles, the nominal power is only

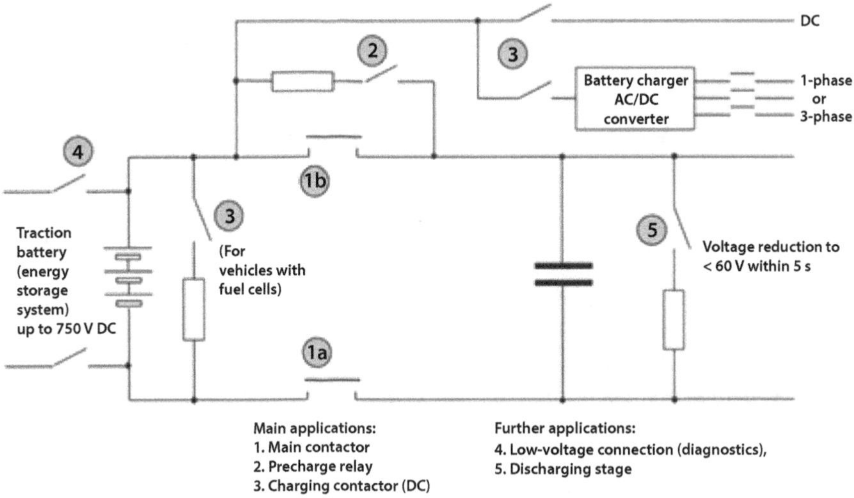

Fig. 12.4 Overview of relays and contactors in on-board HV systems

10 to 20 kW; for high-power sports cars it can reach up to 400 kW. In the medium range, nominal powers of 30 to 80 kW and system voltages of around 200 to 400 V are standard. Therefore, the maximum traction currents are in the low three-digit range (up to ca. 200 A).

Maximum traction current is very rare and usually occurs only in extreme driving situations with maximum acceleration, high towing loads, or a steep incline. The actual traction currents during a journey are usually much lower, fluctuate greatly and quickly, and depend on the situation.

Carrying these load currents primarily results in heat being dissipated from the contactor, caused by the power generated at the contact resistance. This does not cause any wear on the switching contacts and, in turn, does not limit the number of achievable switching cycles. The contact resistance, usually a few 100 microhm for high-voltage contactors, and the cross section of the connected load cables are of vital importance for achieving the maximum continuous current loads. This is because the latter dissipate a considerable proportion of the heat away from the contactor.

Because contactors exhibit very high thermal inertia, short-time current peaks, which reach up as high as many times the continuous current limit, pose no problem whatsoever, not even when they last longer than several seconds. However, the mean load current and the resulting mean temperature rise are important. The limit of the permissible mean load current is reached when the resulting temperatures (in relation to the ambient temperature of the contactor) reach their permissible limit. The weakest point is often the temperature of the load connections. It should not generally exceed 140 to 160 °C. This why heat dissipation by the connected load cables or lead frames plays an important role.

Usually, an ambient temperature of up to 125 °C is factored in for vehicles, depending on the installation location. However, the main contactors in hybrid and electric vehicles are usually situated near the main battery, where the controlled temperature conditions necessary for operating lithium-ion batteries prevail. Therefore, the maximum ambient temperatures typically range from 60 to 70 °C, which is very advantageous for dimensioning the contactors.

On the other hand, the achievable number of switching cycles of a contactor or relay is determined by the electrical load that needs to be switched, taking into consideration load current and voltage, but also inductive and capacitive load characteristics. As mentioned above, the main contactor of the negative path is closed before the precharge relay, without any switching load. But, switch-on inrush currents might occur if Y-capacities in the battery or inverter path are discharged via the ground terminal. Depending on the operating conditions, these current peaks might wear out the switching contacts, resulting in a considerable reduction of service life. The main contactor in the positive path, on the other hand, must switch on the inrush current of the filter capacitors that still remains after precharging. The inverter is usually powered down when the on-board HV system is shut down by the main contactors. Therefore, the contactors do not need to switch off any considerable electrical load here. Overall, the switching loads for the main contactors are comparably low and the established designs cover these without difficulty. Under these conditions, much higher numbers of switching cycles can be achieved than the minimum requirements based on the overall vehicle service life.

If all possible fault scenarios are taken into consideration, however, extremely high switching loads can occur. Different reaction patterns are possible in the event of unexpected overcurrents, depending on the underlying safety philosophy. In many cases, the control system opens both main contactors after a certain waiting period in order to disconnect the high-voltage system. The currents that need to be switched can be much higher than 1,000 A. If the cause of the overcurrent is a "hard" short circuit in the high-voltage system, today's lithium-ion batteries are capable of creating currents of 6,000 A or more within only a few milliseconds. It is to be assumed, however, that, in the event of such a hard short circuit, the high-voltage main fuse would disconnect the short circuit long before the control electronic system opens the main contactors. The main contactors then only have to switch off the system without any contact load. However, they must be able to carry the short-circuit current without damage until the high-voltage fuse melts. Preventing spontaneous contact welding caused by the extreme load current is of particular importance.

An additional concern is so-called levitation: The electromagnetic effect of the load current creates a force that tries to separate the closed contacts [3, 4]. Because this levitation force is essentially determined by the physical phenomenon Lorentz force, it can only be marginally influenced by the technical design. When designing a high-voltage contactor, it is important to dimension the contact forces such that they are higher than the expected levitation force.

The levitation force is very low for regular operational currents up to 300 A. Therefore, it can be neglected completely during dimensioning. It is a quadratic function of the load current, however, and therefore plays a considerable role in the kA range, and at 6 kA it can reach 50 N (Fig. 12.5). The contact force that needs to be realized in the design not only must compensate for this levitation force; it also

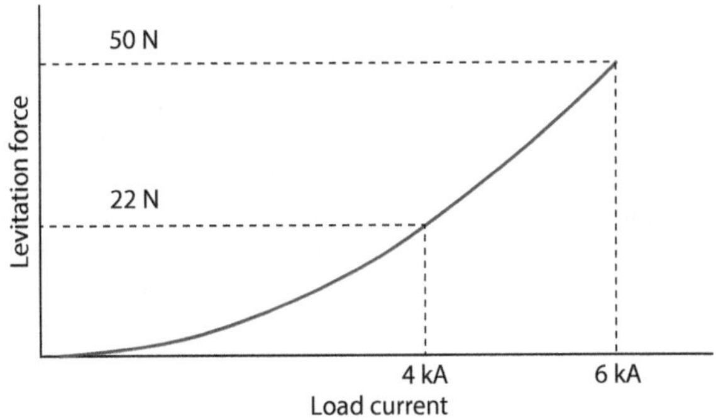

Fig. 12.5 Levitation force as a function of current

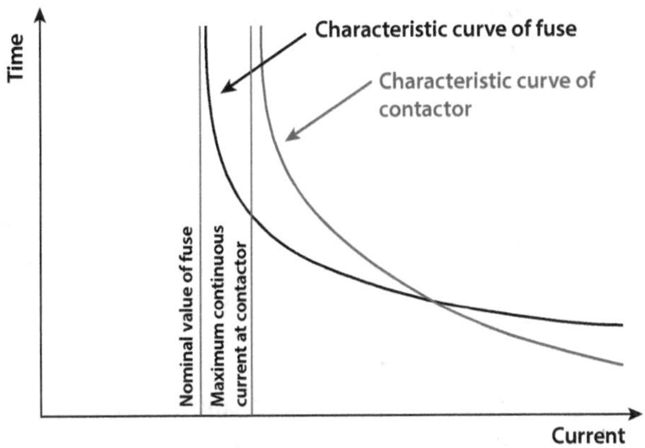

Fig. 12.6 Characteristic curves of fuse and contactor

must provide a sufficient safety margin to cover the necessary vibration and shock resistance of the contactor.

Another scenario that needs to be considered is a "creeping" short circuit. This is the case when the short-circuit current significantly exceeds the regular operation current, but is limited by remaining impedances in the short circuit. This leads to considerably increased reaction times of the fuse, of up to several minutes, depending on the magnitude of the short-circuit current. If the main contactors are switched off by the monitoring system after the above-mentioned waiting period, they must switch off the short-circuit current. The required switching capability can be much higher than 1,000 A, depending on the system.

The employed contactors need to be able to switch these maximally expected overcurrents and their current-carrying capability must also suffice to cover the expected tripping times of the fuse. Fig. 12.6 shows an example of the expected reaction times of a fuse as a function of the overcurrent (solid blue line). It also displays the maximum permissible time for a contactor to carry an overcurrent (solid red line).

If the two curves cross (as shown in Fig. 12.6), this might lead to contact welding, at appropriate currents.

12.4 Design examples

Different design principles have proven of value for optimally fulfilling the above-mentioned main contactor requirements. Controlling the electric arc that occurs during switching is of crucial importance for the switching capacity. Many designs therefore integrate hermetically sealed contact chambers that are filled with a shielding gas. The most common shielding gases are nitrogen and hydrogen. The gas has three main functions: It cools the arc; the gas pressure compresses the arc, and it effectively protects the contact materials from corrosion. Thanks to this corrosion protection, it is not necessary to use precious metals such as silver and pure copper can be used for the contacts instead. Fig. 12.7 shows an example of the basic construction of a nitrogen-filled contactor for continuous currents up to 135 A.

There are also alternatives to gas-filled designs. In these designs, the contact distances are usually larger to reliably extinguish the switch arc. Their contact chambers are not hermetically sealed in general; the atmospheric pressure in these chambers is the same as that of the ambient air. This means that a reliable switching function must be ensured even at expected altitudes.

An additional measure that is used in both gas-filled and regular contact chambers is blowing magnets. Their magnetic field deflects the arc plasma in such a way that the electric arc's length is increased and its diameter reduced. Both effects result in the arc extinguishing more quickly. The blowing magnets are positioned so that their magnetic field vector is perpendicular to the direction of the electric arc. Bridge contacts are used with two fixed contacts and one mobile contact bridge that connects the fixed contacts. This results in two electric arcs, which are then targeted by one magnet pair each. If the polarity is chosen correctly, the electric arcs are deflected by the force F to opposing sides of the contact chamber (Fig. 12.8).

The direction of deflection is dependent on the direction of the load current: In the reversed current direction, the electric arcs are deflected toward the center, instead of toward the outside. This could cause both arcs to merge, generating a short circuit between both fixed contacts. When using this contact type, the preferred direction of the load current must be taken into account, and it must be ensured that the permissible switching current in the reverse direction is never exceeded.

This is of special importance for electric or hybrid vehicles when they are equipped with kinetic energy recuperation systems and if there is the possibility that the contactor is switched during a recuperation phase.

The requirements with regard to the switching and current-carrying capabilities of high-voltage contact systems necessitate that these systems are designed with considerable contact forces and comparably large contact distances. This means that strong coils with high power rating are required for switching the contact systems. Coil designs with lower power rating can only be achieved through very large coils. However, these coils have the disadvantage that they are big, heavy, and expensive.

Therefore, the coil current is often reduced to a holding value after the pull-in so that the full coil power does not have to be continuously maintained during the on-state of the contactor. This reduction is usually achieved by pulsing the coil

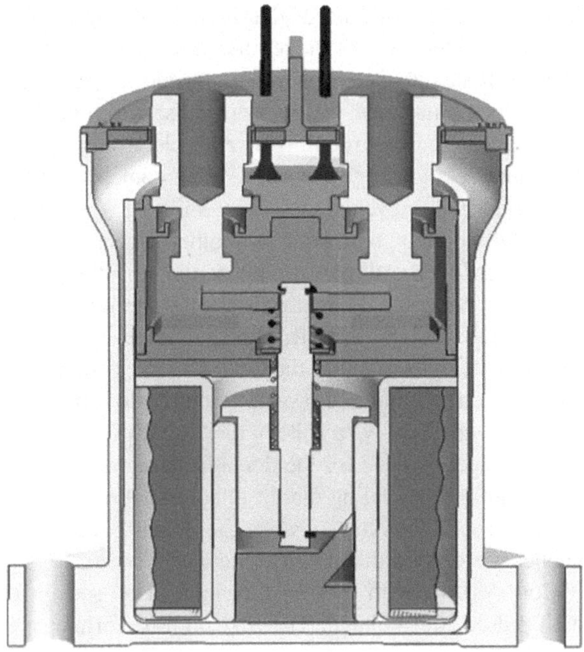

Fig. 12.7 Example of nitrogen-filled contactor

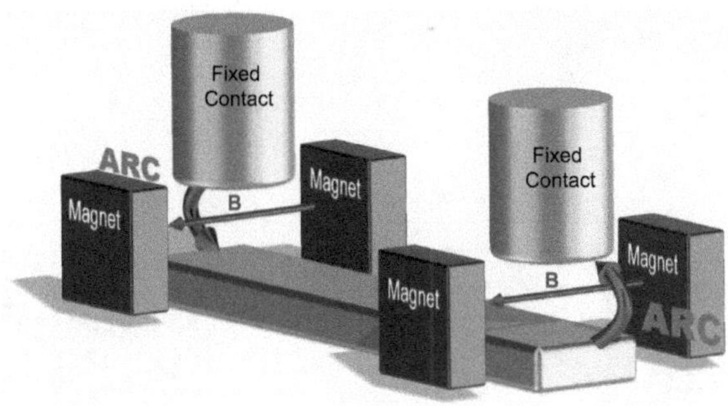

Fig. 12.8 Effect of blowing magnets on bridge contact

voltage with a frequency in the range of 15 to 20 kHz. During the blanking interval, the coil current flows due to self-induction via a free-wheeling circuit with a diode in parallel to the coil. This creates a mildly pulsing coil current, which is only a fraction of the current required to close the contactor (Fig. 12.9).

The pulse frequency must be high enough to prevent a buzzing noise from occurring and, potentially, a micro movement of the switching contacts, which could damage the contactor.

The mean current resulting from the pulsing is dependent on the coil winding's resistance and, therefore, on the temperature that strongly influences the coil resistance. The coil current that is necessary to safely hold the contacts closed is virtually independent of the temperature, however. Therefore, a reduction in current ideally creates a steady mean coil current by adapting the duty cycle to the coil's actual "hot resistance" by means of pulse width modulation (PWM). PWM control is often realized by mounting a small electronic device ("economizer") directly in the contactor housing. External PWM controls are also implemented, which are integrated into the vehicle's control unit.

So-called demagnetization is another aspect that needs to be taken into account with regard to the dimensioning of the coil control. The energy that is stored in the contactor coil's magnetic field must be reduced as quickly as possible when the contactor is powered off. This guarantees that the contact opens at an optimum speed, which ensures that the arc's burning time is minimized. The free-wheeling diode described above, however, would allow the free-wheeling current to decay comparatively slowly, in spite of the "hard" powering off of the coil. To prevent this, a Zener diode is used for the demagnetizing process. This Zener diode is built into a separate free-wheeling circuit in combination with an antiparallel standard diode (Fig. 12.10). The voltage drop at the Zener diode ensures a suitably quick decrease in energy stored in the magnetic field.

Coil systems with two windings are also used instead of the described control with pulsed coil voltage. The first high-performance winding is used for pull-in, and the second, weaker winding is used for holding. It is also possible to combine both windings in the contactor for pull-in and then switch off one of them during holding. This is usually achieved by a small electronic circuit mounted in the contactor housing.

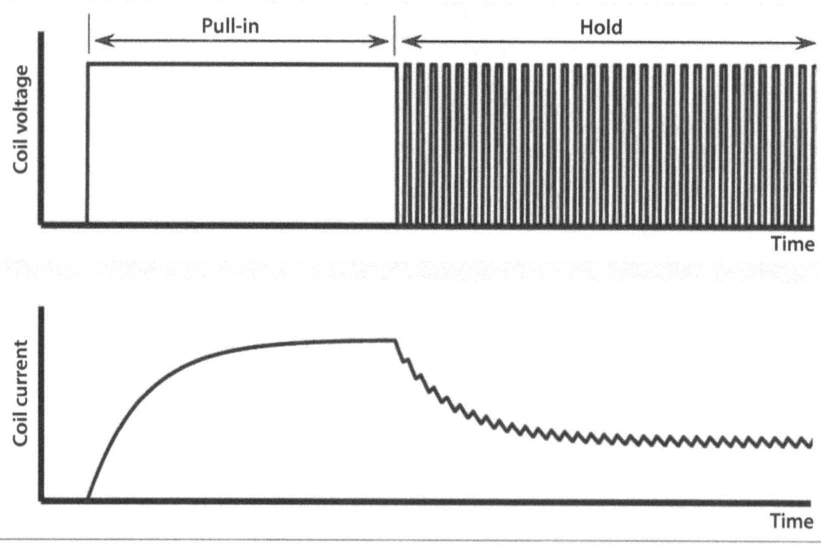

Fig. 12.9 Coil current reduction by means of pulsed voltages

Fig. 12.10 Free-wheeling circuit

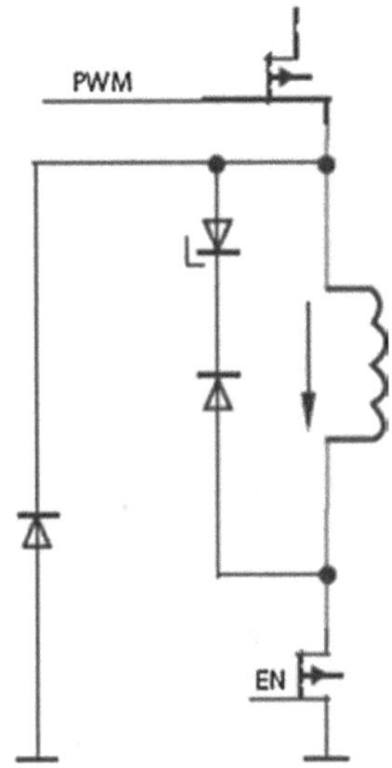

Advantages and disadvantages of different designs Gas-filled contact chambers have proven reliable for most industrial applications of contactors. For the new automotive applications, however, it is necessary to examine the resulting costs. Car manufacturers go to great efforts to reduce the costs for alternative drive systems. Atmospheric contact chambers are advantageous here, because they do not require an elaborate gas-filling process and a hermetically-sealed contact chamber.

Furthermore, the gas-filled contact chambers pose an additional safety and reliability risk because it cannot be entirely ruled out that gas is released during service life, for example, because of prior damage. This could reduce the contactor's switching capabilities, possibly causing a system failure.

Contactors with an integrated economizer or with a two-coil system and integrated electronic coil switch have the advantage that a reduction of coil power is not necessary and thus no PWM control, for example, is required. On the other hand, the circuits and the microprocessor in the control units combined with the existing relay drivers generally allow PWM control just by implementing appropriate software routines. It is cost-effective in such cases to realize PWM control within the control unit. This alternative does not produce additional costs, contrary to the integration of coil power reduction into the contactor.

12.5 Future contactor developments

Developments in the sector of high-voltage contactors for hybrid and electric vehicles will focus on future contactor requirements. In principle, tomorrow's lithium ion-based battery systems are expected to exhibit even lower internal resistances than today's. As a result, the short-circuit currents in the batteries will increase accordingly. On the other hand, implementing smaller battery systems with lower power ratings and reduced energy content in compact hybrid and electric vehicles will reduce the short-circuit current requirements.

Fuel cell-driven vehicles, as opposed to hybrid-only and electric vehicles, will probably provide considerably higher traction power. This means that not only the continuous current requirements will increase, but also that the short-circuit currents will rise accordingly.

The safety requirements for the on-board HV system will have a decisive impact on the employment of high-voltage contactors and their required characteristics [5]. The status quo is that it must be possible to securely shut down both poles of the on-board HV system by means of the main contactors under all conditions. Depending on the expected short-circuit currents, only contactors with a contact system that has been dimensioned accordingly are able to provide this function. An in-depth analysis of all conceivable operational conditions and the corresponding system parameters is of utmost importance for defining the specific requirements for all system components in order to ensure that the system is permanently safe.

Such an analysis therefore needs to be conducted on system level. Thus, in the future, close cooperation between the manufacturers of battery systems, contactors, fuses, and wiring systems will be increasingly important to perfectly balance these components with each other and with the overall system.

12.6 Lithium-ion battery wiring

Lithium-ion batteries for hybrid and electric vehicles are complex systems consisting of many accumulator cells. Alternative drive systems require a voltage level of several hundred volts. To achieve this, the single cells are combined to form accumulator modules, which are then combined to form the final cumulative accumulator. To ensure permanent reliability of different states of charge, the voltages and temperatures need to be monitored meticulously. This is the only way to guarantee early fault detection and reliable warning functions. High amperage must be transmitted within the cells and between the modules. The power provided by the battery is eventually used by the high-voltage components of the system.

Different types of connecting cables are required to fulfill all these tasks: cables with large cable cross-sections for high currents with high voltages and cables with small cable cross-sections for high-voltage signals or low voltages. The requirements related to these cables are very different.

12.7 Cable requirements

Lithium-ion battery wiring usually consists of a low-voltage harness connecting the sensor technology with the control unit and a high-voltage harness connecting the modules and providing power for the electric loads. Signal wiring is similar to that of a conventional on-board electrical system. This is why many low-voltage cables are used that comply with the prevailing standards, e.g., the German delivery specification LV 112 [6] and ISO 6722 [7]. These cables are off-the-shelf standard products that are well-established on the market. Low-voltage cables have the lion's share of the market (90 %) of cables installed in high-voltage batteries. Battery designers have to ensure that the low-voltage harness does not come into contact with high-voltage potentials or cables.

High-voltage wiring consists of special orange signal-colored cables. These cables must fulfill increased requirements:

- designed for voltages of max. 600 V
- suitable for transmission of high amperages
- transmission of short-term current peaks while preventing overheating
- highly flexible
- suitable for small bending radii.

The last two requirements are due to the constricted space within the batteries. Short cables are extremely critical in this respect. Depending on the concept, the cables are welded during production, subjecting them to high short-term temperatures up to 200 °C. This needs to be taken into account when determining the cable materials. Special electromagnetic shielding is not necessary for high-voltage cables used within the lithium-ion battery due to the metal housing. The opposite applies to the connections from the high-voltage battery to the devices inside the vehicle. These need to be additionally shielded to provide protection from electromagnetic radiation.

Contrary to the standard cables, the high-voltage cables used are mainly new or customized products. Current efforts aim to standardize the geometry and requirements for these components.

12.8 Wiring cables

Standard low-voltage cables with typical cross sections of 0.35 to 0.5 mm^2 connect the battery control unit (BCU), the sensor electronics, and the external temperature sensors. Thin-wall PVC-isolated cables are used, designed for voltages up to 60 V DC. The cables are cut to the required length, furnished with connectors or cable lugs, and then combined into a harness as shown in Fig. 12.11. The total length of such a set of cables can reach 30 to 50 meters, depending on the application.

A distinction is made between the high-voltage cables inside a battery depending on their usage: sensor connections or current transmission connections. Special high-voltage relays are used to reliably disconnect the battery from the on-board electrical system when in the passive state or in the event of malfunction. These relays are connected by means of an unshielded high-voltage cable with a cross section of 0.5 mm^2. The cables have very thick walls to cater for applications with

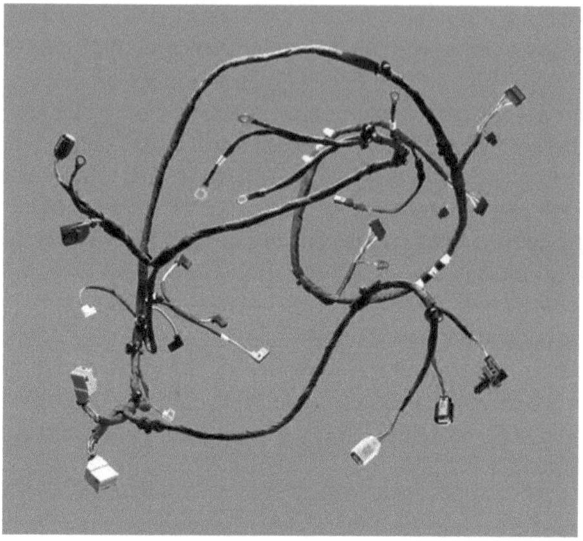

Fig. 12.11 Lithium-ion battery low-voltage harness (*Source* LEONI)

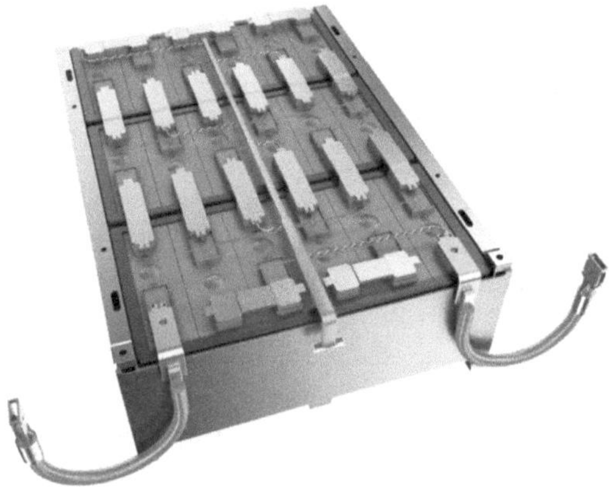

Fig. 12.12 Lithium-ion battery module with high-voltage cable for module connection

voltages up to 600 V because voltage of several hundred volts are present. The small cross section is sufficient because this connection only needs to transmit high-voltage signals and not power.

Similar cables are used to supply the temperature sensors. This is necessary because of the potential voltage that is present. Current-carrying high-voltage cables have much larger cross sections, depending on the power they have to transmit.

Unshielded cables with a cross section of 10 mm^2 or higher are used to connect several modules with each other (Fig. 12.12). This situation also requires isolation

Table 12.1 Overview of isolation materials for high-voltage cables [LEONI]

Continuous use temperature	105 °C	125 °C	150 °C	175 °C	180 °C/200 °C
Isolation material	PVC PP	Cross-linked PE TPE-S TPE-O	Cross-linked PE TPE-E	ETFE	FEP silicone

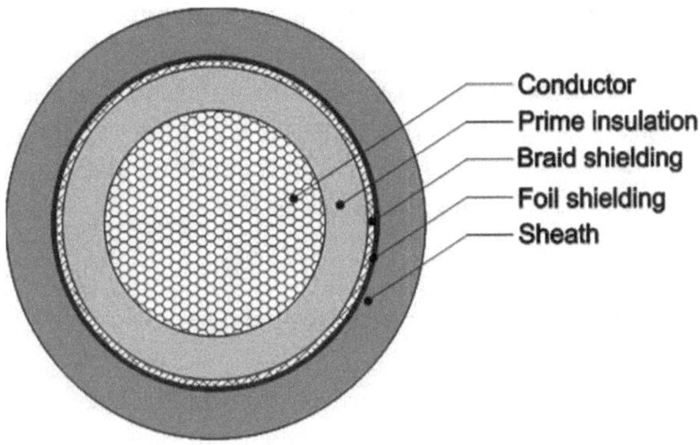

Fig. 12.13 Composition of a shielded high-voltage cable (*Source* LEONI)

that is suitable for 600 V. The very confined space makes the usage of very soft isolation materials and special highly flexible conductor designs a necessity. It is also possible to save space by using narrower cross sections with smaller cable diameters.

However, this leads to self-heating of the cable due to dissipation losses, which needs to be taken into account when determining the materials and dimensioning the temperature management. Table 12.1 provides an overview of high-voltage cable standard isolation materials with different continuous use temperatures.

The current state-of-the-art for connecting cells within a module is metallic lead frames. Compared to connections by means of highly flexible copper meshwork, they need less space and are less expensive. Because they can be welded, very low resistances are achieved as an additional benefit.

The on-board HV system begins outside of the lithium-ion battery and spreads across the vehicle. Two-phase connections are generally accomplished with single-core shielded cables, as shown in Fig. 12.13. Depending on the battery power, the cross sections can range from 16 to 50 mm², sometimes even higher. High thermal stability of the cables is necessary especially in hybrid vehicles, because the cables are placed next to the exhaust line, subjecting them to high external temperatures in addition to self-heating due to dissipation losses.

12.9 Future cable developments

Different harnesses are necessary to integrate the lithium-ion battery into the high-voltage system of the vehicle. Especially components that monitor the system are connected within the battery with a low-voltage harness.

Due to the function and topology of lithium-ion batteries, high-voltage cables are also needed. They are responsible for connecting the sensors and transmitting power. These cables have to fulfill particular requirements, which necessitates special construction of the cables. Another important aspect is the interface to the connectors; it needs to be separately adapted.

Series production and the resulting experience will probably lead to a reduction of high-voltage cable cross sections in lithium-ion batteries, subsequently reducing weight and space requirements. In the future, another option for reducing weight is to replace the current-carrying copper conductors with large cross sections by equivalent aluminum conductors.

12.10 Connectors and terminals

Transmitting electrical power in hybrid, fuel cell, and battery-powered vehicles has led to specific requirements and, in turn, to a new generation of automotive high-voltage connectors. The high-voltage system and 14-V on-board electrical system are kept separate in vehicles. The high-voltage components and their installation conditions define the requirements for the high-voltage connectors. The main components to be connected by means of high-voltage connectors are shown in the overview of the high-voltage on-board electrical system (Fig. 12.14):

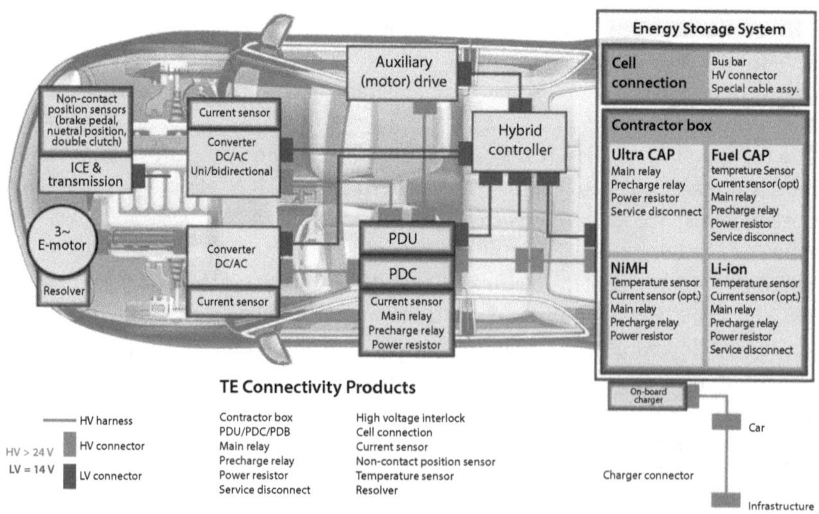

Fig. 12.14 Overview of the high-voltage on-board electrical system

- energy storage systems (battery, supercapacitors, fuel cells, etc.)
- high-voltage switch and distribution units with high-voltage components (relays, power resistors, fuses)
- voltage converter HV-DC/LV-DC for the supply of the 14-V on-board electrical system
- voltage converter HV-DC/HV-DC for the supply of the electric drive
- electric motor
- charging interface to the grid.

The application of the standards to the electrical path of the plug-in vehicle charging sockets is different. The on-board high-voltage circuit is generally separated from the mains power supply in the vehicle's charging unit. Therefore, not only automotive standards need to be heeded but also electrical engineering sector standards.

12.11 Product requirements

If the various product requirements were not 'bundled', there would be a vast range of different high-voltage connectors available for on-board HV systems. These requirements have therefore been defined in LV 215-1 [8] by a working group of the German automobile manufacturers.

This approach has resulted in the standardization of high-voltage connectors for use with different vehicle components. More than 50 product requirements have been specified to enable their use under as many operating conditions as possible.

The most important product characteristics are:

- maximum operating voltage 850 V DC
- ambient operating temperature −40 °C to 140 °C
- touch protection according to IPX2B (VDE test probe) [9], unplugged
- connectors orange or marked with labels
- reverse-polarity and inversion protection
- leading signal contact (interlock) to power off the power side
- shielded connectors only
- shielding by means of transition resistance of > 10 mΩ, entire service life
- current-carrying capability of the shield to handle permanent current loads of 10 A and fault currents of 25 A for 60 s
- power contacts with primary and secondary lock
- unique coding and sufficient connector alignment
- maximum insertion force 100 N
- vibration load up to severity level 3 in accordance with LV 214-1 [10]
- impermeability in accordance with IP6K9K, IPX7, plugged-in [9]
- single-wire and multi-wire shielded cables permissible
- clearance and creepage distance in accordance with IEC 60664-1 [11]

- voltage and isolation stability in accordance with ISO 6469-3 [12]
- screwed-in terminals with power transmission by means of cable lugs
- high-voltage connector and terminal classification into different categories based on cable cross section and current-carrying capability.

Many of these product requirements have a major impact on the design of high-voltage connectors and were not required in this combination up to now. In comparison to a 14-V connector, a maximum operating voltage up to 850 V requires much larger clearance and creepage distances. Optimized isolation designs are also necessary to minimize the space needed for the connectors. Ambient temperatures up to 140 °C most probably occur only for certain applications. However, these specifications have been defined as a standard to ensure universal applicability.

Touch protection according to IP2XB [9] in unplugged condition makes additional design measures necessary for both the connector and the opposing multi-pin connector. These are, for instance, additional walls to prevent penetration (Fig. 12.15). From cable cross sections of around 25 mm^2, the use of a round contact with finger guard is the safest way of preventing contact with live wires.

The leading signal contacts, also referred to as high-voltage interlock (HVIL), are disconnected before the power contacts when the connector is removed from the multi-pin connector. The evaluation unit evaluates the interlock signal and subsequently powers off the power side. Otherwise, an electric arc could occur in presence of direct voltage at the power contacts. This arc could destroy the power contacts and endanger the operator.

Reverse polarity protection and coding with sufficient connector alignment are requirements set out in the specifications in LV 215-1 [8]. Only sealed and shielded connectors are specified in LV 215-1.

The required shielding is determined by the EMC characteristics as well as its direct current contact resistance of < 10 mΩ over the entire service life.

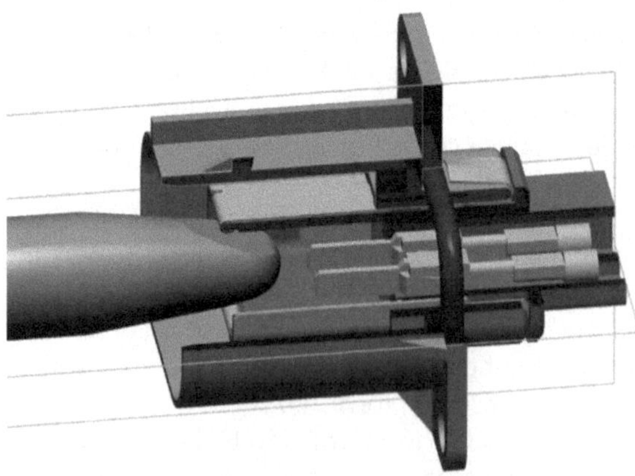

Fig. 12.15 Multi-pin connector with VDE probe

Furthermore, shielding must be able to handle a continuous current of 10 A and a fault current of 25 A for 60 s. Highly conductive copper alloys with good contacting characteristics are needed to fulfill these requirements. The connection between cable shielding and connector shielding, the connection between connector shielding and multi-pin connector, and the shielding between multi-pin connector and devices such as inverters, e-motors, and batteries must all comply with the 10-mΩ requirement. There are two possibilities for cable shielding: shielding of a single wire or of several wires. The connector is designed for one of the concepts.

Only inserting and removing aids can provide the maximum insertion force of 100 N necessary for cable cross sections of, for example, 50 mm^2. Usually, a lever is used to assist with removing and inserting the connectors.

The greatest challenge for high-voltage connectors is vibration loads up to severity level 3 in accordance with LV 214-1 (when installed near the combustion engine). Due to the large mass of the cable and the connector, the resulting resonance frequency of the entire system is much less than that of the lighter 14-V connectors. The vibration profile used can therefore cause considerable wear to the contact points without elaborate measures. Simple countermeasures are to effectively tie the components and to reduce the cable weight by means of an aluminum conductor. Screwed-in terminals have the best vibration characteristics thanks to their design and should be used for high-vibration loads.

The classification of high-voltage connectors and screwed-in terminals into different categories is based on cable cross section and current-carrying capability. Connector and terminal systems designed in accordance with LV 215-1 [15] are discussed below.

12.12 High-voltage connectors and screwed-in terminals

High-voltage connectors and terminals are classified in five classes (Table 12.1):

- Class 1 (25 A): cable diameter 2.5 mm^2 up to and including 4 mm^2 (connector)
- Class 2 (40 A): cable diameter 4 mm^2 up to and including 6 mm^2 (connector)
- Class 3 (80 A): cable diameter 6 mm^2 up to and including 16 mm^2 (connector)
- Class 4 (200 A): cable diameter 16 mm^2 up to and including 50 mm^2 (connector and terminal)
- Class 5 (400 A): cable diameter 70 mm^2 up to and including 120 mm^2 (terminal)

All connectors and terminals shown are able to accommodate shielded high-voltage cables in accordance with LV 216-2 [13]. This high-voltage cable standard defines the dimensions as well as the approval tests for shielded copper and aluminum cables (Fig. 12.16).

Two-pole or three-pole high-voltage angle connectors (90 or 180 degrees) are suitable depending on the application. Mostly, the latter are used.

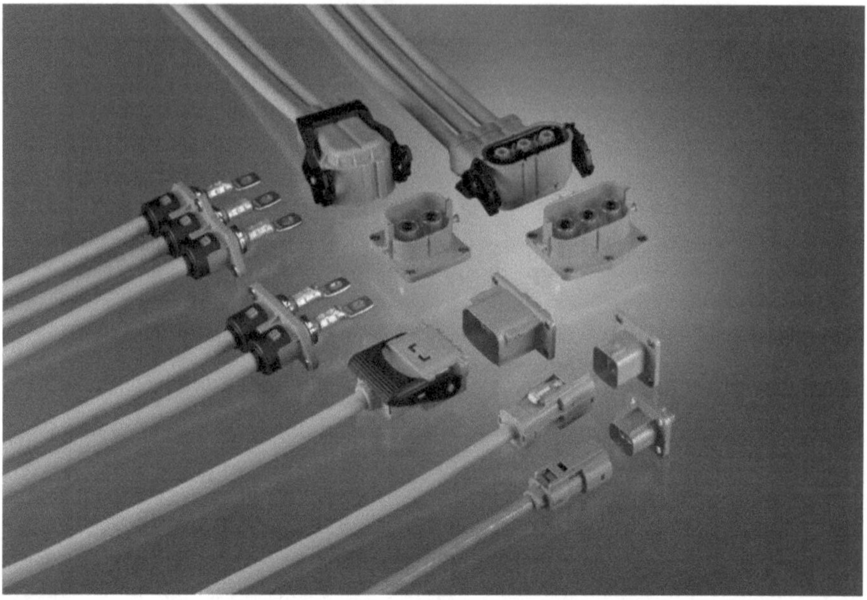

Fig. 12.16 Connectors and terminals of Classes 1, 2 and 4

12.13 Charging sockets

The charging sockets in accordance with IEC 62196-1/-2 [14, 15] are categorized into three classes. All types are tested with 10,000 cycles of operation under contamination. The maximum ambient temperature during charging is specified at 50 °C. To ensure heating wire resistance of the plastic housing, special additives, which decrease processability and impair the material's mechanical characteristics, are required. The high-voltage connectors in the vehicle are not subject to this requirement.

The maximum voltage for a charging socket type 2 (Fig. 12.17 [left]) is 500 V. The design shown enables transmissions of 70 A (one-phase) or 63 A (three-phase). The three-phase connector (Fig. 12.17 [right]) connects the type 2 charging socket with the internal charger and, therefore, the on-board HV system with the mains (Fig. 12.14).

12.14 Future connector and terminal developments

Transmission of electrical power plays an important role in the overall system of hybrid, fuel cell, and battery-driven vehicles. High-voltage connectors and terminals have to comply with different requirements that depend on the system used. The modular five-class system has led to a reduction in the number of variants while at the same time covering almost all applications.

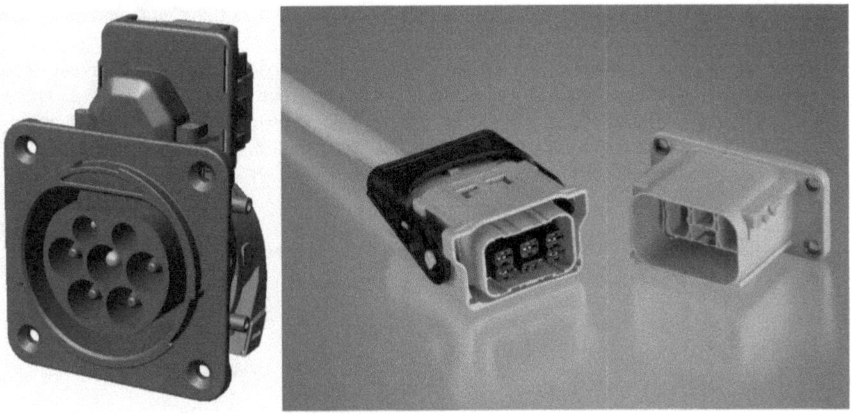

Fig. 12.17 Type 2 charging socket (*left*), three-phase connector (*right*)

A holistic approach is necessary to further optimize high-voltage connectors and terminals for the requirements in relation to the on-board HV system. With an approach of this kind, the connector would no longer be considered merely as a separate part but as part of the whole. Today, derating graphs in accordance with LV214-1 [10] provide test parameters that are only an indication of the current-carrying capability under real operating conditions.

For example, advanced approaches integrate lower cable cross sections, thus reducing the costs and weight of the overall system.

Bibliography

1. ISO 26262 "Road vehicles – Functional safety"
2. OEM white paper "Elektrische Eigenschaften und elektrische Sicherheit von Hochvolt-Komponenten in Kraftfahrzeugen – Anforderungen und Prüfungen" (Edition 2009 – 05)
3. Kroeker M, Faul J LV 123 Besondere Anforderungen an HV–Schütze in Hybrid– und Elektrofahrzeugen, TE Connectivity
4. Kroeker M (2011) Untersuchungen der Stromtragfähigkeit und des Schaltvermögens von Kontaktanordnungen in nicht hermetisch gedichteten Schaltkammern bei 400 V, 21. Albert Keil Kontaktseminar Karlsruhe, Germany, VDE Fachbericht 67
5. ISO 6469 Electrically propelled road vehicles – Safety specifications
6. LV 112 (2006) Elektrische Leitungen für Kraftfahrzeuge einadrig, ungeschirmt, November 2006
7. International Standard ISO 6722–1 (2011) Road vehicles – 60 V and 600 V single-core cables – Part 1, Dimensions, test methods and requirements for copper conductor cables, 4th edition, October 15, 2011
8. LV 215-1 (2011) Elektrik/Elektronik Anforderung an HV-Kontaktierungen
9. IEC 60529 (1991) (DIN 40050-9) Schutzarten durch Gehäuse (IP-Code)
10. LV 214-1 (2010) KFZ Steckverbinder – Prüfvorschrift
11. IEC 60664-1 (2008) Isolationskoordinaten für elektrische Betriebsmittel in Niederspannungsanlagen
12. ISO 6469-3 (2010) Electrically propelled road vehicles – Safety specification – Part 3: Protection of persons against electric shock

13. LV 216-2 (2011) Hochvolt – Mantelleitungen geschirmt für Kraftfahrzeuge und deren elektrische Antriebe
14. IEC 62196-1 (2011) Stecker, Steckdosen, Fahrzeugsteckvorrichtungen und Fahrzeugstecker – Konduktives Laden von Elektrofahrzeugen Teil 1: Generelle Anforderungen
15. IEC 62196-2 (2010) Stecker, Steckdosen, Fahrzeugsteckvorrichtungen und Fahrzeugstecker – Konduktives Laden von Elektrofahrzeugen Teil 2: Anforderung an und Hauptmaße für die Austauschbarkeit von Stift und Buchsenvorrichtungen

Battery thermal management 13

Achim Wiebelt and Michael Guenther Zeyen

Contents

13.1 Introduction

A lithium-ion battery for mobile applications needs efficient thermal management to guarantee the required service life of more than 10 years as well as full performance and availability under all operating and environmental conditions. Thermal management comprises cooling as well as heating. Lithium-ion batteries are available in a multitude of variants, with considerably differing cell chemistry and design. Thermal management therefore must be adapted to the respective variant's requirements. This also applies to the large number of different vehicle applications, ranging from the subcompact car to the sports car and from the low-level hybrid vehicle to the entirely electricity-driven vehicle. These different types impose different thermal management requirements on the drive battery. Integration into the vehicle's thermal management also plays an important role. The car manufacturers pursue approaches that vary immensely.

A. Wiebelt (✉)
Behr GmbH & Co. KG, Heilbronner Strasse 393, 70469 Stuttgart, Germany
e-mail: achim.wiebelt@behrgroup.com

M. G. Zeyen
vancom GmbH & Co. KG, Marie-Curie-Strasse 5, 76829 Landau, Germany
e-mail: m.zeyen@vancom.de

R. Korthauer (ed.), *Lithium-Ion Batteries: Basics and Applications*,
https://doi.org/10.1007/978-3-662-53071-9_13

In the following Chapter, the necessity of efficient thermal management will be explained in detail, and concrete solutions will be presented and assessed.

13.2 Requirements

The optimum operating temperature for a lithium-ion battery ranges from 20 °C to 40 °C (Fig. 13.1). Its performance rates peak in this range while showing acceptable aging characteristics. If a longer service life is required, cooling should be configured closer to the low temperature limit because cell aging increases with temperature.

Operating temperatures below 20 °C raise internal resistance disproportionately as the temperature decreases. This negatively influences battery performance and thus decreases the power available to drive the vehicle.

Below 0 °C, this performance deficit can be as high as 30 % and even higher, below −20 °C. In addition, special aging mechanisms occur at temperatures below zero that can lead to irreversible cell damage. The most important of these mechanisms is so-called lithium plating, which is when pure lithium is deposited on the anode during charging. This reduces cell capacity, causing an internal short circuit at worst, when the plated lithium forms dendrites stretching from the anode to the cathode. Of course, a suitable operational strategy can prevent lithium plating and, therefore, aging by preventing energy intake at battery temperatures below zero, because it only occurs in these conditions. However, kinetic energy recovery would also not be possible, reducing the overall efficiency of the hybrid vehicle and the range of the electric vehicle in a way that would not be acceptable. Thermal management's task here is to quickly and efficiently heat the cells up to temperatures above zero. This enables full availability and battery performance.

Aging also plays a role at operating temperatures above 40 °C at which lithium-ion cells age disproportionately quickly. These mechanisms can be explained with the Arrhenius equation.

As a rule of thumb, service life decreases by 50 % when the operating temperature is raised by 10 K. Assuming a cell operating at a permanent operating temperature

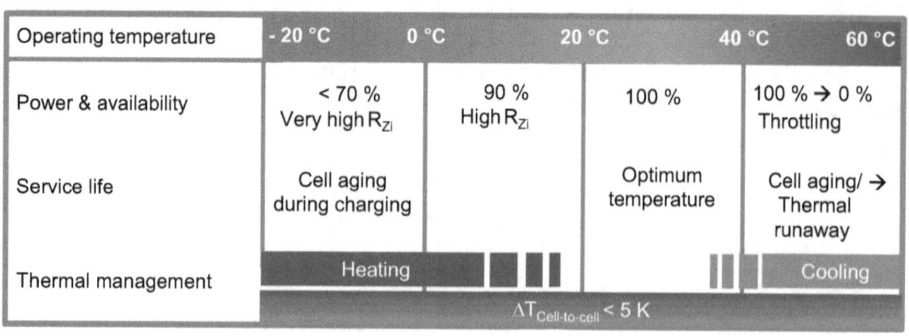

Fig. 13.1 Correlation of operating temperature, thermal management, and service life

of 40 °C has a service life of 10 years, this service life would be reduced to five years at a permanent operating temperature of 50 °C. At even higher temperatures, it is possible that the electrolyte thermally degrades and sets the cell on fire. This is why the cells need to be cooled in such a way that the maximum permitted temperature is never exceeded in any environmental or operating conditions. If this temperature limit is exceeded, the battery management system must decrease the battery output.

However, keeping the battery temperature from exceeding the specified maximum temperature is not the thermal management's only task. It also keeps the battery's temperature as homogeneous as possible. Cell temperatures, measured at the same location of each cell, should not differ by more than 5 K. This poses a particularly difficult challenge in large-scale batteries in which the distance between cells can reach more than one meter. If a homogeneous cell temperature is not maintained, the cells age differently, which in turn increases the balancing outlay required to keep the cells in the same state of charge. This reduces the battery's usable energy content.

13.3 Cell types and temperature balancing methods

The foundation for efficient thermal management is an efficient cell cooling concept on cell level. This concept depends on the cell's type, outer dimensions, inner design, and on the amount of heat that needs to be dissipated. Fig. 9.4 shows the currently available cell types. The active material of the round cell and of the prismatic hardcase cell consists of electrodes and separators, is generally wound up and installed in a stable aluminum housing. In pouch cells, however, the individual layers of active material are stacked or folded and packed in a flexible aluminum composite foil. With regard to cooling, round cells have a geometrical disadvantage compared to prismatic or pouch cells. They have a relatively low surface-volume ratio, and heat transport is impeded from the inside to the surface, thus increasing radial temperature gradients inside the cell. The rounded outer surface also prevents a thermally perfect connection to the structures that dissipate the waste heat to a heat sink. Round cells are nevertheless often deployed nowadays because cooling is not the only factor that influences the choice of cell type; other factors are availability, maturity, safety, service life, and, last but not least, costs.

All three cell types show very high thermal conductance (λ_{II}) laterally to the electrodes (20 $Wm^{-1}K^{-1} < \lambda_{II} <$ 50 $Wm^{-1}K^{-1}$). This is due to the physical property that good electrical conductivity is always accompanied by good thermal conductivity. In the perpendicular direction to the layers, thermal conductance $\lambda_{\perp}$ is lower by two orders of magnitude (0.5 $Wm^{-1}K^{-1} < \lambda_{\perp} <$ 2 $Wm^{-1}K^{-1}$). The range of thermal conductivity specified is caused by the different active materials used by different cell manufacturers. Another reason is the cell's electrical performance requirements. Electric vehicle cells have a slightly different internal design and therefore different thermal characteristics to hybrid vehicle cells.

Fig. 13.2 shows a diagram of cell cooling principles.

Air cooling (Fig. 13.2a): Cooling air moves around the cell and cools the freely accessible surfaces. This type of cooling does not require thermal contact to the cell, facilitating design of the cooling system interface. For this reason, such an interface is often the first choice for pragmatic reasons. This cooling concept requires a large amount of space between the cells to install the cooling ducts, however. Its cooling efficiency and homogeneity are frequently insufficient.

Other cooling concepts, which integrate a thermal connection to the cell and are based on thermal conductance, are advantageous because they require less space and are more effective. Designing thermal connections, however, poses new challenges. The interface with the cooling system is more complex because there is a direct connection between the cooling device and the battery's electrical components. For low-height cells with sufficiently thick walls, a cooling plate in direct contact with the cell's bottom is adequate (Fig. 13.2b). If more cooling is required, additional thermally conducting elements (Fig. 13.2c), which support heat dissipation to the cooling plate, can be installed between the cells. However, as a result, the cooling system is heavier and more expensive.

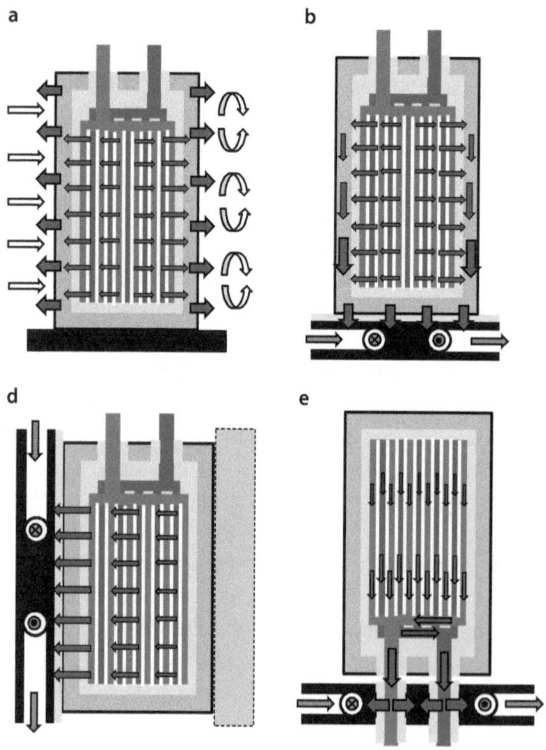

Fig. 13.2 Cell cooling principles: **(a)** air cooling, **(b)** bottom cooling, **(c)** side cooling (passive), **(d)** side cooling (active), **(e)** conductor cooling

Thermodynamically, the best solution is cooling plates carrying coolant that are located between the cells (Fig. 13.2d), as the thermal conductance path from the cell's interior to the coolant is very short and, therefore, high temperature homogeneity can be achieved inside the cell. This solution is, however, the heaviest and the most expensive.

Conductor cooling is an exceptionally effective cooling technology (Fig. 13.2e). The electrodes are used to directly cool the cell interior. Pouch cells are particularly suited for such systems because they have flat conductors which enable a good connection. All of the above-mentioned technologies dissipate the cells' waste heat to a cooling plate. The heat is either absorbed by a coolant or by an evaporating refrigerant and is discharged into the environment via the vehicle's cooling system.

13.3.1 Battery cooling

Especially in summer, the surrounding air is not suitable for cooling lithium-ion batteries. The difference between the surrounding temperature and the maximum permitted battery temperature is not large enough to effectively dissipate the battery's waste heat. There is only one controllable heat dissipation technology in the vehicle that can ensure full battery availability in warm conditions: the refrigerant circuit of the air conditioning system. There are three ways of dissipating a battery's waste heat: with the conditioned air, with the refrigerant from the air conditioning system's refrigerant circuit, or with a separate cooling circuit (Fig. 13.3). The method

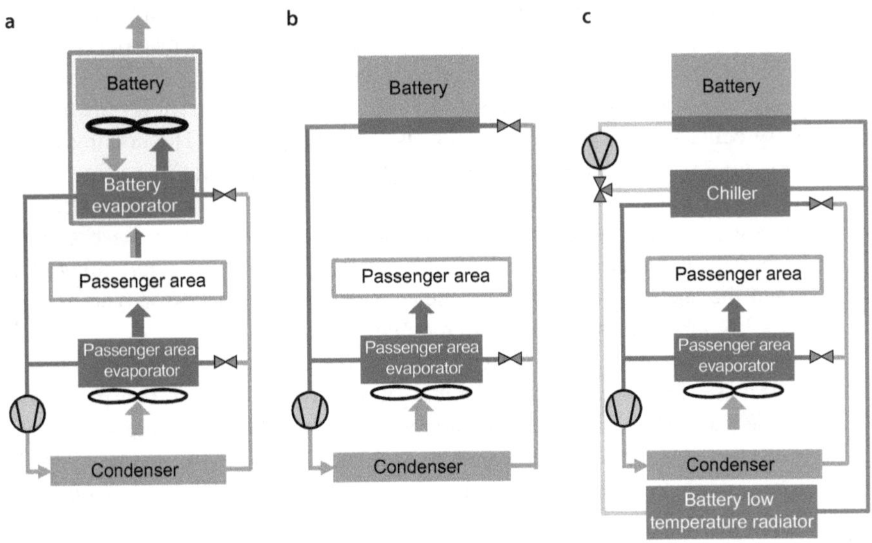

Fig. 13.3 Battery cooling concepts: **(a)** with air, **(b)** directly with refrigerant, **(c)** with coolant (secondary circuit)

deployed depends on the usage profile of the respective vehicle type. Each method has its own specific advantages and disadvantages; all methods are currently being used in series production and will be presented below.

Air cooling The cooling air is guided to the battery through large ducts. As the air moves through the battery and along the cells, it is heated up and usually discharged into the environment. This is a very simple system with several built-in disadvantages such as the large air ducts to and from the battery, the fan's weight, possible unwanted noise in the vehicle interior, and safety issues if air from the passenger area is used and if there is a direct connection between the battery and the passenger area. The air must be filtered to prevent it from contaminating the battery's interior, as contamination can cause creepage currents or impede heat transfer.

If air from the passenger area is used for cooling, the battery can only be cooled at the same time as the passenger area is being cooled. This can result in a conflict of interest between cooling the battery and maintaining comfort in the passenger area. This issue can be solved by installing a small separate air conditioning system for the battery, similar to the systems at the rear of luxury-class sedans, and connecting it in parallel to the passenger area's air conditioning system. Such a system is displayed in part (a) of Fig. 13.3. The battery evaporator cools the battery with airflow from a fan. Its disadvantage is that it requires additional weight and space, further reducing the system's energy density. Its advantage is that battery cooling operates in a recirculated-air mode, and a filter system is not needed. This technology is mainly used in large vehicles that have sufficient space for it, such as SUV's.

Direct cooling with refrigerant Direct cooling with refrigerant is the most compact battery cooling method. This method uses a compact evaporator for a battery cooling plate, which is mounted inside the battery. This evaporator has a thermoconducting connection with the lithium-ion cells (this method is displayed in part (b) of Fig. 13.3). The heat necessary to evaporate the refrigerant is drawn from the battery cells, very effectively cooling them in the process. The cooling ducts must be configured and arranged in such a way that the evaporating refrigerant is available in any place and at any time. This ensures that the temperature remains homogeneous, as required. Only two additional refrigerant pipes are necessary to connect the refrigerant circuit with the battery cooling plate: the pressure line to the battery and the suction line from the battery back to the compressor. The battery evaporator is connected in parallel with the main evaporator.

Cooling the battery only slightly decreases the lithium-ion battery's energy density because there are very few devices outside of the battery. The passenger area and the battery have different cooling requirements. This is why the overall circuitry must be balanced carefully, which is facilitated by deploying an electrical air conditioning compressor with a variable revolution speed instead of a conventional belt-driven version.

A refrigerant compressor is always necessary for battery cooling. Compared with the energy needed for air conditioning the passenger area, it does not consume much additional power, however. Battery cooling systems with refrigerants are mainly used in vehicles in which there is little space for battery cooling and the additional compressor output does not negatively influence overall efficiency. The refrigerant circuit only operates up to a temperature of around – 7 °C and therefore limits battery cooling at lower temperatures, which is acceptable in most situations.

Cooling with coolant Cooling with coolant is the most flexible battery cooling method. It is also very efficient if an additional battery low temperature radiator is used (part [c] of Fig. 13.3). The battery also has a cooling plate, but the cooling medium flowing through it is a mixture of water and Glysantin® in a so-called secondary circuit. A well-balanced and well-designed cooling duct layout ensures optimum temperature homogeneity across the plate and compensates for increases in temperature of the coolant. The secondary circuit's temperature depends on the vehicle's operational strategy and the battery cells' characteristics. Typically, it ranges from + 15 °C to + 30 °C. Recooling is achieved with a so-called chiller that connects the refrigerant circuit with the secondary circuit. This chiller evaporates the refrigerant and draws the heat necessary for this process from the cooling medium of the secondary circuit. The currently available chillers have a high power density, which is an advantage. A disadvantage is that they require more space outside of the battery because not only the chiller needs to be installed, but also a pump, pipes, and, if bivalent operation is desired, an additional low temperature radiator and a valve.

The battery cooling's energy efficiency plays an important role in electric vehicles and in plug-in hybrid vehicles which frequently drive in electric mode. If a secondary circuit is used, an additional low-temperature radiator can be implemented in the cooling module to prevent the electric compressor from operating unnecessarily because this would decrease the vehicle's range. A switching valve enables the driver to choose between chiller operation in the summer and low-temperature radiator operation at lower ambient temperatures.

13.3.2 Battery heating

From a technical point of view, there are several similarities between heating and cooling a battery. However, some solutions are completely independent of each other. Both aspects need to be taken into account when designing the battery and the battery management system.

In general, battery heating must be considered as a means to an end. It is technically required because of the cells' characteristics. However, its design influences the applications and, therefore, the acceptance of electric vehicles on the market. Heating should be safe, quick, lightweight, cost-efficient, and it should not unnecessarily reduce the electric vehicle's range. The latter leads to restrictions, especially for electrical heating systems.

Standard cell types require a battery temperature of at least + 5 °C for a moderate electricity output. Although the minimum operating temperature can vary depending on the cell type, it must be ensured. Furthermore, it needs to be raised to a "comfortable" temperature of at least 20 °C by means of suitable operational strategies and devices.

Before providing an overview of the types of battery heating systems, the question of energy sources needs to be answered. As opposed to combustion-engine vehicles, heat in electric vehicles comes at a price. The electrical devices' losses during operation considerably depend on the drive state, are a little below 1 kW on average, and are not available at the beginning of the journey.

They can reduce the range of a pure electric vehicle by as much as 50 % during winter. The main reasons are reduced battery capacity and the power required to heat the passenger area electrically, if available. The energy source and its efficient use are of paramount importance.

Electricity is the most common energy source for heating batteries. There are some vehicles equipped with fossil fuel heating systems. They use standard fuels (some of them use ethanol) and need very little electricity and, therefore, only negligibly influence the vehicle's range. Their market acceptance depends on several factors and registration regulations, so that most battery heating systems in pure electric vehicles are powered directly (electric heater) or indirectly (heat pump) by electricity.

Direct heating with air works similarly to air cooling; only the heat is transferred in the opposite direction.

The advantages and disadvantages described above apply accordingly. The heat source can be the passenger area heating system, which is generally an electrical air heating system as part of the centralized heating, ventilation, and air condition unit (HVAC) in electric vehicles. Such a battery heating system can, if required, be supplied by a decentralized electrical air heating system situated directly in the battery's intake air current. Another possibility is the recirculated-air mode, which further increases the outlay, however. High-voltage heating systems operated with direct voltages above 60 V are much more expensive than low-voltage heating systems. This is due to the high safety requirements.

As mentioned, fossil-fuel heating systems are also used. They can (pre-)heat the passenger area and also provide battery conditioning. This technology is of interest especially for hybrid vehicles with a range extender as well as for commercial electric vehicles. A simple and cost-efficient solution is to guide centrally heated air from the passenger area to the battery cells and dispense with a separate battery heating device.

Heating the battery with coolants is a particularly interesting technology, especially in respect to the vehicle's overall thermal management. This technology uses the same heat transfer elements in the battery as in the coolant conditioning system and can also be very flexibly integrated into more complex temperature circuits.

Many electric vehicles have a high-voltage liquid heating device in the passenger area, which can be used to introduce the energy needed to heat the cooling circuit. Exact dosing of the energy volume introduced and battery preconditioning before the start are advantages of this technology. If preconditioning is performed while the plug-in vehicle is still plugged in, this does not reduce the range.

Advanced temperature circuits increasingly use heat pumps and waste heat technologies. The heat energy obtained in this process can be used to heat both the passenger area and the battery. Depending on the operational state, it is also possible to introduce the battery's heat loss into the medium circuit.

Direct cell heating by inert liquids Direct cell heating by inert liquids is rarely used nowadays in spite of the fact that it is an interesting approach. In terms of heat transfer, most other technologies have to deal with material transition areas, some of them with air pockets. Direct liquid cooling, however, has very high heat transfer rates and ensures a very homogeneous temperature distribution in the battery pack if designed and configured correctly. However, because the requirements related to the heat transfer liquid are high, this technology is used very rarely.

The liquids should have a specific heat capacity, similar to that of a water-Glykol® mixture. They must also be entirely inert and non-toxic in all circumstances. In the event of a thermal incident in the cell, the medium should not add to the existing hazards.

Moreover, the system outlay is comparably high, and maintenance questions need to be answered if such media are used in large series productions. It remains to be seen what future developments will bring.

Direct cell heating by electric heating elements Direct cell heating by electric heating elements is a widely used method to quickly and effectively heat battery cells. Large-surface thermo foils with resistance heating are installed in the cell pack near the cells, if necessary also between the cells. These foils heat up very quickly if suitably electrically controlled and, therefore, can increase the cell temperature with precision. Some heating foils do not have an inherent temperature-limiting PTC effect. If this is the case, their control and local thermal management have to fulfill very high functional safety requirements.

Indirect cell heating by electric heating elements Indirect cell heating by electric heating elements is another promising technology. Similar to side and bottom cooling, the adjacent surfaces are electrically heated. Heat transfer occurs indirectly via the surfaces and, therefore, is less effective than foil heating in the vicinity of the cells.

As a rule, all battery heating technologies described above must be evaluated together with the thermal isolation of the battery housing. Depending on the design, it is possible to use the residual heat of the prior drive cycle for a certain amount of time to keep the battery in the acceptable range.

13.4 Outlook

Effective drive battery thermal management is very important for service life, availability, and safety. Several technologies are available for cooling and heating and are often combined with each other.

Depending on the vehicle's requirements and its power requirements, refrigerant battery cooling is feasible with cooling media or air. When operating under very high specific loads and high environmental temperatures, both systems may require support from the refrigerant circuit. Air cooling systems are cheaper and less heavy. Coolant cooling is much more efficient because it uses just one medium to cool and heat the battery and is also easier to integrate into more complex thermal management circuits.

This is similar for battery heating. Depending on the requirement profile, heating methods that directly heat the cells with near-cell heating elements or indirectly with a fluid that is often the same as the cooling fluid, i.e., coolant or air, are also a possibility. The advantages and disadvantages of these systems are similar to those of the cooling systems.

There is a trend toward smaller batteries, such as those used in hybrid vehicles, needing only cooling. Their own waste heat heats the cold batteries very quickly during operation. Medium-sized batteries such as those used in plug-in hybrid vehicles require cooling and heating. Pure electric vehicles have such large batteries that they do not have to be actively cooled because their thermal mass suffices to buffer any operational waste heat. If these batteries are capable of quick charging, however, active cooling is still be necessary. For electric vehicles, it is all the more relevant that the cold battery is efficiently heated in winter to achieve the required temperature.

The future outlay for the thermal management of drive batteries will depend on developments in the field of cell chemistry. If internal resistances can be reduced and permitted maximum cell temperatures can be raised, cooling requirements will decrease. However, as developers are simultaneously striving to raise power and energy density, the overall cooling requirements could remain the same or even rise. The same applies to the outlay for heating: Reducing internal resistances at low temperatures and decreasing the sensitivity of cell chemistry at temperatures below zero will lower heating requirements. On the other hand, efficient and high-performance heating is still required to fulfill the demands related to the battery's service life and efficiency. This involves reliably keeping the battery's temperature in the perfect range at all times.

For the first generation of vehicles with fully or partly electrified drive, development mainly focused on the components. In the coming years, this will change, as optimizing thermal management and improving the interaction of all components become more and more important. Developments such as heat pumps open the door to new technologies and increase efficiency and range. Market demand and specific usage profiles have determined the vehicle types on the market, and this will also be the case for battery cooling and heating solutions. The goal is to find the perfect compromise between cost, weight, safety, efficiency, and space requirement.

Battery management system 14

Roland Dorn, Reiner Schwartz, and Bjoern Steurich

Contents

14.1 Introduction

The battery management system's most important task is to protect the drive battery's individual cells and to increase service life as well as cycle numbers. This is especially important for lithium-ion technology because the batteries must be protected from overcharging and excess temperature (Fig. 14.1) to prevent cell destruction.

R. Dorn (✉)
Texas Instruments Deutschland GmbH, Haggertystr. 1, 85356 Freising, Germany
e-mail: Roland.Dorn@ti.com

R. Schwartz
STMicroelectronics Application GmbH, Bahnhofstrasse 18, 85609 Aschheim-Dornach, Germany
e-mail: reiner.schwartz@st.com

B. Steurich
Infineon Technologies AG, Am Campeon 1-12, 85579 Neubiberg, Germany
e-mail: bjoern.steurich@infineon.com

R. Korthauer (ed.), *Lithium-Ion Batteries: Basics and Applications*,
https://doi.org/10.1007/978-3-662-53071-9_14

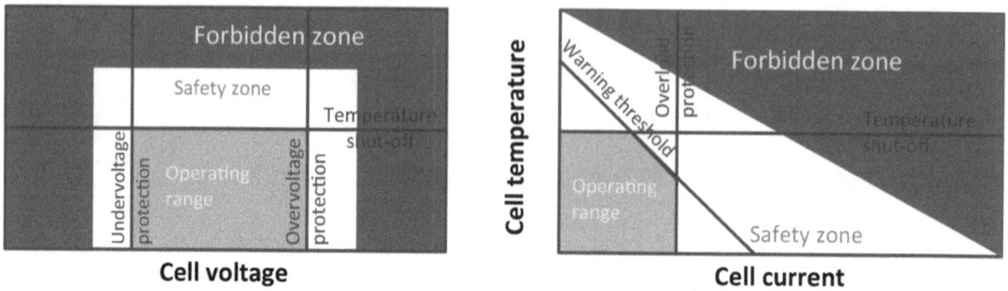

Fig. 14.1 Lithium-ion battery cell area of operation

14.2 Battery management system tasks

The battery management system (BMS) measures the control parameters cell voltage, temperature, and battery current. A typical battery cell has a nominal voltage of 3.6 V at a maximum end-of-charging voltage of 4.2 V and a minimum end-of-discharge voltage of 2.5 V. High discharging (< 2.5 V) causes irreversible damage such as capacity loss and increased self-discharging. Overvoltage (> 4.2 V) results in spontaneous self-ignition and thus poses a safety risk.

Capacity loss is mainly caused during charging when temperatures and voltages are too high. If properly used, a standard battery is good for around 500 to 1,000 cycles before it loses around 20 % of its initial capacity.

Monitoring cell voltage, current, and temperature enables forecasting of a battery's state of charge (SOC) and state of health (SOH). SOC describes the current state of charge compared to the battery's maximum capacity. SOH describes the current state of health compared to that of a new battery.

Both parameters are important for ensuring the vehicle's state of function (SOF) (Fig. 14.2). Ultimately, this is the essential information for the driver: whether the vehicle will reach its destination or the battery needs to be charged beforehand. There are three ways of calculating these parameters.

- Electrical model
  The battery is simulated by an analogous electrical model. The individual components' parameters are adapted to the material characteristics in terms of capacitive, inductive, and purely ohmic properties, while taking into consideration age-related changes.
- Electrochemical model
  This simulation is based on the battery cells' chemical characteristics in order to model the electrical properties.
- Kalman filter method
  This method constantly adapts the filtering parameters, which should reflect the battery's properties, to the current conditions.

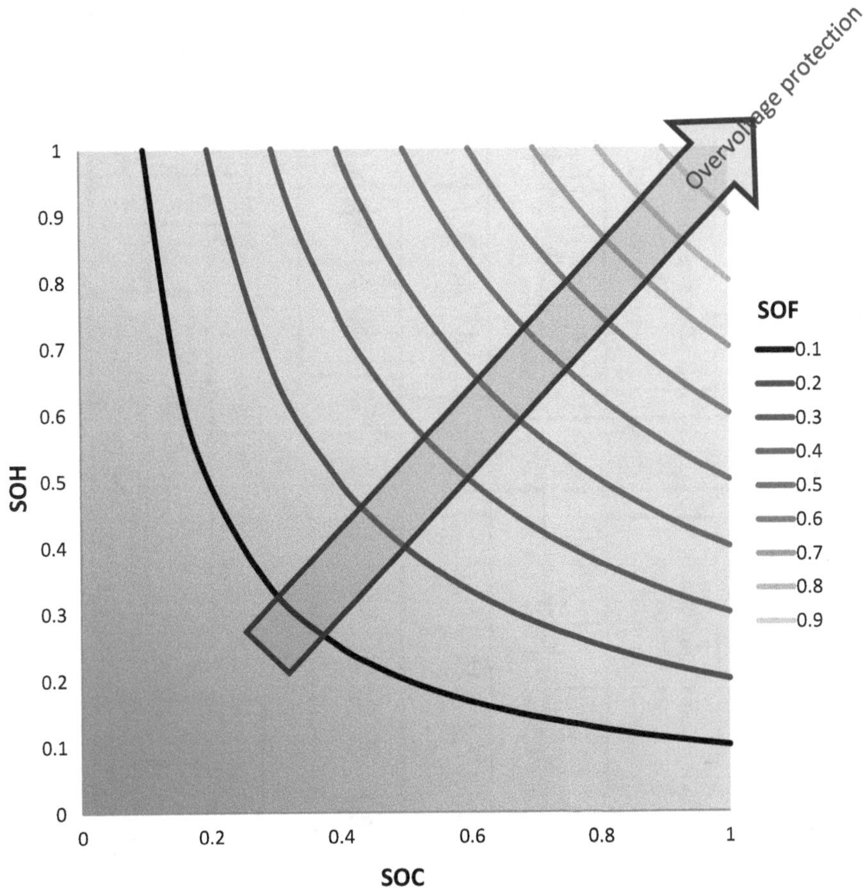

Fig. 14.2 Relationship between SOC, SOH, and SOF

14.3 Battery management system components

In general, a drive battery consists of five components (Fig. 14.3):

- Typically, battery modules comprise several stacked cells. Cell voltage and cell temperature are monitored in these modules, and the values are transmitted to the control unit. Also, a charge equalization between the individual cells is performed in the modules to reduce battery wiring outlay. Charge equalization and cell supervision are mostly controlled by an ASIC, the cell supervisory circuit (CSC). Drive batteries comprise several modules connected in series and have an output voltage of several hundred volts.
- The control unit calculates SOC and SOH and controls charge equalization. Standard automobile interfaces such as the CAN bus or FlexRay bus are used

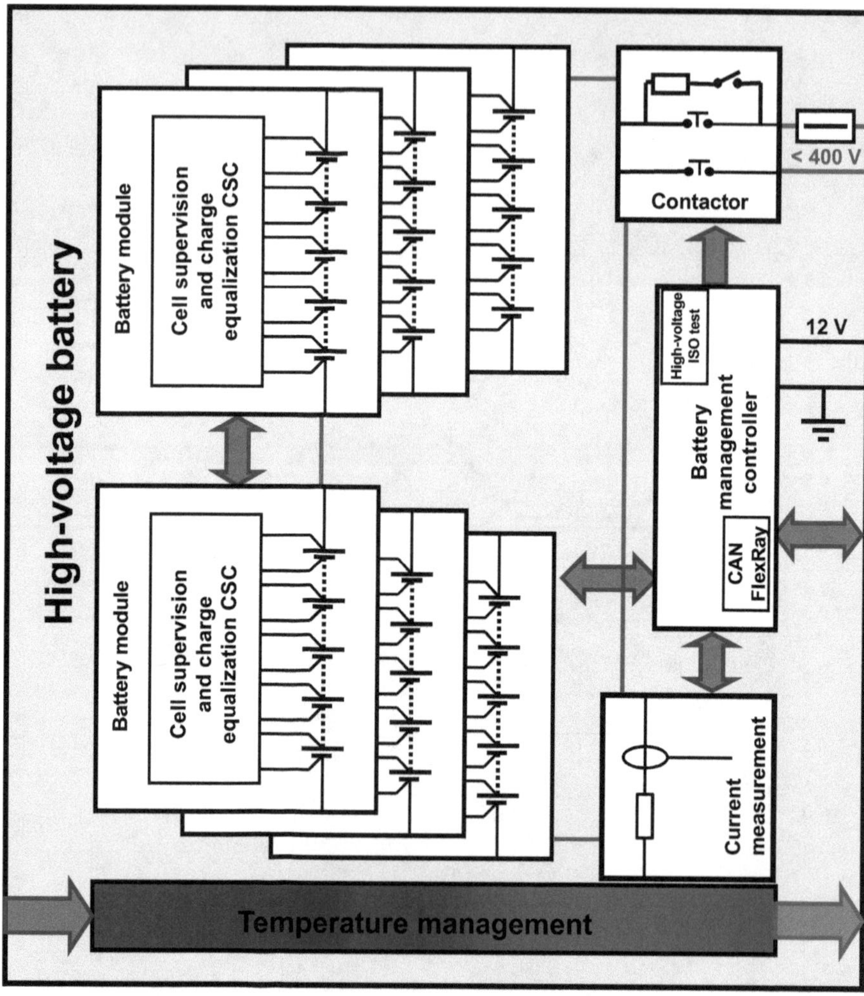

Fig. 14.3 Block diagram of BMS components

to communicate with the vehicle to determine the SOF. The interface also controls battery charging from the power grid. This is why the control unit must also control the battery's performance management and should reduce its power requirement in passive state to a minimum.

- The HS contactor isolates the battery cells from the vehicle in passive state to prevent unnecessary losses or hazards. It can also isolate the system in the event of extreme malfunctions such as short circuits, excess temperature, or accidents. The battery is also protected by a fuse in the event of short circuits.
- The current is generally measured with a special sensor directly at the battery. Two independent systems are used for safety reasons. State-of-the-art systems use a measuring resistor as sensor or they use the electromagnetic field.

- Temperature management ensures that the drive battery operates at optimum temperatures. This is especially important to ensure that the cells age evenly. Service life, availability, and safety considerably depend on this.

14.4 Cell supervision and charge equalization

Charging a large number of cells always poses a challenge; overcharging individual cells must be prevented. The individual cells show residual charge variations. This is inherent to such a system. Thus, individual cells can reach their maximum voltage during charging earlier than others (Fig. 14.4), resulting in overvoltages or premature interruption of the charging process.

This leads to capacity loss caused by incomplete charging of all cells. The weakest cell determines the battery system's characteristics.

High-precision measurement is used for each battery cell. The CSC with its accuracy of 2 mV plays an important role for determining SOC and SOH. Current BMSs use different cell voltage monitoring approaches, which each have advantages and disadvantages. Cell voltage is measured with analog-to-digital converters (ADC) as a rule. They are connected to the battery pack cells directly or via a multiplexer (MUX).

Three converter topologies have currently established themselves on the market (Fig. 14.5). A successive approximation register (SAR) converter is used if cells can be measured sequentially. Sigma-delta converters are used for permanently monitoring cell voltage, if desired. SAR converters store the samples in a sample-and-hold (SH) circuit. Dedicated SH circuits can be used for each cell and their outputs can be interfaced with the ADC by means of an analog voltage multiplexer, or the cell voltage can be transmitted directly to the ADC's SH circuit by means of a multiplexer.

Topology choice depends on whether the measured values of all cells should be measured synchronously or whether sequential measurement suffices. The SH circuit greatly influences measurement accuracy and performance. Both values directly determine the chip surface and the costs of monitoring circuits. To detect

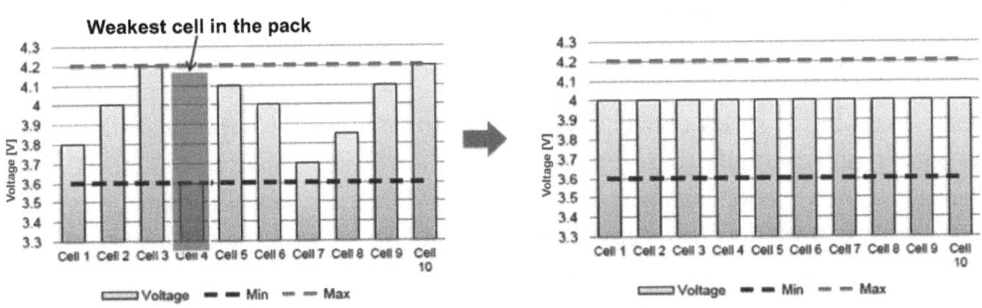

Fig. 14.4 Cell pack variations during charging and discharging (*Source* TI)

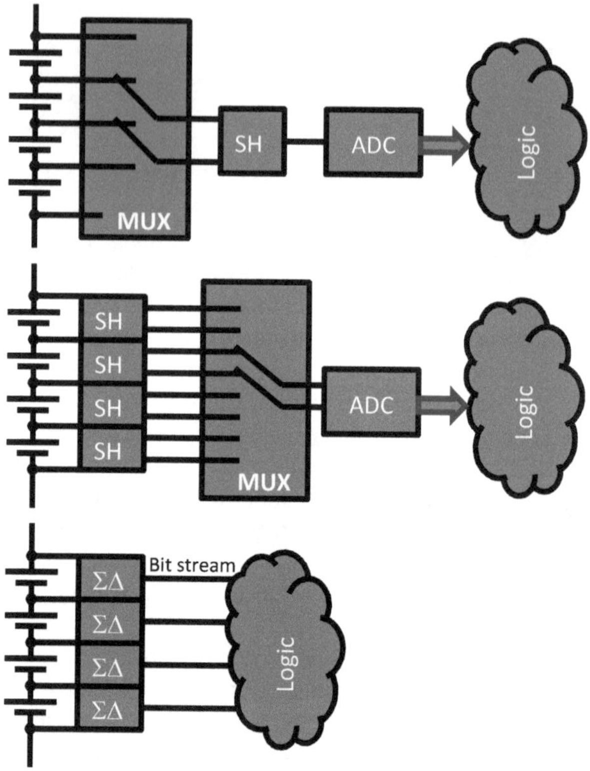

Fig. 14.5 Converter topologies: SAR converter topologies (*top, center*), sigma-delta converter (*bottom*)

errors, systems with just one SH circuit are advantageous. However, these solutions pose high demands on the sample rate and input signal filtering. They generally use higher-order input filters than solutions with sigma-delta converters. First-order filters suffice for the latter. The voltage offset caused by battery cell stacking can be adapted to the logic by a digital signal converter. A cascaded reference concept is used here for the individual input comparators.

14.5 Charge equalization

The length of the charging and discharging processes is generally determined by the weakest cell in the series connection. This is why the BMS charges and discharges the battery only partially (e.g., from 30 % to 80 % instead of from 0 % to 100 %). Thus, the usable energy density is decreased, and valuable resources remain untapped. A method has been developed that channels away the "surplus" electricity to charge all cells as evenly as possible. This passive charge equalization (Fig. 14.6) prevents cell overcharging by converting the surplus energy into heat by means of resistors. The resulting heat limits the maximum current. During discharging, on the other hand, the available energy cannot be used to prevent deep-discharging of the weakest

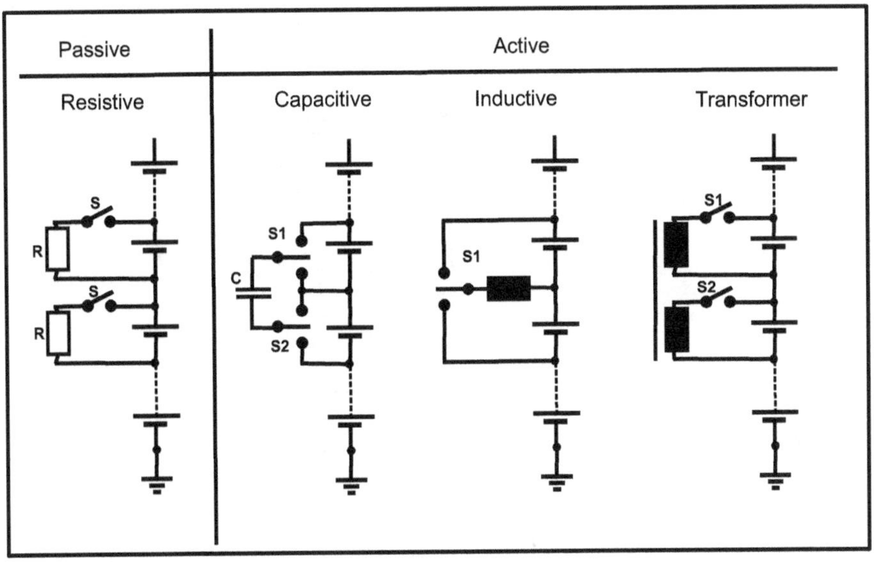

Fig. 14.6 Passive and active charge equalization circuitry

cells. The stronger cells are left with residual energy. Fig. 14.4 shows that two cells reach the upper and lower limits at different times during a charging and discharging cycle.

Passive balancing does not improve the discharging process. The capacity discrepancy of cells will increase over time and further decrease maximum capacity utilization.

Active charge equalization approaches are an alternative to today's passive technologies. DC-DC converters transfer charges between the cells. This is possible during charging and discharging as well as in passive state. More effective charge transfers enable higher compensating currents. This means that larger cells with higher capacities can be used and cell equalization is quicker.

Capacitive and inductive approaches load the cell charge into a storage element and transfer it to an insufficiently charged cell. A switching matrix consisting of transistors connects the storage element with several cells. To maintain high efficiency, this technology is mainly used for lower cell numbers. If cell voltages differ only slightly, the efficiency of capacitive "shuffling" technology decreases. A transformer-isolated charge transfer remedies this situation. Battery cells are connected via switches with transformer windings that enable energy transfers between the battery cells.

The bidirectional transformer (Fig. 14.7) renders two equalization types possible. The first (during discharging) is called "bottom balancing". The first step of this approach is to identify the weakest battery cell of a battery block. Then the charge is redistributed from the entire battery block. The second method (during charging) is called "top balancing". This involves evenly spreading the energy surplus of the

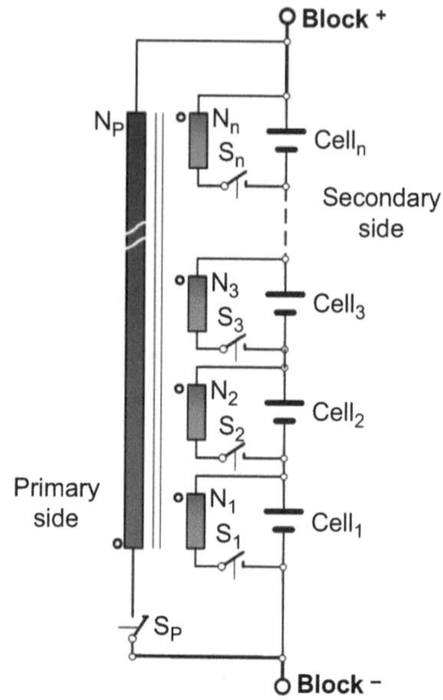

Fig. 14.7 Bidirectional transformer (*Source* Infineon)

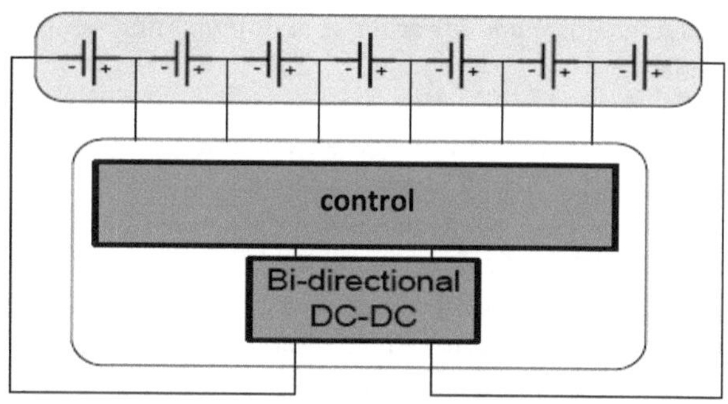

Fig. 14.8 Block diagram of a bidirectional DC-DC converter (*Source* TI)

strongest battery across the entire battery block. Balancing losses for one battery block are around 1.5 W and therefore are considerably lower than the standard loss of 18 W for a comparable passive system.

The inductive components can be reduced by a modified transformer concept which integrates bidirectional switching matrices and DC-DC converters (Fig. 14.8). Any cell can be directly charged or discharged thanks to the switching matrix between the cells and the converter. The charges are transferred between an individual cell and the cell group.

All active processes also open up potential for cutting cooling costs. Battery modules of up to 12 battery cells are generally compatible with bidirectional transformer concepts.

ASICs are mostly used to control charge equalization. The procedure described above can easily be extended to charge equalization between battery blocks. In this case, an additional transformer selectively feeds energy into the 12-V on-board electrical system.

14.6 Internal battery communication bus

Measurements in the cell set are synchronized and all errors such as increased cell voltage, communication disruption, disconnected sampling lines, and increased cell temperature are recorded and transmitted to the main controller. Today's drive batteries essentially use two approaches to exchange data between individual modules and the central control unit.

Star wiring (Fig. 14.9a) involves fitting each module with a galvanically isolated data transmitter. Many systems use the CAN bus for this purpose. The main advantage of this system is the potential disconnection of the communication bus from the battery cells. Thus, a short circuit between the communication lines and cells does not automatically destroy individual electrical components. In addition, thanks to the distribution in parallel bus systems, the data rate can be reduced. The expensive galvanic isolation required in each module offsets these advantages, however.

A vertical communication interface is an alternative to a large number of cell supervisory circuits (CSC) (Fig. 14.9b). Its main task is to transmit voltage and temperature data from the stacked modules to the lowest module, which then communicates with the control unit by means of galvanic isolation. Commands for cell voltage balancing are transmitted in the reverse direction. The data rate of such a

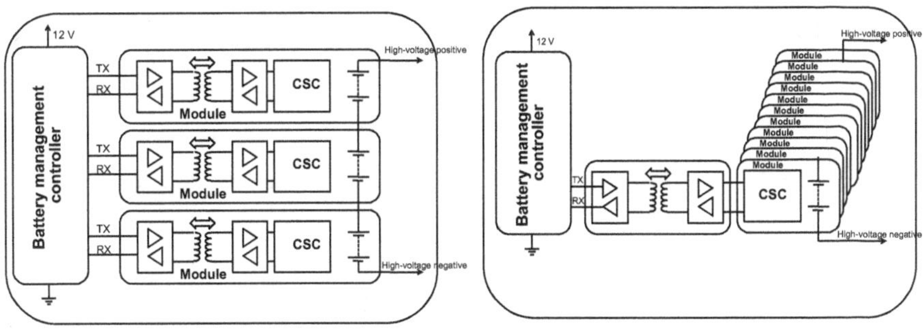

Fig. 14.9 Internal battery system communication bus, star wiring *(a)*, and vertical communication interface *(b)*

system is considerable. The data protocol consists of the address, the measured values for each cell of each module, a cyclic redundancy check (CRC) for serial data transmission, and module programming.

Data transmission must have a certain guaranteed redundancy to ensure a limp-home mode for the battery-powered vehicle. This can be achieved by means of a differential signal, which should be able to maintain communication in the event of an error. The reduced spectrum of radiation is another advantage of differential data transmission. In general, twisted-pair lines without shielding suffice for bus wiring. Failure analysis of a vertical interface must, similar to that of star wiring, take into account high-voltage short circuits. Erroneous measurement results or destruction of the CSC must be prevented.

This is why the communication lines are mostly connected capacitively or via inductive transformers. However, such measures reduce the cost advantages derived from vertical communication.

Radio transmission is another possibility for exchanging data within the battery pack. These wireless systems are currently being researched.

14.7 Battery control unit

The battery control unit (Fig. 14.10) evaluates all measurement data and exchanges battery-specific information with a higher-level control unit or a battery charger. A CAN bus or FlexRay bus is used in today's OEM batteries. The control unit is connected with the 12-V power supply. In passive state, the current consumption should be as low as possible to prevent discharging the low-voltage battery.

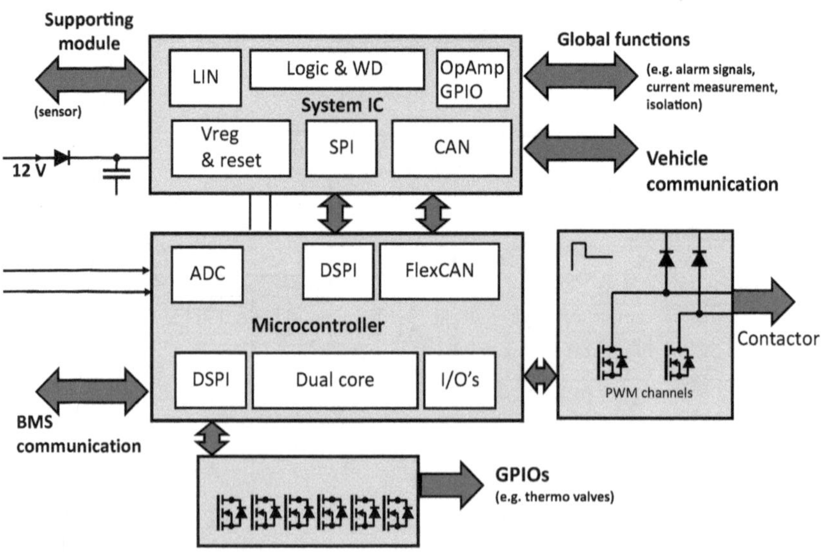

Fig. 14.10 Battery control unit block diagram

On the other hand, cell supervision and battery cell charge equalization in passive state are desirable. This is why so-called power management system circuitry is used to "wake up" the battery electronics in due time. As a rule, the power supply, communication interface, and control unit functions are implemented on a system circuit. The circuit also monitors the main CPU, an additional safety feature. This so-called watchdog monitors CPU functions and, in the event of failure, implements the necessary measures to safely switch off the battery.

This is especially important for the central CPU, whose task is to protect the entire system. Modern microcontroller designs are perfect for this task. They are equipped with two processor cores, monitor each other in the so-called lock-step mode, and integrate error correction and a comprehensive BIST (built-in self-test). These systems have already been tried and tested in safety-relevant automobile applications such as brakes, steering, and airbag.

The main CPU also calculates the battery model in order to determine SOC and SOH. It processes battery cell voltage, current, and temperature data. The computer used should have a floating-point function to fulfill the requirements of the monitoring circuitry. It also controls cell charge equalization and ensures that the battery temperature profile is as homogeneous as possible, providing the battery has controllable temperature management.

Software

15

Timo Schuff

Contents

15.1 Introduction

The growing number of control units in cars and their increasing networking make car electronics more and more complex. Also, car manufacturers face the challenge of having to cut design cycles and reduce costs. As a result, quick and cost-efficient software development methodology is becoming essential. The software for lithium-ion batteries is located in distributed systems and it also has elements that are critical for safety, which generates additional requirements.

This Chapter explains how a new function can be efficiently implemented in a software, how it is integrated into the overall system, and how to ensure that the software operates flawlessly.

15.2 Software development challenges

When lithium-ion batteries are used in cars, several devices such as temperature management components, one or more control units, generators, and DC and AC

T. Schuff (✉)
ITK Engineering GmbH, Im Speyerer Tal 6, 76761 Ruelzheim, Germany
e-mail: timo.schuff@itk-engineering.de

R. Korthauer (ed.), *Lithium-Ion Batteries: Basics and Applications*,
https://doi.org/10.1007/978-3-662-53071-9_15

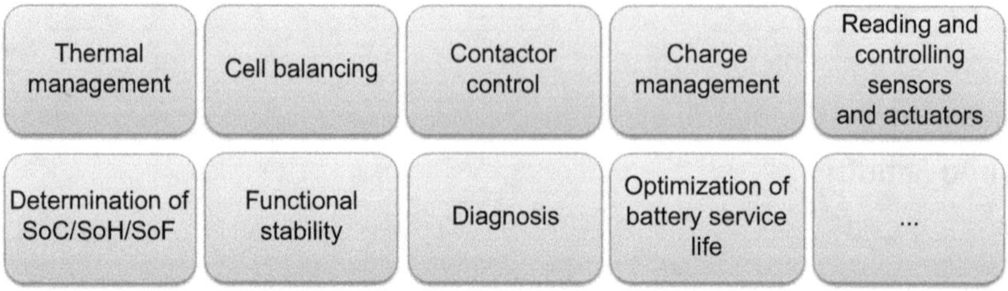

Fig. 15.1 Typical lithium-ion battery applications

converters have to interact with the battery. Specific software applications are necessary to enable secure lithium-ion battery operation (Fig. 15.1). Their number and scope depend on the overall system's complexity, which in turn depends on the battery voltage level, its load type, and size.

A traction battery that is used continuously and operates at high voltage levels has to fulfill much more challenging requirements than a 12-V starter battery.

If the complexity of the system is low, certain applications such as diagnosis functions or actuator control are not required and can be transferred to the peripheral devices (Fig. 15.2a). In this way, internal battery measurement technology can be relocated to the exterior, and cumbersome communication between the battery and control unit can be dispensed with. However, the control unit must still guarantee that the battery operates safely by means of charge management and external sensor technology.

A system of medium complexity such as a starter battery also does not need temperature management, due to its small size (Fig. 15.2b). Charge and temperature management merge because thermal energy, aside from the drive train's waste heat, only stems from electrical losses during charging and discharging. While the battery measures and transmits its internal temperature and cell voltages, the charging current and pack voltages are measured by external sensor technology.

In contrast, Fig. 15.2c shows a highly complex system. The battery, which is actively cooled and heated, has internal measurement points for parameters such as temperature and voltage. It is charged by a generator, by the electric engine in generator mode, or by a voltage converter, and is controlled by a control unit outside the battery. This control unit and the battery ensure that the overall system operates safely.

It can therefore be concluded that the more complex the overall system becomes, the higher the number of total and internal applications required. It is important to define interfaces early in the development cycle because the applications are generally developed by different suppliers. To increase efficiency, there is a need to guarantee that the individual software modules used are interchangeable and can be used on different platforms. This can be achieved by complying with the AUTOSAR standards.

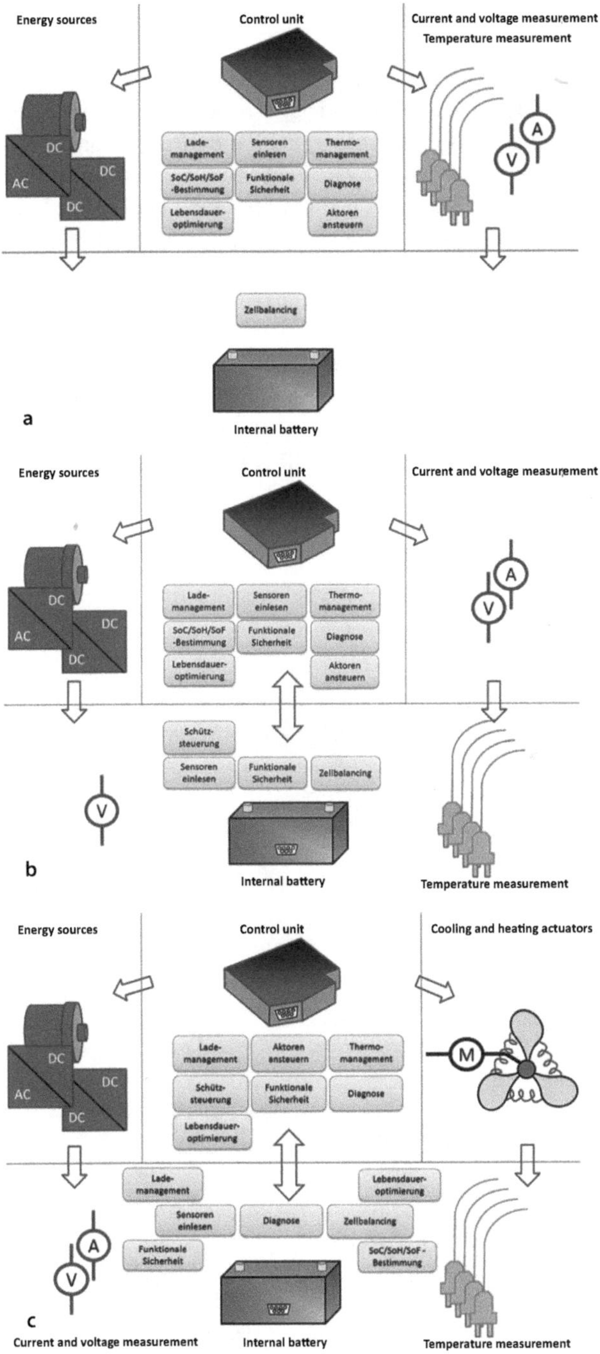

Fig. 15.2 Software modules in systems with low (*a*), medium (*b*) and high (*c*) complexity levels

15.3 AUTOSAR – a standardized interface

The Automotive Open System Architecture (AUTOSAR) was developed and introduced by car manufacturers and suppliers. The objective of this extended operating system for control units is to provide a foundation for platform-independent applications to facilitate functional software exchange, reuse, scaling, and integration in a vehicle's control units. The architecture is especially suited for large systems with numerous components as well as for systems where applications are expected to be transferred from one control unit to another.

Its virtual functional bus has several standardized interfaces and it enables communication between other applications even without any information on signal carriers such as internal signals or buses. An interface description, e.g., an XML document, is created with the application and can be used to automatically create an interface for the application. Signals that need to be transferred internally via bus systems such as CAN, LIN, and FlexRay are created automatically (Fig. 15.3).

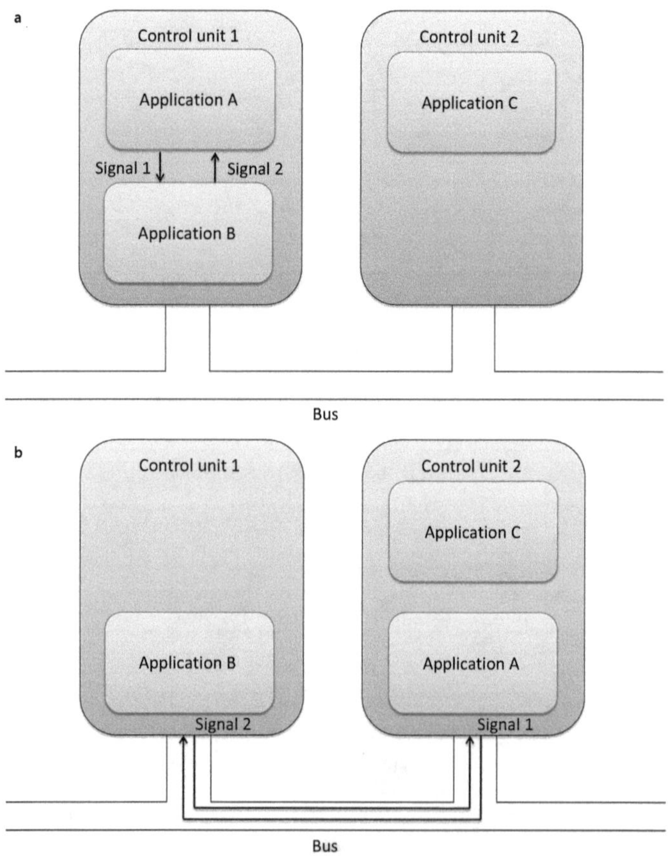

Fig. 15.3 Applications before transfer (*a*), applications after transfer (*b*)

The example shows application A, which is being transferred from control unit 1 to control unit 2. In the original control unit, application A communicated by means of internal signals with application B. After the transfer, direct communication by means of internal signals is no longer possible, and signals 1 and 2 are changed into bus signals.

This can cause different errors: interrupted bus lines, bus lines with short circuits, or the transfer of erroneous signals for a variety of reasons. To prevent such errors, correct communication between the applications must be ensured and an additional fallback system must be available to ensure safe operation even in case of malfunction. This can be achieved by detecting a hardware-related failure of the bus driver by performing cyclic redundancy checks (CRC) of the most important bus messages, or by so called "alive" signals such as message counters or other cyclically alternating signals. These signals enable the receiver to determine whether the sender is still active and sending correct signals.

AUTOSAR establishes a run time environment (RTE) on the control unit. A basic software, a default bus driver, and an operating system on different abstraction levels provide connection to carrier hardware such as a processor. A processor must have a certain number of interfaces and meet minimum performance requirements to be suitable for an AUTOSAR-compliant development.

15.4 Quick and cost-efficient model-based development

To reduce development times for embedded systems and simultaneously optimize software functionality, it is recommended to combine AUTOSAR-compliant development with model-based development. The objective is to ensure that algorithms are correctly implemented and robust without straining development times.

The model-based method, combined with automatic code generation, has been widely adopted not only in the automotive industry but also in other branches. Various tools are combined to create a graphically programmable abstraction level, which can be used for automatically generating C code.

There are various advantages of such a development method over conventional manual coding. The easy-to-understand display facilitates work with the model for all project members. By modularizing and creating hierarchies, complex systems become clear, and individual and project-independent modules can be reused if necessary.

The model is continually refined until the executable source code can be generated automatically. This means that the architecture and detailed design are executed by the same model. Models that have been verified in the respective development stage can be used up to implementation. Although there is a high abstraction level at the beginning, the model becomes more and more detailed during development. The development process is geared to the German "V-Modell", which requires that the results are validated and verified in each development phase. Various analytical and test methods such as static code analysis using code review, model-in-the-loop (MiL), software-in-the-loop (SiL), hardware-in-the-loop (HiL), and the test of the entire system in the car are used for testing in the various phases (module test, integration test, system test).

The models can be immediately embedded in MiL simulation environments and their functionality tested very quickly. It is no longer necessary to implement the application on the destination hardware. This enables automated testing during early development stages and significantly reduces the amount of hardware required. Many tool chains available from different suppliers provide extensive automated testing functions. Test cases that are described in scripts contain the prerequisites, test steps, and expected results. The automatically generated code can also be embedded in the test environments. Scaling errors, interface errors, and overruns can be detected by means of software-in-the-loop tests, which use the correct data types for their fixed-point computations.

Additionally, model-based development can be used for simulations. Tests to verify that the control and feedback algorithms function correctly are performed before a prototype of the real system exists. With this rapid prototyping, finite state machines can be tested already during the design phase. It is possible to determine whether the simulation algorithm works correctly very early, and the model's functionality and quality are continuously improved. Flexible real-time hardware with the respective interfaces is required; at best, the models can be directly implemented on this hardware.

A development computer monitors the internal model parameters, and the real-time system is substituted for the control unit or is added to the vehicle network. Usually, these rapid prototyping systems have enough interfaces to be used as a gateway (Fig. 15.4). Signals can be intercepted, changed, or added. Various systems are available on the market that can be equipped with multiple bus controllers and I/O modules.

Rapid prototyping enables developing and testing of new functions from the outset by means of best-practice approaches. Connected bus systems as well as digital and analog inputs and outputs can generally be accessed easily. Models can be computed in floating-point format thanks to powerful microprocessors and sufficient memory, and the developers do not need to make allowances for scalability and memory requirements. Thus, not only new functions but also extended existing functions can be tested in advance.

The possibility of testing models and automatically-generated code in the early development stages improves design quality, as deviations from the specifications can be detected earlier and more easily. This also results in a sound and sustainable design. A comprehensive changeover from manual coding to model-based development is not necessary because manual code can be integrated. By selecting special targets such as simulation targets, DLLs, and microcontrollers, the generated code is optimized for the application. If the destination target is a microcontroller, for example, processor-optimized code can be generated.

Of course, creating a model-based development tool chain requires certain investments of time and money. Due to the complexity of a transition from manual coding

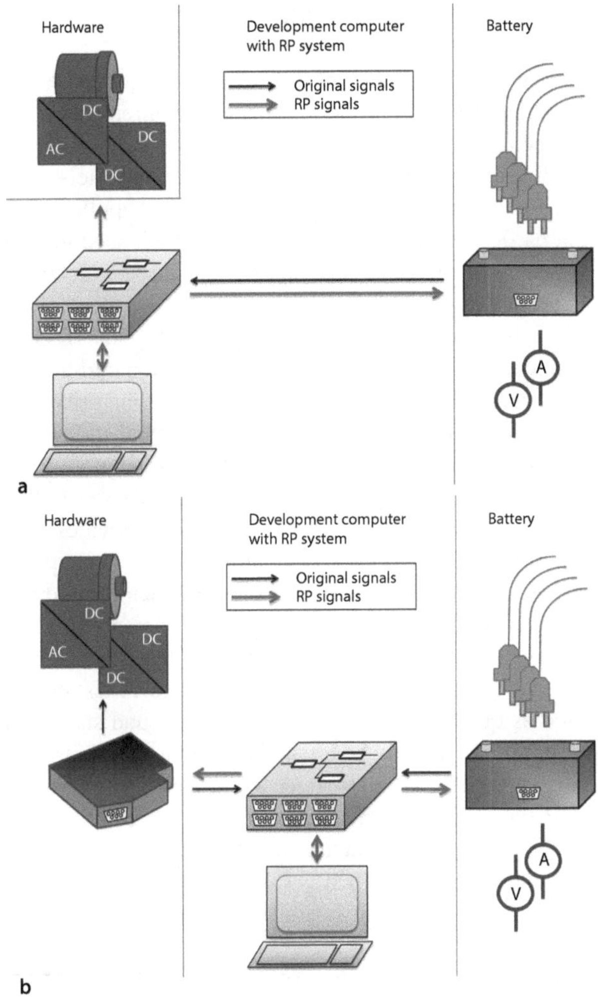

Fig. 15.4 RP system in replacement mode (*a*), RP system in gateway mode (*b*)

to model-based development, it is recommended to obtain support from experts for this process.

Furthermore, when implementing complex functions, certain modeling guidelines must be rigorously observed to keep track of all changes implemented. Furthermore, automatically generated code is not always easily readable or only partially readable. In the medium term, however, the reduction in development time will pay off. Implementing a model-based method across the entire organization leads to further benefits, such as cross-organizational collaboration.

15.5 Requirements engineering

In distributed systems, such as lithium-ion batteries, requirements engineering is responsible for tracing, implementing, and testing requirements. All requirements for the system to be developed are defined on the specification sheet. It is important to describe all requirements constructively and completely because detailed specifications are the foundation for efficient testing. All applications, often developed by different parties, must function according to their specifications when integrated into the overall system. Change management is used to document all requirement amendments such as changes or additions.

A unique ID is assigned to each individual requirement. This ID can be used by all software development participants to track the status of a requirement (e.g., "implemented", "under review", "tested"), and thus its level of maturity, at any time. The implementing team, for example, can change a requirement status accordingly after implementing it. The test team then either creates one or more test cases for the implemented requirements or assigns already existing test cases to the requirement. Having separate teams for implementation and testing provides a certain control mechanism that, in some cases, is even mandatory.

Generally, one or more test cases are assigned to each individual requirement. This may not always be possible in the early development stages, however. The ratio of the number of requirements that already have test cases assigned to them to the total number of requirements is called test coverage. A test coverage of 100 % should be realized as quickly as possible to obtain a sound statement on the level of maturity. Test cases can already be created during or even before the implementation phase. Once created and automated, a test case can be used for all future software levels. If the requirements change, the associated test case must also be changed. Ideally, a new software version is automatically tested on MiL, SiL, or HiL test benches with the existing test cases. Once all test cases have been completed successfully, the requirement is considered successfully tested and can be released.

When creating test cases, the test platform needs to be specified. It depends on whether the test is a pure software test or an algorithm test or a test in which the application needs to interact with other applications or with the hardware.

15.6 An example of requirements engineering

In this example, the task is to develop the function "behavior in case of undervoltage". The developer has finished the IDs A1 to A3, but not ID A4 (Fig. 15.5). The code or the model is tested for correct implementation. ID A3 did not pass the inspection, so its implementation needs to be amended. Test cases already exist for all requirements and the level of maturity can be determined from the test result. In this case, only one of four requirements has been correctly implemented and tested.

ID	Requirement	Finished	Review	Test	Test result
A1	If cell voltage goes below 2.6 V and if the message counter does not show an error, send signal "undervoltage" for 1 s.	Yes	OK	Yes	OK
A2	If signal "undervoltage" changes from 1 to 0, request "Charge" from DC/DC converter.	Yes	OK	Yes	NOK
A3	If DC/DC converter is inactive or if there is no charging current (I > 0.2 A) 5 s after requesting, send "Open_Contactor = 1".	Yes	No	Yes	NOK
A4	Check 1 s after "Open_Contactor" whether the contactor has opened. If it has not opened, save error code "Contactor does not open" in the error log.	No		Yes	NOK

Fig. 15.5 Sample requirements for "behavior in case of undervoltage"

The requirements from ID A1 to A4 could also be fulfilled with the text "ID B1: The battery must be able to protect itself against undervoltage."

In principle, ID B1 could describe the same functional behavior as the IDs A1 to A4 but software developers have a great amount of developing freedom. The signal names, voltage and current thresholds, and the decision as to whether the battery should be protected against undervoltage by a contactor and/or by charging is at the discretion of the implementing team. Such a description does not guarantee that functional behavior is implemented as intended by the requester. Automated testing based on ambiguous requirements is difficult or even impossible.

Because the IDs A1 and A2 are just internal signal and algorithm tests, they can be tested in a MiL environment. Successful testing of IDs A3 and A4, however, is only possible if external hardware such as DC/DC converters and contactors can interact with the system. A test bench with real hardware connected to it or a test bench that enables a simulation of such hardware (hardware-in-the-loop) is therefore suitable for this purpose.

15.7 Outlook

The increasing complexity of electronics in cars, especially due to lithium-ion battery use, creates considerable requirements for software development. The required software functionality, robustness, and safety can only be achieved if defined development processes and methods are adhered to. To meet these demands, AUTOSAR-compliant development, combined with model-based development and automatic code generation, is a suitable approach which consumes little time and money. In the automotive sector, this method is already established as a standard process. In the future, model-based development is also expected to spread to other domains such as industrial plants, agricultural and construction machinery, or household appliances.

Next generation technologies 16

Juergen Janek and Philipp Adelhelm

Contents

16.1 Introduction

Rechargeable lithium-ion batteries have been continually developed since their introduction by Sony in 1991. Energy density is one of the key parameters for

J. Janek (Prof.) (✉)
Physikalisch Chemisches Institut & Laboratory for Materials Research (LaMa)
Justus-Liebig-Universitat Giessen,
Heinrich-Buff-Ring 17, 35392 Giessen, Germany
e-mail: juergen.janek@phys.chemie.uni-giessen.de

P. Adelhelm (Prof.)
Institut fuer Technische Chemie und Umweltchemie, Center for Energy and Environmental Chemistry (CEEC Jena), Friedrich-Schiller-Universitaet Jena, Philosophenweg 7a, 07743 Jena, Germany
e-mail: philipp.adelhelm@uni-jena.de

R. Korthauer (ed.), *Lithium-Ion Batteries: Basics and Applications*,
https://doi.org/10.1007/978-3-662-53071-9_16

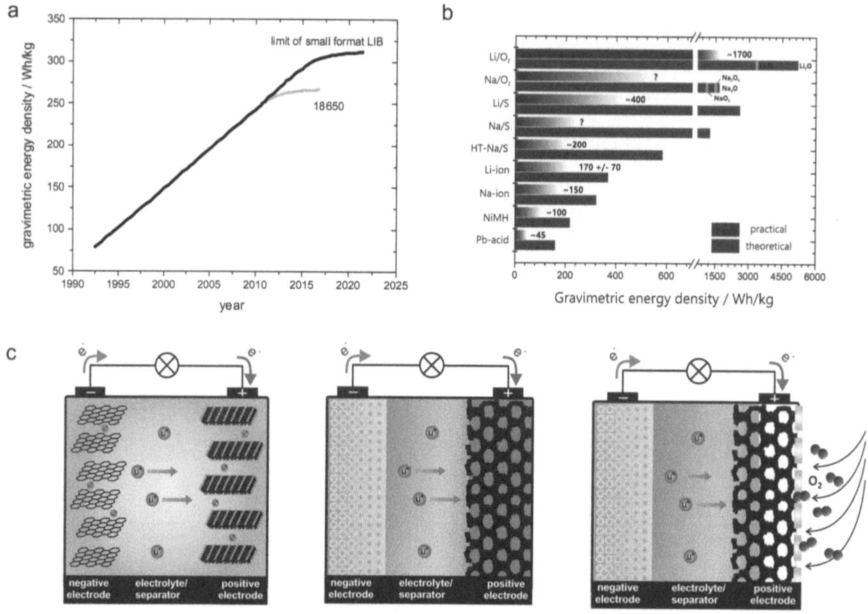

Fig. 16.1 (**a**) Chronological development of lithium-ion cells' mean practical energy densities (high energy design). Diagram after ref [1]. (**b**) Theoretical and (forecast) practical energy densities for different rechargeable cell systems (secondary elements). Practical energy density values are reference values only and strongly vary with the respective cell design (size, geometry, high energy, high power): lead – lead battery (car battery, 12 V), NiMH – nickel metal hydride (cell level, AA), lithium-ion – mean value across different types (cell level), HT-Na/S – high temperature sodium/sulfur (cell level) Li/S – data from by Sion Power (cell level, pouch), Li/O2 – data from by Polyplus (cell level, primary element)) (**c**) Comparison between the different cell concepts of conventional lithium-ion cells, metal/sulfur cells and metal/oxygen cells. [1] Gesamt-Roadmap Energiespeicher für die Elektromobilität 2030, Fraunhofer-Institut für System und Innovationsforschung ISI, Karlsruhe, Dezember 2015

lithium-ion batteries. It was steadily increased by optimizing battery components such as electrode materials or electrolyte as well as by improving the cell construction technologies. The cell level progress during recent years is shown in Fig. 16.1. Both gravimetric (specific) and volumetric energy density were more than doubled.

It is obvious however, that such development is not to be expected in the coming years since technology increasingly reaches its natural limits. Eventually, the electrode materials used will be the limiting parameters. Currently, carbon materials (mainly graphite) are used in anodes, while cathodes contain oxidic transition metal compounds. These are for example lithium cobalt oxide ($LiCoO_2$), Li-NCM ($Li(Ni_{1-x-y}Mn_xCo_y)O_2$), lithium iron phosphate ($LiFePO_4$), or lithium manganese oxide ($LiMn_2O_4$), whose lithium content can be modified electrochemically. The theoretical gravimetric energy density[1] for these cell types typically ranges from 350 to 400 Wh/kg.

[1]The theoretical (gravimetric) energy density is the stored chemical energy based on the pure electrode materials' mass.

If further cell components such as electrolyte, separator, current collector, additives, and housing are taken into account, energy density typically decreases by more than 50 percent. The transition from single cell to battery[2] causes additional losses, resulting in values of 80 to 120 Wh/kg in electric vehicles. With today's commercial cells' values in mind, it can be predicted that a considerable energy density increase is not achievable with the conventional approaches (Fig. 16.1a and Table 16.1).

Table 16.1 Practical energy densities of a choice of lithium-ion cells compared to Li/S and Li/O2 systems (manufacturer's information, cell level). Depending on size, cell geometry, and application (high-power or high-energy), energy densities show a certain range for cells with an identical cell chemistry

Cell type	Type/manufacturer	E° [V]	Useful spec. energy density [Wh/kg]	Useful energy density [Wh/l]
C/$LiCoO_2$	Cylindrical; VL 34570 – SAFT	3.7	160	380
C/$LiCoO_2$	Prismatic; MP 144350 – SAFT	3.75	143	344
C/$LiCoO_2$	Prismatic; MP 174565 – SAFT	3.75	175	423
C/$LiCoO_2$-based	Cylindrical; ICR18650-26F – Samsung	3.7	209	581
C/$LiCoO_2$-based	Prismatic; ICP103450 – Samsung	3.7	185	415
C/$LiFePO_4$	Cylindrical; VL 45E Fe – SAFT (high energy)	3.3	156	292
C/$LiFePO_4$	Cylindrical; VL 10 V Fe – SAFT (high power)	3.3	55	122
C/$LiFePO_4$	Cylindrical; IFR18650-11P – Samsung	3.2	82	213
C/$LiFePO_4$	Cylindrical; ANR 26650 – A123 Systems	3.3	109	239
Li/S_8^*	Prismatic; – Sion Power Corp.	2.15	350	320
Li/S_8^*	n/a – Oxis Energy Ltd.	-	300	-
Li/O_2^*	n/a; primary element – Polyplus	-	> 700	-

[2]Initially, the terms "cell" and "battery" had strictly different definitions. Thus, an electrochemical cell is the smallest battery unit and consists of anode, cathode, electrolyte, separator, current collector, and housing. As opposed to that, a battery consists of at least two cells connected in series or in parallel. A 12-V lead battery for instance is made of 6 2-V cells. Nowadays however, a cell is often called battery. The electrochemical processes do not differ from cell to battery and this is why the present Chapter does not differentiate between those two terms. Specifying the practical energy densities however calls for a differentiation. All practical energy density values (with the exception of lead batteries) in this Chapter refer to cells.

Currently, important approaches to maximize energy and power density by improving electrode materials are in particular (a) partially or completely substituting lithium alloys containing tin and silicon for carbon in the negative electrode and (b) increasing the positive electrode's nickel content.

Lithium-sulfur and lithium-air systems are much discussed next generation technologies and use cell chemistries considerably different from those of standard lithium-ion batteries. Developing these battery types to marketable systems in regards to gravimetric energy density would mean a great step forward when compared with the current lithium-ion technology.

Theoretical and provisional practical energy densities of both technologies are compared with conventional systems in Fig. 16.1b and Table 16.2. A lithium-air cell has a theoretical energy density of several thousand Wh/kg. This explains the fascination sparked by this cell system.

The most important reasons for the high energy density of both cell concepts are

(1) the substitution of the light elements sulfur and oxygen for the cathode's comparatively heavy transition metal compounds.
(2) the storing of more lithium per formula unit. Conventional cathode materials change the transition metal's oxidation state during the lithium's intercalation and deintercalation. Hence, maximally one lithium ion per formula unit can be stored with the redox pairs Co^{4+}/Co^{3+}, Fe^{3+}/Fe^{2+} and Mn^{4+}/Mn^{3+}. Effectively, only 0.8 lithium ions ($LiMn_2O_4$) and 0.5 lithium ions ($LiCoO_2$) respectively can be utilized per formula unit. In contrast to that, sulfur and oxygen can incorporate 2 lithium ions each per formula unit during a complete reaction.
(3) the replacement of the anode's graphite with metallic lithium in the future. However, using a pure metal anode poses challenges that have yet to be resolved, most notably the dangerous dendrite formation. The continuous electrolyte consumption induced by the reaction with the metal anode during full charging and discharging is a decisive factor for cycling stability.

Developers hope on good grounds that the usage of sulfur or oxygen instead of the comparatively expensive transition metal compounds enables the production of low-cost batteries.

This Chapter takes an in-depth look at the cell chemistries of next generation technologies using Li/S and Li/O_2. It also discusses advantages and disadvantages as well as solutions for the current problems. "All-solid state batteries" become more and more interesting and are reviewed at the end of this Chapter.

16.2 The lithium-sulfur battery

16.2.1 Basic principle

The lithium-sulfur cell chemistry has been examined for several decades [2–4]. The reasons for this are the high energy density, the almost unlimited availability

Table 16.2 Theoretical gravimetric (Wh/kg) and volumetric (Wh/l) energy densities w_{th} for different cell reactions

Cell type and reaction (lithium - anode)	E° [V]	w_{th} [Wh/kg]	w_{th} [Wh/l]	Cell type and reaction (graphite - anode)	E° [V]	w_{th} [Wh/kg]	w_{th} [Wh/l]
$\frac{1}{2}\,Li + Li_{0.5}CoO_2 \underset{charge}{\overset{discharge}{\rightleftarrows}} LiCoO_2$	3.9	534	2,723	$\frac{1}{2}\,LiC_6 + Li_{0.5}CoO_2 \underset{charge}{\overset{discharge}{\rightleftarrows}} 3\,C + LiCoO_2$	3.75	375	1,432
$Li + FePO_4 \underset{charge}{\overset{discharge}{\rightleftarrows}} LiFePO_4$	3.45	586	2,110	$LiC_6 + FePO_4 \underset{charge}{\overset{discharge}{\rightleftarrows}} 6\,C + LiFePO_4$	3.3	385	1,169
$0.8\,Li + Li_{0.2}Mn_2O_4 \underset{charge}{\overset{discharge}{\rightleftarrows}} LiMn_2O_4$	4.00	474	2,044	$0.8\,LiC_6 + Li_{0.2}Mn_2O_4 \underset{charge}{\overset{discharge}{\rightleftarrows}} 4.8\,C + LiMn_2O_4$	3.85	346	1,225
$2\,Li + \frac{1}{8}\,S_8 \underset{charge}{\overset{discharge}{\rightleftarrows}} Li_2S$	2.2	2,613	4,286	$2\,LiC_6 + \frac{1}{8}\,S_8 \underset{charge}{\overset{discharge}{\rightleftarrows}} 12\,C + Li_2S$	2.09	589	1,222
$2\,Li + \frac{1}{2}\,O_2 \underset{charge}{\overset{discharge}{\rightleftarrows}} Li_2O$	2.91	5,220	10,508	$2\,LiC_6 + \frac{1}{2}\,O_2 \underset{charge}{\overset{discharge}{\rightleftarrows}} 12\,C + Li_2O$	2.76	850	1,885
$2\,Li + O_2 \underset{charge}{\overset{discharge}{\rightleftarrows}} Li_2O_2$	2.96	3,458	7,989	$2\,LiC_6 + O_2 \underset{charge}{\overset{discharge}{\rightleftarrows}} 12\,C + Li_2O_2$	2.81	793	1,804

Conventional lithium ion cells are based on a graphite anode which forms an intercalation compound LiC_6. Metallic lithium should be the main substance used in cells with Li/S and Li/O_2 as the anode material. An improved comparability of energy densities is achieved by calculating densities with graphite and a lithium anode. When graphite is used, the cell voltage decreases about 0.15 V compared with metallic lithium. The volumetric energy densities are calculated for cells in a discharged state, i.e. densities and mass fraction of the charged products are included in the calculation.

of sulfur, and its toxicological safety when used as active material. There have been great improvements during the last 10 or 15 years. But the technology still faces great challenges, even now. On first glance, a lithium-sulfur cell is based on reversible transformation of lithium by means of sulfur according to the following principle:

$$\text{Anode reaction}: 16\,\text{Li} \rightarrow 16\,\text{Li}^{+} + 16\,\text{e}^{-}$$

$$\text{Cathode reaction}: \text{S}_8 + 16\,\text{e}^{-} \rightarrow 8\,\text{S}^{2-}$$

$$\text{Overall reaction}: 16\,\text{Li} + \text{S}_8 \rightarrow 8\,\text{Li}_2\text{S}\ (\text{E}^{\circ} = 2.24\ \text{V})$$

The theoretical cell voltage is calculated as E° = 2.24 V from the reaction enthalpy of this reaction ($\Delta_r G^{\circ}_{25\,^{\circ}C} = -432.57$ kJ/mol [Li_2S]). Combined with the theoretical capacity of 1,167 mAh/g (Li_2S), the resulting theoretical energy density is 2,613 Wh/kg (Li_2S). This value is several times higher than that of conventional systems. Literature often describes the cathode reaction by itself, thus the capacity mentioned there is usually referring to sulfur. It is 1,672 mAh/g (S).

Fig. 16.2a shows a schematic diagram of a cell. A suitable cathode structure must be generated since both sulfur and the discharge product are not electrically conducting. Usually, porous carbon particles with large surfaces are used as carrier material. They provide electronic contact and also ensure sufficient electrolyte accessibility. The considerable volume changes are characteristic for cell reactions. Li_2 (ρ = 1.66 g/cm^3, V_m = 28.0 ml/mol) has a lower density than sulfur (ρ = 2.07 g/cm$_3$, V^m = 15.5 ml/mol). Therefore, the cathode must provide enough space to compensate for a volume increase of around 80 %. In general, a sulfur cathode contains between 50 and 70 % weight percent sulfur. The remaining weight is divided among the carbon carry material and the binding agent (small amount). Carbon also needs to be added to standard cathode materials as a conducting additive, but with lower percentages. The energy densities of a lithium-sulfur cell attainable in practice are much lower than the theoretical energy densities. This is mainly due to the high carbon content and the required high porosity.

The electrolyte is made of a mixture of organic solvents and the applicable conducting salt. As opposed to standard carbonate-based solvents such as ethylene carbonate (EC)/dimethyl carbonate(DMC) with the conducting salt $LiPF_6$, the Li/S_8 cell in general features a mixture of dimethoxyethane (DME, $C_4H_{10}O_2$), 1,3-dioxolane (DOL, $C_3H_6O_2$), and lithium bis(trifluoromethylsulfonyl)imide ($LiN(SO_2CF_3)_2$, LiTFSI), because, at this stage, this mixture seems to be best compatible with the metallic lithium anodes.

This cell reaction seems quite simple at first glance, however if one takes a closer look it becomes very complex. Several intermediate steps for the reduction of sulfur to the sulfide ion (S^{2-}) are the reason:

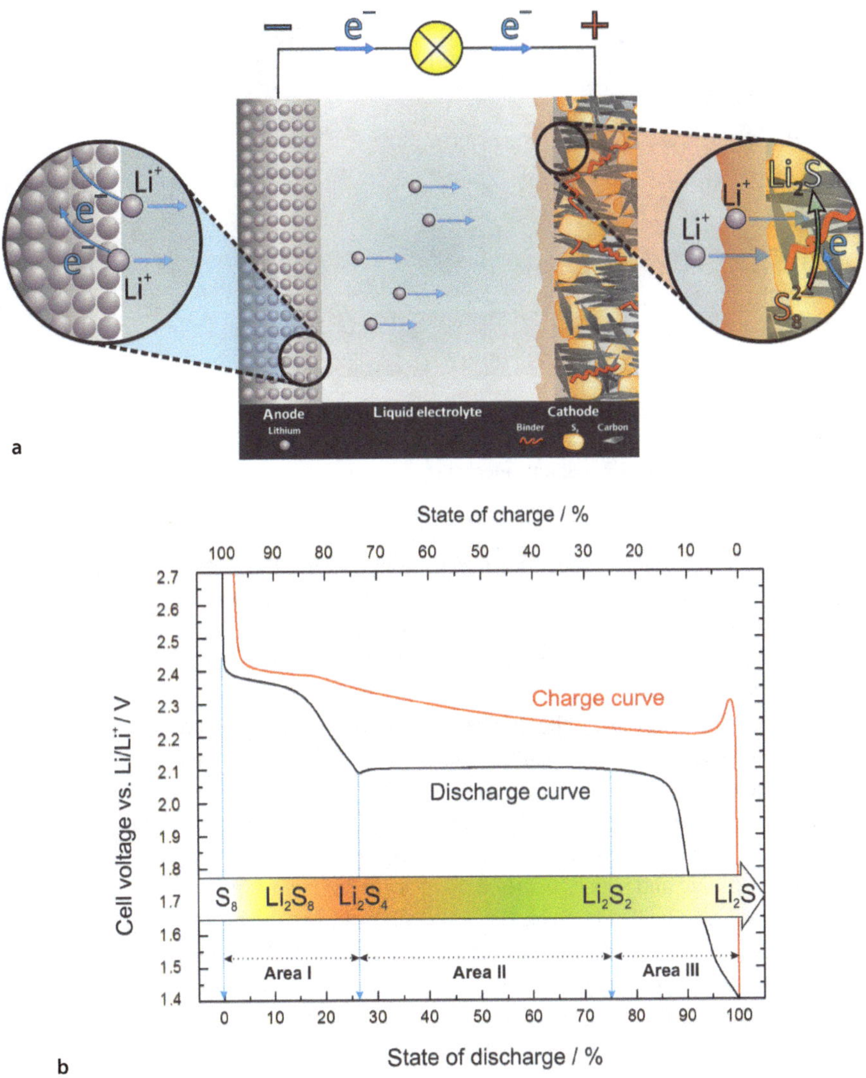

Fig. 16.2 **(a)** Schematic diagram of a lithium-sulfur cell. Metallic lithium is the anode material. The cathode consists of a mixture of sulfur and carbon particles, which is mechanically stabilized by means of a binding agent. **(b)** Typical voltage profile (charging and discharging cycle) of a lithium-sulfur cell

Most polysulfides dissolve very well in the electrolyte, hence the reaction mechanism is considerably different to that of conventional lithium-ion batteries where the reactions are pure solid state reactions. Fig. 16.3 shows polysulfide solubility in the electrolyte with the help of a demonstration. The electrolyte takes on color immediately after the beginning of the discharge reaction because of the dissolved species.

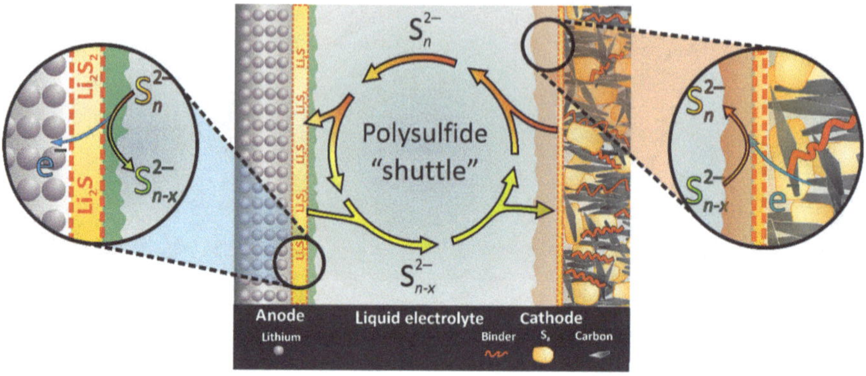

Fig. 16.3 First discharging of a lithium-sulfur cell in a glass cell. In the beginning of the reaction, soluble polysulfides are formed at the cathode from sulfur and lithium. They diffuse to the lithium anode (Section 16.2.2 – Shuttle mechanism)

It is still unsure which species emerge, how quickly the individual partial reactions take place, and how high the respective concentrations are at a specific point in time. The influence of polysulfide formation on the cell reaction directly correlates with the discharging cycle of the cell (Fig. 16.2b). The discharging process can be divided into three areas.

In the beginning (area 1), the discharge reactions starts with the reduction of elementary sulfur S8. Higher polysulfides such as Li_2S_8 and Li_2S_6 are formed. They dissolve in the electrolyte, which causes the discharge voltage to continually decrease. A minimum characterizes the transition to area 2. The incipient formation of solid Li_2S (Li_2S_2) phase is the reason. The required nucleation enthalpy causes an additional overvoltage and subsequently a discharging cycle minimum. Sulfur reduction continues in area 2. Aside from the solid phases, low-order soluble polysulfides (Li_2S_4, Li_2S_3) exist. Ideally, a full transition from Li_2S_2 to Li_2S takes place at the end of discharging. However, this is usually not achieved in cells. In general, 2/3 to 3/4 of theoretical capacity are reached. Reasons for this could be: insufficiently quick solid-state diffusion during the transition from Li_2S_2 to Li_2S, low electrical conductivity from Li_2S_2 to Li_2S, blocking of the porous electrode by the growing Li_2S particles, or the shuttle mechanism (Section 16.2.2). The cell reaction is complex and it has two plateaus. Hence, the average registered cell potential of around 2.15 V is lower than the calculated thermodynamic value of 2.24 V for the direct transition from lithium and sulfur to Li_2S.

There is less knowledge about charging process mechanics. This process features a continually increasing potential and appears to be simpler, namely by means of multiple electron transfer. Peled et al. [5] used cyclic voltammetry during the discharging process to show an incremental reduction, i.e., the voltammogramm displays several maximums. However, there was only a single maximum for the charging process (oxidation). This correlates with the S8 forming.

Fig. 16.4 Li/S cell shuttle mechanism during charging. The polysulfide solubility in the electrolyte causes a sulfur loss at the cathode. This reduces capacity ("fading") and eventually causes reduced cell service life. Li_2S/Li_2S_2 depositions form at the anode and cathode surface during cell cycling

Oxidation to S8 at the end of the charging process apparently is incomplete [6, 7]. It is still unclear how much of the solid phases (S_8, Li_2S_2, Li_2S) eventually form during cell cycling and how high the share of the respective polysulfides is. Only gradually, elaborate in situ experiments shed light on cell chemistry [8, 9].

16.2.2 Shuttle mechanism

Polysulfides dissolve very well in electrolytes. This causes another lithium-sulfur cell characteristic, the so-called shuttle mechanism (Fig. 16.4). The polysulfides $S_{n2}-$ that form and dissolve at the cathode, diffuse to the lithium anode and are reduced to Li_2S_2 and Li_2S.[3]

Subsequent high-order polysulfide species react with these compounds and form low-order polysulfides $S_{(n-x)}$.

This means that the desired electrochemical reaction of sulfur at the cathode partly also takes place at the anode in uncontrolled fashion (chemical or electrochemical reactions both are conceivable), which negatively influences cell characteristics.

If low-order polysulfide species form near the anode, they diffuse to the cathode. When the cell is then discharged, these diffused species are further reduced to Li_2S_2 or Li_2S. Simply put, the cathode reaction partly takes place at the anode during the discharging process or, rather, the cell self-discharges. Both are undesirable effects decreasing capacity. In contrast to that, the diffusion to the cathode during the charging process is followed by a re-oxidation of the polysulfide species from low order to higher order. These polysulfides then diffuse to the anode again. This cycle is generally known as the shuttle mechanism. If the shuttle mechanism is very pronounced, it is possible that a cell can accept an unlimited charge, it is "chemically short-circuited". Kumaresan et al. [10] gave a mathematical description of the discharging and charging reaction and the shuttle mechanism.

In general, the shuttle mechanism causes a parasitic sulfur active matter loss. This is due to the uncontrolled separation of Li_2S_2 and Li_2S outside of the cathode

[3]The polysulfide species S_n2^- that form at the cathode during discharging dissolve in the electrolyte there. A concentration gradient versus the anode develops, which causes the polysulfides to diffuse toward the anode. Step by step, the polysulfides are distributed in the electrolyte.

area and it eventually causes a considerable decrease in cell cycling capability and service life. Further aging mechanisms can be an inhomogeneous separation of Li_2S_2 or Li_2S on the cathode or a mechanical cathode structure breakup due to volume changes during cell reaction.

16.2.3 Long-term stabilization concepts

Currently, several concepts are tested to improve Li/S cell cycling characteristics. These concepts concern the cathode (carbon/sulfur), the electrolyte, and the anode (lithium metal). The goals especially are to limit the active mass loss, to suppress the shuttle mechanism, and to keep the cathode structures mechanically stable over a longer period of time. Hence, special carbon materials with a defined porosity are used for the cathode. They ensure sufficient electrical sulfur contact and also immobilize the polysulfide species to limit their loss in the electrolyte. A great many number of scientific papers treating these issues have been published during recent years. Nazar et al. [11] for example melt-infiltrated a special carbon (CMK-3) with a defined pore size of d = 3 nm with sulfur at 155 °C in their highly regarded paper (Fig. 16.5). The material showed a considerably less pronounced shuttle mechanism and a comparably stable cycling behavior at high capacities when compared to not-infiltrated samples. This is due to the material's special nanostructure.

Modifying the carbon surface with polyethylene glycol (PEG) resulted in additional improvement. Other researched carbon materials include activated carbon, carbon black, graphene, and carbon nanotubes.

Cathode characteristics strongly depend on the used carbon material and on the electrode preparation. Electrode sulfur surface load, layer thickness, binder content, and binder type are among significant parameters in this respect [12, 13]. Decrease in electrode layer thickness or decrease in sulfur surface load also are simple measures to improve cycling behavior. Research is also looking into polysulfide loss reduction by adding cathode additives such as Al_2O_3 or SiO_2 (polysulfide trap). Currently, a high surplus of electrolyte and lithium is still required to guarantee sufficient cycling stability. These measures however defeat the purpose of creating a high energy density battery. Therefore, there must be a balanced approach between improving individual parameters and decreasing energy density.

Another approach is to prevent the diffusion of polysulfide species to the anode. To that end, solid or gel polymer electrolytes are used as barriers [14]. Compared to liquid electrolytes, however, this material has a decreased conductivity and contacting capability that cause higher overvoltages and thus lower cell energy efficiency. Aurbach et al. [15] researched lithium anode passivation by means of the electrolyte additive $LiNO_3$. Cell reaction remains unchanged since the resulting protective layer is lithium-ion permeable. However, the layer's electron conductivity is negligible, a fact that effectively decreases parasitic high-order polysulfide reduction at the lithium anode. Meanwhile, $LiNO_3$ has established itself as standard additive in lithium-sulfur battery research. However, $LiNO_3$ is

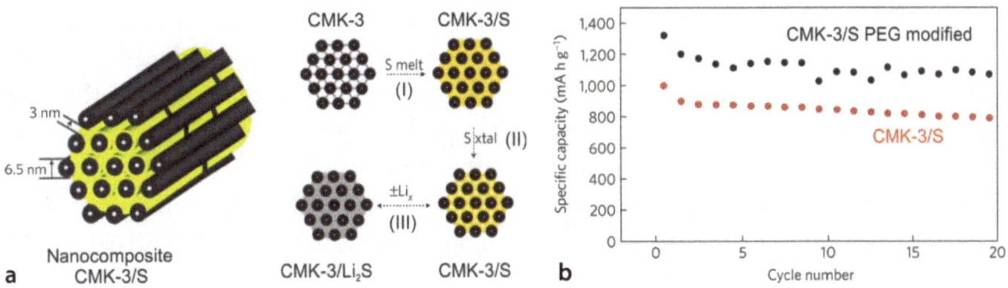

Fig. 16.5 **(a)** Production of a nano-structured cathode from mesoporous carbon (CMK3) and sulfur. (I) Carbon melt infiltration with liquid sulfur. (II) Sulfur cooling and crystallization (30 wt% CMK-3, 70 wt% S). (III) The volume increase that results from Li_2S formation is compensated by the carbon that provides enough free space. **(b)** Nanocomposite (CMK-3/S) or PEG-modified material cycling stability
(Material reproduced and modified by permission of Macmillan Publishers Ltd: Nature Materials [11], Copyright 2009)

continually broken down in the cell and therefore does not prevent aging, it only slows it down.

In addition to all of the above-mentioned cell concepts, additional approaches to create an electrochemical storage device based on lithium and sulfur compounds are currently being researched or rediscovered. It seems as if sulfur can be reversibly implemented in novel, Li^+-conducting compounds such as lithium polysulfidophosphates [16]. A different approach substitutes a liquid catholyte for the solid C/S cathode, a concept similar to that of redox-flow batteries [17]. Other approaches are: cells based on dissolved polysulfides (an approach that had been looked into several decades ago and that currently enjoys a renaissance) and the opposite, namely all-solid state batteries in which a solid electrolyte replaces the liquid electrolyte. It remains to be proven whether the latter approach is viable, however.

16.2.4 Status quo

Up to now and in spite of many different approaches, Li/S cells could not be sufficiently improved to ensure satisfying cycling behavior with respect to all important application parameters. For example, it is generally not possible to cycle Li/S cells till now while ensuring sufficiently high capacity. Also, compared to lithium-ion batteries, cell reaction speed (kinetics) is still insufficient due to the usage of nonconducting species such as S_8, Li_2S_2, and Li_2S. Increased research and development efforts are therefore required to develop a mass-marketable lithium-sulfur cell. There exist however novel applications relying on high gravimetric energy densities (e.g., aerial drones). The intensified research efforts produce a high publication rate of 5 to 10 per week. In addition to the lithium-sulfur battery discussed here, a sodium-sulfur battery that operates at ambient temperatures is also conceivable. The number of research results on this cell type is very limited, especially when compared to lithium-sulfur cells.

Sion Power Corp. (Tucson, USA) [18] and Oxis Energy Limited (Oxfordshire, UK) [19] are two of the few companies having produced marketable (to a certain degree) Li/S cells. These cells however are currently not available on the free market. Their gravimetric energy density values starting at 350 Wh/kg (at 320 Wh/l) are much higher than those of standard lithium-ion batteries. According to Sion Power Corp. 600 Wh/kg (at 600 Wh/l) can be achieved within the next years. Problems still are cycling stability and increased operating temperatures, both caused by the shuttle mechanism mentioned above.

16.3 The lithium-air battery

16.3.1 Basic principle

Abraham et al. in 1996 were the first to describe the function of lithium-air batteries (more precisely lithium-O_2 batteries) using non-aqueous electrolytes [20]. The main difference to standard batteries is that this cell type is an open system because – as with fuel cells – atmospheric oxygen is changed over at the cathode (Fig. 16.6). One might assume that the discharge product of an Li/O_2 cell is Li_2O, similar to fuel cells where H_2O is formed from hydrogen and oxygen. However, it is lithium peroxide (Li_2O_2) that is generally formed as a discharge product for thermodynamic reasons. Research has shown that potentially formed Li_2O is difficult to oxidize again. Metastable lithium superoxide probably develops as interstage product.

Anode reaction: $2Li \rightarrow 2Li^+ + 2e^-$	Anode reaction: $2Li \rightarrow 2Li^+ + 2e^-$
Cathode reaction: $0.5O_2 + 2e^- \rightarrow O^{2-}$	Cathode reaction: $O_2 + 2e^- \rightarrow O_2^{2-}$
Overall reaction: $2Li + 0.5O_2 \rightarrow Li_2O$ ($E^\circ = 2.91$ V)	Overall reaction: $2Li + O_2 \rightarrow Li_2O$ ($E^\circ = 2.96$ V)

Battery mass increases with an increasing discharge state because oxygen is introduced from the outside during discharging. This is why the described theoretical energy density of this system varies considerably. If the oxygen mass is included in the calculation, the free reaction enthalpy $\Delta_r G^\circ_{25\,°C} = -439.08$ kJ/mol (Li_2O) combined with a capacity of 1,793.88 mAh/g (Li_2O) results in a theoretical energy density of 5,220.16 Wh/kg (Li_2O) (or 1,168.33 mAh/g and 3,458.26 Wh/kg for Li_2O_2 as discharge product). Without the oxygen mass, the energy density for both reaction products is above 11,000 Wh/kg. This is similar to theoretical energy densities of standard fuels where oxygen mass is also not included in energy density calculations.

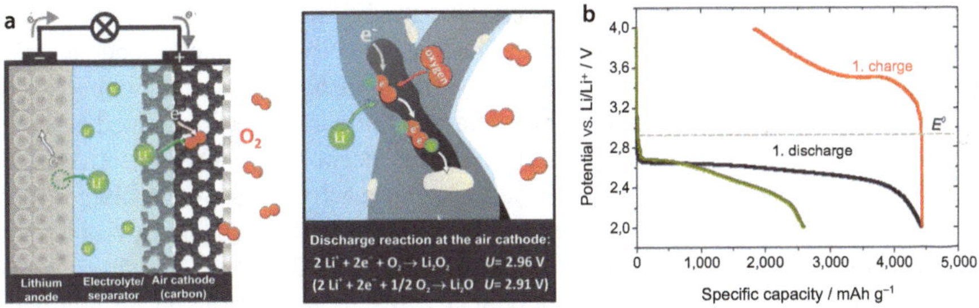

Fig. 16.6 **(a)** Schematic diagram of a lithium-air cell and electrode reaction for Li_2O_2 as the discharge product (oxygen reduction reaction, ORR). **(b)** Typical charging and discharging cycle. The displayed capacity relates to the mass of the carbon material used. The thermodynamically calculated voltage is E° = 2.6 V

Fig. 16.6 shows that a suitable cathode structure (conductive carbon matrix covered with catalytic particles, if necessary) is required to enable cell reaction and to provide enough free space for the resulting solid reaction products. The reaction products have a very low conductivity. This is why large-surface carbon materials (> 50 m^2/g) are generally used to ensure a homogeneous distribution of product particles that are as small as possible. Since it is difficult to specify the theoretical cathode capacity, literature specifies measured capacity as absolute (mAh) or sets it in reference to the carbon carrier material's weight (mAh/g[C]).

For an operation of a lithium-oxygen cell containing air to function, several undesired secondary reactions with air components must be prevented by using suitable membranes. Examples are: unwanted introduction of N_2 (nitride formation, Li_3N), H_2O (hydroxide formation, LiOH), and CO_2 (carbonate formation, Li_2CO_3). Currently, no simple solutions have been found in this respect. Hence, cell chemistry research is not conducted with air, but with oxygen. Also, cell drying-out, i.e., electrolyte evaporation from the cathode that is in contact with the atmosphere, must be prevented during operation. At the same time, lithium anode corrosion by dissolved oxygen in the electrolyte must be inhibited. In practice, this also calls for an additional protective layer. Because of the additionally required components, the attainable energy density will be much lower in practice than the theoretical values. However,the energy density values expected in practice from companies such as IBM and Polyplus are 1,700 Wh/kg and > 700 Wh/kg, respectively, and are therefore still much higher than those of current batteries [21, 22].

16.3.2 Electrolyte stability, efficiency, and reversibility

Fig. 16.6b shows a typical charging and discharging cycle of a Li/O_2 cell. The big difference between the measured discharge voltage (~2.6 V) and the charging voltage (> 3.5 V) is representative for this cell type and results in pronounced hysteresis. This is caused by high overvoltages during cell reactions, especially during the charging process (oxygen oxidation). Hysteresis values directly correlate with the cell reaction's energetic inefficiency. Therefore, cell reaction efficiency is only around 60 to 70 % for current aprotic Li/O_2 cells. This observation led to increased catalyst usage (MnO_2, Pt, Au, etc.) and better cathode materials (activated carbon, carbon black, graphene, nanotubes, etc.) during recent years. The objective is to improve system kinetics and therefore enable cell reactions with overvoltages that are as low as possible. Reaction reversibility improvement especially was hoped for in addition to the above. However, it generally was less then 20 cycles, combined with high-capacity losses. Sporadically, research mentions cycle numbers of > 100 [23].

Early research results indicated complex cell reactions [24–26]. However, it was recently proven that carbonate-based electrolytes (i.e., 1 M $LiPF_6$ in propylene carbonate) that had been used until then irreversibly degrade during cell reaction [23, 27]. Instead of the desired product (Li_2O_2), a multitude of degradation products (Li_2CO_3, CO_2, H_2O, $C_3H_6[OCO_2Li]_2$, …) were accounted for. Partially, these degradation products are further degraded during charging. This is due to the superoxide radical's (O^{2-}, forming during oxygen reduction) chemical reactivity. Thus, capacity values achieved in experiments were based not only on the reversible formation and degradation of Li_2O_2, they also were based on the irreversible electrolyte degradation. The examined catalysts merely accelerated this degradation process even more [28].

Research focus has changed due to these results and currently is directed toward developing electrolytes with a sufficient stability. It was not possible to identify a suitable system for practical applications till now. However, cell reactions, which still are accompanied by undesired secondary reactions, can at least be better examined in ether or DMSO. For example, the discharge product Li_2O_2 generally occurs as nanoscopic particles shaped as a torus (Fig. 16.7). Using gold instead of carbon as electrode material at least seems to enable easier scientific examinations of cell reactions [29].

16.3.3 Status quo

The current Li/O_2 cell status quo without a doubt still requires more fundamental research. At present, there is no cell concept for aprotic electrolytes that is able to prove reversible forming and degradation of Li_2O_2 (or Li_2O) across several cycles without simultaneous electrolyte degradation in practice.

Recent research results make it obvious that currently the greatest challenge for Li/O_2 cell production is the development of chemically stable electrolytes and electrode materials. In terms of the electrolyte, additional requirements need to be fulfilled aside from chemical stability. These are: lithium-ion conductivity, oxygen solubility and diffusivity as well as suitable electrode wetting. To better determine and understand the many various secondary reactions, current research uses a great variety of analysis methods. Mass spectrometer testing or pressure measuring during cell operation for example provide clear evidence of undesired secondary reactions.

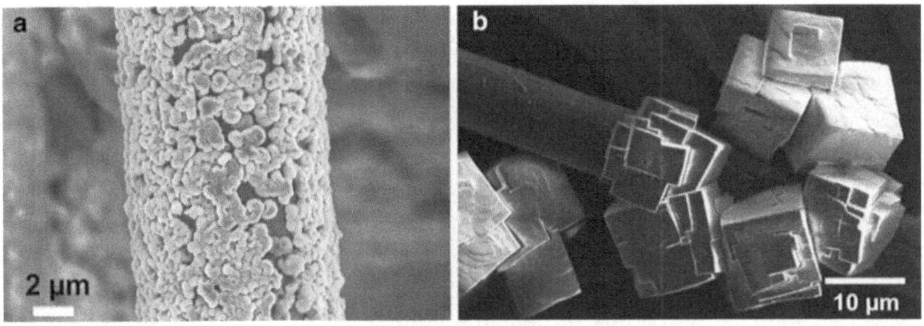

Fig. 16.7 SEM micrographs of an Li/O_2 cell cathode (*left*) and an Na/O_2cell (*right*) after discharging. Carbon fibers were the conducting carrier material in both cases. It is clearly discernible that Li_2O_2 is present as nanoparticles, while NaO_2 forms cubic crystallites in the µm range

A novel approach to reducing charging overvoltages uses redox mediators such as tetrathiafulvalene (TTF), lithium iodide, or tetramethylpiperidinoxyl (TEMPO). Oxidation is attained through dissolved compounds in this case. During charging, these compounds themselves are electrochemically oxidized at first, then they are reduced by Li_2O_2, which is degraded in the process. Also currently in research focus is the influence of contaminated water on cell chemistry.

Surprisingly, it could be proven that a substitution of sodium for lithium makes the cell reaction much more reversible. The discharge products in this case are not nanoscopic Na_2O_2 particles however, but large sodium superoxide (NaO_2) crystallites (Fig. 16.7) [30]. A catalyst may not be required because overvoltages are very low. Compared to an Li/O_2 cell, energy density is lower (2,643 Wh/kg[Na] and 1,105 Wh/kg[NaO_2], respectively).

Li/O_2 cells not only are developed by academic institutions, but also by companies [27, 31]. Li/O_2 batteries with an aqueous electrolyte are an alternative to the non-aqueous Li/O_2 batteries discussed here [32]. In such a system, oxygen from the ambient air reacts with lithium and water to lithium hydroxide (LiOH). Very high energy densities are to be expected with this cell concept. It is a prerequisite however, to effectively protect the lithium anode against the aqueous electrolyte.

16.4 Challenges when using metallic lithium in the anode

Maximum energy densities can better be achieved if metallic lithium is used in the anode instead of intercalation materials such as graphite. This is why metallic lithium is already being used successfully in primary elements.

Reversible dissolution and re-plating of lithium that are required for rechargeable batteries harbor technical difficulties however. This is why metallic lithium has as of yet not been used for safety reasons. This is due especially to the chemical reactivity of metallic lithium with the electrolyte as well as to lithium plating and subsequent forming of needle-like dendrites during charging.

In general, metallic lithium reacts with all known electrolytes for thermodynamic reasons. Suitability of an electrolyte for usage in a cell with a lithium metal electrode primarily depends on whether a passivating layer is formed that is self-inhibiting and prevents further parasitic reactions at the metallic anode. The layer must be conductive to lithium ions and block electrons to achieve this. It should be mentioned that all lithium-ion batteries with graphite anode are also operated outside of the electrolyte's stability range. Caused by a reaction with the carbonate-based electrolyte (i.e., $LiPF_6$ in EC/DMC and additives), a very stable SEI is formed that has a thickness of only a few nanometers. Contrary to that, ether-based solvents such as dioxolane or glyme are preferable for metallic lithium anodes [33]. This in spite of the fact that they are irreversibly degraded over time. Creeping corrosion of the lithium electrode requires that lithium anode cell systems have to be operated with excess lithium and liquid electrolyte. This negatively influences achievable energy density. It should be pointed out that especially when lithium-sulfur cells are researched in a laboratory environment, this is often done with a great excess of liquid electrolyte and lithium. This excess must be considerably reduced to achieve a marketable cell concept. Therefore, there is great need to develop suitable approaches to increase lithium anode and electrolyte compatibility.

Metal anodes typically form dendrites, a kinetic phenomenon that is often described for metal plating. This is another challenge for researchers that is more pronounced with higher charging currents. If the dendrites grow all the way to the cathode, they would cause a short circuit in the cell resulting in thermal runaway. It is possible that the dendrites fully become detached from the anode during cycling. Then they would not be available for the electrode reaction ("dead lithium"). Cell capacity would continually decline. An excess of lithium could balance out this loss. The amount of added lithium strongly depends on the system researched and the degree of cell optimization. Earlier research with lithium anodes (MoS_2 cathode) could give some pointers in this respect. A threefold lithium excess resulted in a service life of 300 cycles [34].

The following requirements must be fulfilled to use metallic lithium in rechargeable cells: parasitic reactions must be kept as low as possible and lithium separation should be as planar as possible, i.e., dendrite formation needs to be prevented. It might be possible to achieve these goals if a suitable electrolyte composition is developed that promotes formation of a pertinent SEI in the contact with lithium. Separately manufactured layers or membranes made from lithium-ion conducting solid electrolytes could be an alternative to the protective layers which are formed in the cell. These layers or membranes could be applied directly to the anode and cathode, or – as with separators – could be installed in the space between the electrodes. Another approach to prevent dendrite formation subjects the cell to mechanical pressure above the lithium flow limit [35]. There is also research looking into different additives to improve lithium electrode reversibility.

If metallic lithium cannot be used, conventional anode materials such as graphite also lend themselves to this purpose. This would however considerably reduce the theoretically achievable energy density (Table 16.2). This is why another approach favors silicon as anode material in Li/S cells [36, 37]. Based on the assumption that cell voltage decreases by 0.2 V vs. a metallic lithium anode because of the different anode reaction (formation of $Li_{4\cdot4}Si$), cell voltage is calculated as E° = 2.04 V and theoretical energy density as 1,862.45 Wh/kg and 3,299.25 Wh/l, respectively. Another alternative is tin (E° = 1.72 V, 922.84 Wh/kg and 2,628.14 Wh/l, respectively) [14].

16.5 All-solid state batteries

A very recent trend is the development of rechargeable solid state batteries (SSB), often also noted as "all-solid state batteries" (ASSB) or "solid state lithium batteries" (SSLB) in order to highlight the fact that this type of battery is constructed without the typical liquid electrolyte [38–41]. The current rate of development is fast, and only some of the most important topics related to the development of SSB shall be discussed here briefly (see [39–40] for a discussion of the essential challenges).

Two major types of SSLB have to be distinguished on the basis of the cell geometry and cell charge capacity: *Thin film* SSLB (tf-SSLB) have a relatively low area capacity in the order of 0.1 mAh/cm^2 and are typically prepared by gas phase deposition. Thin film SSB are already commercialized, but the corresponding market is yet small. Commercial tf-SSLB are usually based on single cells of the type Li/"LiPON"/LCO, i.e. contain a thin film cathode (a few µm thick) of $LiCoO_2$, a thin film solid electrolyte separator made of "LiPON" (an amorphous Li_3PO_4 film containing a few percent of nitrogen) and a lithium metal anode. Once well-functioning, tf-SSB can be run reversibly for thousands of cycles without significant degradation. Production of large scale batteries on the basis of multi-layer tf-SSB will not be cost-efficient, and therefore, the further development and use will be restricted to low-energy mobile applications (smart cards, flexible batteries for textiles, etc.).

Thick film or *large scale* SSB are still in the state of research [39–42–43,]. Here, two general trends can be identified: In "solidified" lithium ion batteries (SE-LIB; Solid Electrolyte LIB) the liquid electrolyte is substituted by either a solid polymer or a glass-ceramic electrolyte, or a combination of both. As solid electrolytes have usually a higher density than liquid (organic) electrolytes, SE-LIB will only gain improved energy density if the use of high capacity electrodes or of high voltage cathodes are enabled. One particular development in the direction of high energy SSB would then be LiM-SSB (Lithium Metal SSB) in which a thick lithium metal anode is utilized. Due to the high specific capacity of the lithium metal anode, LiM-SSB could offer a significant jump in energy density (up to +70%), but both severe mechanical effects during metal dissolution and deposition, as well as the risk of dendrite growth through the solid electrolyte separator have to be overcome.

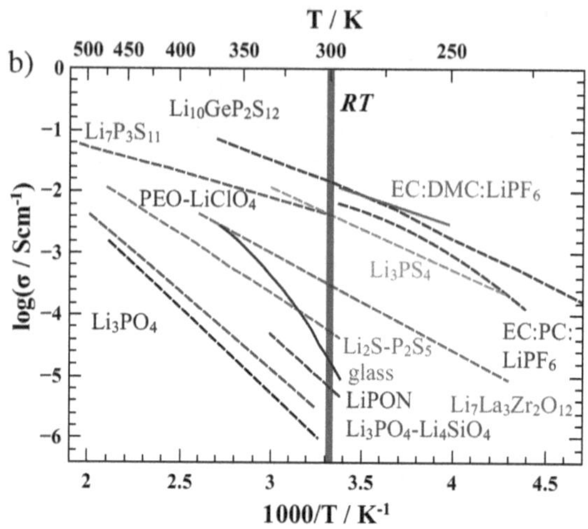

Fig. 16.8 Ionic conductivity of various solid electrolytes (by permission of …)

Power density is considered as "Achilles heel" of SSB, as solid electrolytes are considered as comparably poor ionic conductors. This is not correct, and today already quite a number of inorganic solid electrolytes with lithium ion conductivity higher than that of liquid electrolytes have been reported, see Fig. 16.8 for comparison [44–45]. In the case of SE-LIB, authors of a recent report even suggest higher rate capability than in the case of conventional LIB [42]. This can be of great relevance for fast charging and will surely drive further research.

It is generally accepted that interface kinetics is key to SSB with sufficient performance [39–40]. Once the lithium metal anode is employed, SEI formation with the solid electrolyte will take place, as the best solid electrolytes are all thermodynamically unstable against reduction [46–47]. If the forming SEI is highly resistive, the cell impedance will increase and will reduce the energy efficiency. If the forming SEI is highly conductive it may stabilize the cell, as in the case of SEI in conventional LIB. At the cathode/solid electrolyte interface also degrading interfacial reaction may take place and may also increase the cell impedance.

In summary, solid state batteries may open an attractive avenue toward high energy and high power lithium batteries. It is too early to judge the potential commercial success, as still some critical issues have to be solved.

16.6 Outlook

Lithium-sulfur and lithium-air batteries are two of the few systems with which it seems feasible to achieve a significant gravimetric energy density increase compared to lithium-ion batteries. Increase in volumetric energy density is less significant however. When sulfur or oxygen are used instead of transition metal compounds, there is also hope for producing low-cost batteries provided that the need for special carbons (or alternative electrode materials) or catalysts does not increase costs.

Both systems still pose great challenges from a technological point of view. This is especially true for lithium-air systems, because no suitable aprotic electrolyte could be found up till now. Both systems require protection of the lithium metal anode from parasitic secondary reactions. Special electrolytes or protective layers are used to prevent the shuttle mechanism in Li/S systems and the anode's oxygen corrosion in Li/O_2 cells. Furthermore, dendrite formation must be averted. Also, isolating and solid reaction products develop in both cell types during cycling (S, Li_2S_2, Li_2S and Li_2O_2, Li_2O). They may block the electrode and considerably impair the cell reaction's kinetics. Therefore, it is probably necessary to use cathode structures with a suitable porosity to promote a fine (nanometer range) reaction product distribution. A lithium-air cell also requires an effective and low-cost membrane, which specifically enables oxygen transport into and from the cell. This calls for innovations on both material and cell level.

It is hard to forecast whether and, if possible, when the next generation technologies for secondary elements described here can actually be put into practice. This is due to the current development status and the very diverse requirements for different application areas. A study conducted by the Fraunhofer Institute for Systems and Innovation Research expects marketable Li/S_8 cells by 2020. For lithium-air systems however, the prediction is the end of 2030. This shows that there are still fundamental problems to be solved. In terms of primary elements, lithium-air and similar systems are expected to be on the market earlier than that.

Bibliography

1. Broussely M, Archdale G (2004) Li-ion batteries and portable power source prospects for the next 5–10 years. J Power Sources 136(2):386–394
2. Herbert D, Ulam J (1962) Inventors; electric dry cells and storage batteries
3. Nole DA, Moss V, Cordova R (1970) Inventors; battery employing lithium-sulphur electrodes with nonaqueous electrolyte
4. Abraham KM (1981) Status of rechargeable positive electrodes for ambient-temperature lithium batteries. J Power Sources 7(1):1–43
5. Yamin H, Penciner J, Gorenshtain A, Elam M, Peled E (1985) The electrochemical-behavior of polysulfides in tetrahydrofuran. J Power Sources 14(1–3):129–134
6. Akridge JR, Mikhaylik YV, White N (2004) Li/S fundamental chemistry and application to hig-performance rechargeable batteries. Solid State Ionics 175(1–4):243–245
7. Mikhaylik YV, Akridge JR (2004) Polysulfide shuttle study in the Li/S battery system. J Electrochem Soc 151(11):A76–A1969
8. Nelson J, Misra S, Yang Y, Jackson A, Liu Y, Wang H et al (2012) In operando x-ray diffraction and transmission x-ray microscopy of lithium sulfur batteries. J Am Chem Soc 134(14):6337–6343
9. Dominko R, Demir-Cakan R, Morcrette M, Tarascon J-M (2011) Analytical detection of soluble polysulphides in a modified Swagelok cell. Electrochem Commun 13(2):117–120
10. Kumaresan K, Mikhaylik Y, White RE (.2008) A mathematical model for a lithium-sulfur cell. J Electrochem Soc 155(8):A576–A582
11. Ji X, Lee KT, Nazar LF (2009) A highly ordered nanostructured carbon-sulphur cathode for lithium-sulphur batteries. Nat Mater 8(6):500–506

12. Schneider H, Garsuch A, Panchenko A, Gronwald O, Janssen N, Novak P (2012) Influence of different electrode compositions and binder materials on the performance of lithium-sulfur batteries. J Power Sources 205:420–425
13. Cheon SE, Ko KS, Cho JH, Kim SW, Chin EY, Kim HT (2003) Rechargeable lithium sulfur battery – II. Rate capability and cycle characteristics. J Electrochem Soc 150(6):A800–A805
14. Hassoun J, Scrosati B (2010) A high-performance polymer tin sulfur lithium ion battery. Angewandte Chemie Int Edition 49(13):2371–2374
15. Aurbach D, Pollak E, Elazari R, Salitra G, Kelley CS, Affinito J (2009) On the surface chemical aspects of very high energy density, rechargeable li–sulfur batteries. J Electrochem Soc 156(8):A694–A702
16. Lin Z, Liu Z, Fu W, Dudney NJ, Liang C (2013) Lithium Polysulfidophosphates: A Family of Lithium-Conducting Sulfur-Rich Compounds for Lithium-Sulfur Batteries. Angewandte Chemie. 125(29):7608–11
17. Yang Y, Zheng G, Cui Y (2013) A membrane-free lithium/polysulfide semi-liquid battery for large-scale energy storage. Energy & Environmental Science 6(5):1552–8
18. www.sionpower.com
19. http://www.oxisenergy.com/
20. Abraham KM, Jiang Z (1996) A polymer electrolyte-based rechargeable lithium/oxygen battery. J Electrochem Soc 143(1):1–5
21. Girishkumar G, McCloskey B, Luntz AC, Swanson S, Wilcke W (2010) Lithium–Air Battery: Promise and Challenges. The Journal of Physical Chemistry Letters. 1(14):2193–203
22. http://www.polyplus.com
23. Mizuno F, Nakanishi S, Kotani Y, Yokoishi S, Iba H (2010) Rechargeable Li-air batteries with carbonate-based liquid electrolytes. Electrochem 78(5):403–405
24. Read J (2002) Characterization of the lithium/oxygen organic electrolyte battery. J Electrochem Soc 149(9):A1190–A1195
25. Sawyer DT, Valentine JS (1981) How super is superoxide. Acc Chem Res 14(12):393–400
26. Aurbach D, Daroux M, Faguy P, Yeager E (1991) The electrochemistry of noble-metal electrodes in aprotic organic-solvents containing lithium-salts. J Electroanal Chem 297(1):225–244
27. Freunberger SA, Chen Y, Peng Z, Griffin JM, Hardwick LJ, Barde F et al (2011) Reactions in the rechargeable Li-O_2 battery with alkyl carbonate electrolytes. J Am Chem Soc 133(20):8040–8047
28. McCloskey BD, Scheffler R, Speidel A, Bethune DS, Shelby RM, Luntz AC (2011) On the efficacy of electrocatalysis in nonaqueous Li-O_2 batteries. J Am Chem Soc 133(45):18038–18041
29. Peng ZQ, Freunberger SA, Chen YH, Bruce PG (2012) A Reversible and Higher-Rate Li-O_2 Battery. Science. 337(6094):563–6.
30. Hartmann P, Bender CL, Vracar M, Dürr AK, Garsuch A, Janek J, Adelhelm P (2013) A rechargeable room-temperature sodium superoxide (NaO_2) battery. Nat Mater 12: 228–232
31. http://www.ibm.com/smarterplanet/us/en/smart_grid/article/battery500.html
32. de Jonghe LC et al (2007) inventors; protected active metal electrode and battery cell structures with non-aqueous interlayer architecture
33. Aurbach D et al (2009) On the surface chemical aspects of very high energy density, rechargeable Li-sulfur batteries. J Electrochem Soc 156(8):A694–A702
34. Brandt K (1994) Historical development of secondary lithium batteries. Solid State Ionics.69(3–4):173–183
35. Monroe C, Newman J (2005) The impact of elastic deformation on deposition kinetics at lithium/polymer interfaces. J Electrochem Soc 152(2):A396–A404
36. Yang Y, McDowell MT, Jackson A, Cha JJ, Hong SS, Cui Y (2010) New nanostructured Li_2S/Silicon rechargeable battery with high specific energy. Nano Lett 10(4):1486–1491
37. Elazari R, Salitra G, Gershinsky G, Garsuch A, Panchenko A, Aurbach D (2012) Rechargeable lithiated silicon–sulfur (SLS) battery prototypes. Electrochem Commun 14(1):21–24
38. *Handbook of Solid State Batteries*, 2nd ed., Dudney N J, West W C, Nanda J (Eds.), World Scientific 2015

39. Janek J, Zeier W (2016), Nat. Energy. 1(9):16141
40. Luntz A C, Voss J, Reuter K (2015), J. Phys. Chem. Lett. 6:4599–4604
41. Robinson A L, Janek J (2014), MRS Bulletin 39:1046
42. Kato, Y. et al. (2016), Nat. Energy 1:16030
43. Oh G, Hirayama M, Kwon O, Suzuki K, Kanno R (2016), Chem. Mater. 28:2634–2640
44. Bachman JC, Muy S, Grimaud A, Chang HH, Pour N, Lux SF, Paschos O, Maglia F, Lupart S, Lamp P, Giordano L, Shao-Horn Y (2016), Chem. Rev. 116(1):140–162
45. Minami T, Hayashi A, Tatsumisago M (2006), Solid State Ionics 177:2715–2720
46. Wenzel S, Weber D, Leichtweiss T, Sann J, Janek J (2016), Solid State Ionics 286:24–33
47. Zhu Y, He X, Mo Y (2015), ACS Appl. Mater. Interface 7:23685–23693

Part III

Battery Production – Resources and Processes

Lithium-ion cell and battery production processes

17

Karl-Heinz Pettinger, Achim Kampker, Claus-Rupert Hohenthanner, Christoph Deutskens, Heiner Heimes, and Ansgar vom Hemdt

Contents

K.-H. Pettinger (✉)
Technologiezentrum Energie, Hochschule Landshut, Am Lurzenhof 1,
84036 Landshut, Germany
e-mail: karl-heinz.pettinger@haw-landshut.de

A. Kampker (✉) · H. Heimes · A. vom Hemdt
Chair of Production Engineering of E-Mobility Components, Campus Boulevard 30,
52074 Aachen, Germany
e-mail: A.Kampker@pem.rwth-aachen.de; H.Heimes@pem.rwth-aachen.de;
A.Hemdt@pem.rwth-aachen.de

C.-R. Hohenthanner
Evonik Technology & Infrastructure GmbH, Rodenbacher Chaussee 4,
63457 Hanau, Germany
e-mail: claus-rupert.hohenthanner@evonik.com

C. Deutskens
PEM Aachen GmbH, Karl-Friedrich-Strasse 60,
52072 Aachen, Germany
e-mail: c.deutskens@pem-aachen.de

R. Korthauer (ed.), *Lithium-Ion Batteries: Basics and Applications*,
https://doi.org/10.1007/978-3-662-53071-9_17

17.1 Introduction

Lithium-ion batteries for electric mobility applications consist of battery modules made up of many individual battery cells (Fig. 17.1). The number of battery modules depends on the application. The modules are installed in a lithium-ion battery together with a battery management system, a cooling system, temperature management, and power electronics. Different cell types can be used in battery modules; they include round cells, prismatic hardcase cells, or flat cells such as pouch cells (more detailed information available in Chapter 9).

This Chapter describes battery cell production processes as well as battery module and battery pack assembly processes.

17.2 Battery cell production processes and design rules

Lithium-ion cell production can be divided into three main process steps:

- electrode production
- cell assembly
- forming, aging, and testing.

Cell design is the number one criterion when setting up a cell production facility. For all designs, four basic requirements must be fulfilled:

1. Each cathode sheet must face an anode sheet with the same or higher capacity. The distance between the sheets must be uniform.
2. Cathodes and anodes must be electrically isolated from each other at all times.
3. Electrical conductivity must be implemented at the best possible rate in all cell parts. Good contacting between the conductors as well as between the active materials and the conductors must be ensured in all parts of the system.
4. Ionic conduction must be ensured in all cell body parts by means of homogeneous electrolyte wetting.

These basic rules must be implemented throughout the process, starting with the design through to the production of the individual cells. Exceptions are not

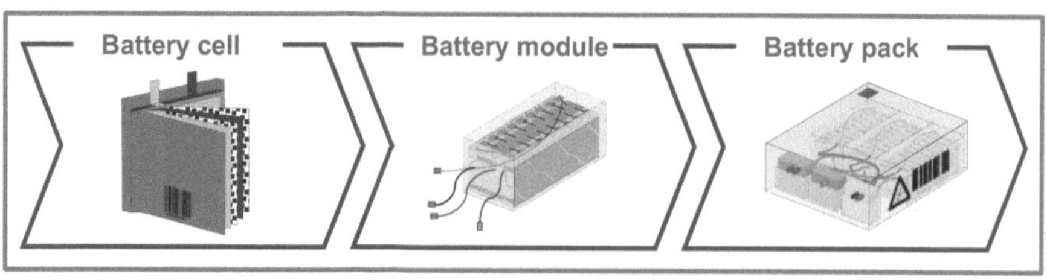

Fig. 17.1 From battery cell to battery pack

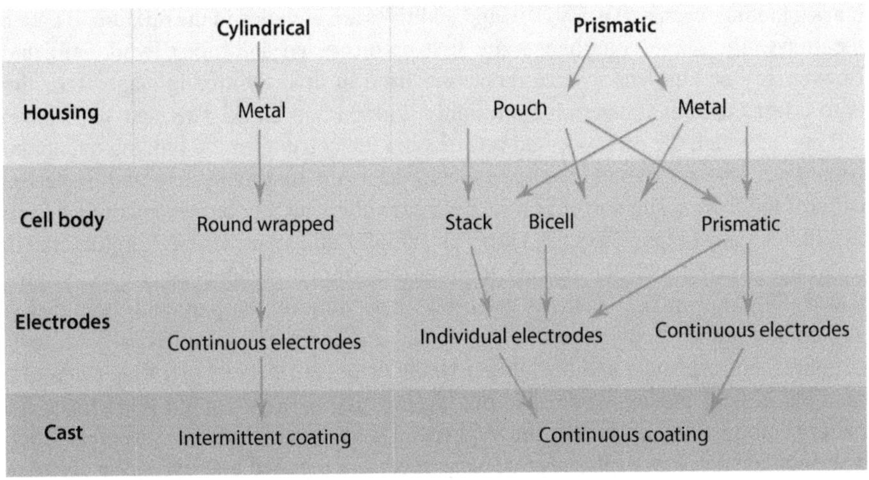

Fig. 17.2 Lithium-ion cell design

permitted and would cause functional defects. The rules must be heeded especially when defining and verifying production tolerances. This includes coating errors.

Taking into account these rules, cells can have different designs (Fig. 17.2). Process planning must define each cell's design because it is the foundation for all subsequent process steps. This not only applies to individual machines during the process, but also to handling and transport devices. The need to have a complete cell design two to three years before the actual start of production makes process planning especially difficult. During the process set-up, slight cell design changes are still possible. Multiple variants and larger design changes disproportionately increase the price of the production plant. Cell standardization seems necessary and would increase the security of investment in plants. Major design deviations can strongly influence the process set-up, causing considerable additional costs and delays.

17.2.1 Electrode production

All cell manufacturers use their own formulations and processes to produce electrodes. Separators, on the other hand, are freely available on the market in sufficient quality and amounts. Cell manufacturers make their own electrodes for several reasons: Electrodes govern the majority of a cell's properties (terminal voltage, capacity, C-rate stability, cycling behavior, aging, impedance, etc.). Most of the know-how and intellectual property in the field of cell production is in the electrodes.

The first sub-process in lithium-ion cell production involves mixing the active materials. It combines different components and results in a coating mass known as slurry [1]. Besides the active materials, electrical conductive agents (e.g. carbon black), binding agents (e.g. PVDF), and additives are included in the mixture to create the slurry [2]. These components are first preprocessed and mixed while still dry. Low-energy or high-energy treatments are used in this dry-mixing stage. The aim is to cover the active materials thoroughly with carbon black. The next step is wet mixing, in which the preprocessed powder and the solvent for the binding agent and sometimes also for the additives are conveyed to the mixing system and dispersed to form the slurry. The solvent on its own cannot create a homogeneous mass from the individual components; the slurry is initially still considerably agglomerated. Thorough stirring with the mixing tools ensures that a homogeneous mass is produced. Different mixing systems are used, depending on the particular application and quality requirements. These systems may use different mixing techniques, temperatures, atmospheres, and batching. Mixing errors are irreversible; they cannot be rectified in later process steps. The mixing process is of paramount importance for battery cell quality. This is why the requirements for the mixture are extremely strict. The individual components must be very precisely metered and exceptionally pure with a low residual water content. The slurry must fulfill certain parameters regarding homogeneity and viscosity to ensure a safe and continuous coating process. Changes in slurry viscosity and homogeneity over time must be taken into consideration and require that the slurry is processed rapidly.

Coating involves applying the slurry from the mixing process onto thin metallic foils. Copper foil with a typical thickness of 6 to 15 μm is used for the anode. Aluminum foil is used for the cathode; its typical thickness is 15 to 25 μm. The metallic foils form the electrode base and the current conductor. Like the other raw materials, they are extremely pure (> 99.8 %).

The additives' task is to increase the slurry's conductivity. Binding agents ensure a cohesive electrode structure and the slurry's adhesion to the foil. During mixing, these components are combined with a solvent and processed into a homogeneous paste. The solvent only dissolves the binding agent and some of the additives. It changes the rheologic characteristics so that the backing foil can be coated with the slurry [2–4]. The flat cell's capacity is mainly determined by the thickness of the active material coating on the metal foils. High-energy cells usually have an electrode thickness up to 200 μm [3].

Different application procedures are used to coat the metal foil with the paste-like or almost liquid slurry. These are doctor-blade coating, slot die coating, and reverse roll coating. The procedure itself can be continuous or intermittent, single-pass or double-pass, which involves coating both sides simultaneously. High precision is required here as well, especially with regard to the coat thickness and surface weight. Optimum-precision web runs and web tension are essential in order to produce a high-quality result. The foils must be dried after coating. Air jets and different dryers are used, comprising airborne sheets, suction jets, and roll conveyors. The temperature profile set in the dryer determines subsequent coat adhesion to the metallic foils and the distribution of the binding agent in the active material layer.

The solvent expelled during drying is extracted from the dryer's exhaust air, condensed, cleaned, and reused in the mixing process.

Coating and drying are followed by calendering (French: calandre = roll). Several top and bottom rollers gradually reduce the thickness. In addition, using several rollers provides better compacting control to improve active material cohesion to the foil [5]. Calendering is a continuous process, which can be set at speeds of 20 m/min and more. Linear loads can reach 40 to 60 t. The gaps between the rolls can be adjusted in the micrometer range by motors [6]. If the load is set too high during calendering, the foils may break or become brittle. Linear loads should be kept constant during the process. Another factor that influences quality is the coating of the calendering roller.

17.2.2 Cell assembly

The first step, irrespective of the cell type, involves cutting the cathode and anode coils to a certain width. This process is called slitting. The standard width of master rolls is around 600 mm. During slitting, they are cut into several slave rolls. They are then rolled up again and transported in vacuum boxes to the next process step. Like coating and calendering, slitting is a continuous process. The slitting machines use either blade or laser cutting with speeds of more than 100 m/min. The slitting blades are subject to considerable mechanical wear which has an impact on the quality of the cutting edge. If cutting is performed with lasers, it is important to take the heat-affected zone into account. The cutting remnants must be removed completely.

Cell assembly is then continued in a dry room with a dew-point temperature of −55 °C or less. Before the electrode rolls can be conveyed into the dry room, they must be dried again, leaving only little residual water. They can be dried as coils in a vacuum drying cabinet or as sheets during throughput. The dryer also serves as an airlock for the electrodes before the dry room.

"Decollation" and "stacking" are subsequent processes in flat-cell production. Decollation produces the individual cathode and anode sheets with several different technologies: laser cutting, rotating cutting, die-board cutting, and cutting by punchs. Decollation can take place in an integrated assembly line or through interim storage of several sheets in magazines.

Mechanical cutting procedures pose the risk of producing uneven cutting edges bent in one direction. This can be due to incorrectly set tools, tool wear, or smearing of active material over the cutting edges. Special care needs to be taken with the heat-affected zone during laser cutting, because the binding agent may froth up, causing a higher edge. Meticulous laser focusing in the electrode thickness mean is a prerequisite for exact cuts and a small heat-affected zone.

After cutting, the subsequent cleaning process removes the chipped-off particles from the edges of the electrode sheets produced during decollation as well as the particles from the electrode sheets. These particles can damage the separator and cause a short circuit if they are electroconductive. The input components for the stacking process comprise individual cathode and anode sheets as well as the separator foil (Fig. 17.3). Two different stacking processes can be used: Z-folding or

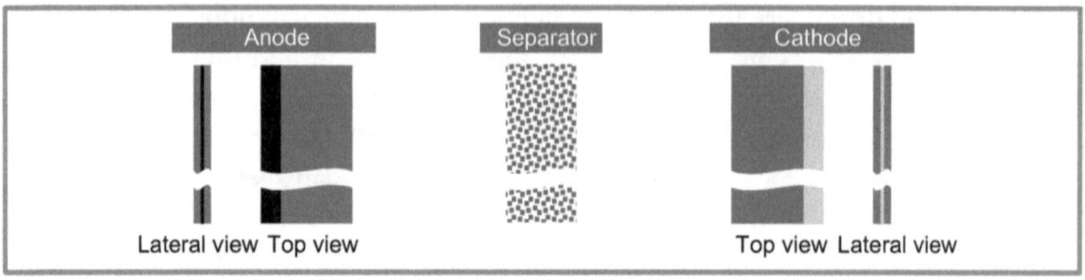

Fig. 17.3 Input materials for cell assembly

single-sheet stacking. Z-folding involves inserting the individual anode and cathode sheets sideways into the Z-shaped separator web. Single-sheet stacking involves decollating the separator, and the resulting sheets are stacked alternately with the cathode and anode sheets. The sequence is as follows: anode, separator, cathode, separator, and so on. Special challenges for the stacking process are handling, position recognition, and position alignment of the separator, anode, and cathode sheets with a vacuum gripper. Adhesive tape is used to secure the newly created electrode stack and prevent the individual sheets from sliding.

Prismatic or round cells are not manufactured using decollation and stacking, but with a winding process that follows the slitting process instead (Fig. 17.4). Winding is a process during which the different webs are wound around a prismatic body (separator). As with stacking, the sequence is cathode web, separator web, anode web, separator web, and so on. To prevent short circuits between the electrodes, the webs are wound in such a way that the separator reaches beyond the coated anode and cathode areas. The coated anode band is wider than the cathode band and always covers the cathode coating. The end product is called a jelly roll and is held together with adhesive tape.

The main production parameters that determine quality are the web tension, web guidance, winding speed for the individual webs, and the geometry or radius of the wraps. To finish off, the isolator foil is folded into a pocket and also fastened with adhesive tape.

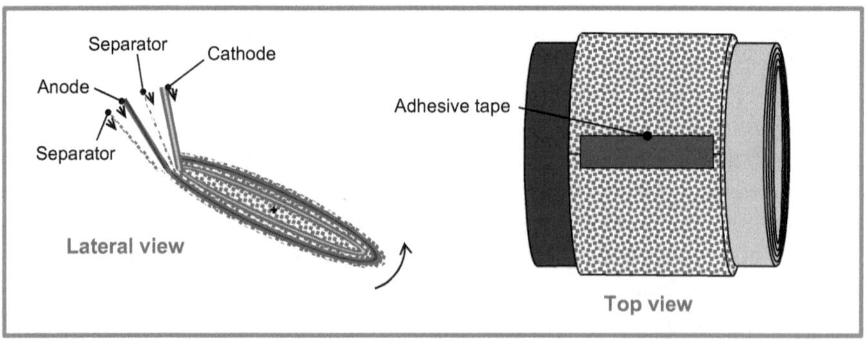

Fig. 17.4 Winding of a prismatic cell

The folded pouch is then slipped over the jelly roll. This ensures that the jelly roll is isolated from the inside of the metallic cell housing. After this, the uncoated electrode contact tabs are welded together with the conductors.

This process step for stacked cells includes the parallel switching of the individual, electrically isolated anode and cathode sheets. It is important to select the right welding technology because the parts are thin and susceptible to mechanical loads and temperature. Not every welding process is sufficiently accurate or it creates excessive heat so it can not be used for cell assembly. The processes that are suitable for welding the electrodes' uncoated contact tabs with the conductors are resistance welding, plasma-arc welding, ultrasonic welding, and laser welding. Industrial production normally uses ultrasonic welding and, more recently, laser welding.

17.2.3 Housing design

Lithium cells must be protected from moisture ingress. This is achieved by means of metal barriers. The materials used most commonly are steel and aluminum, which is easily machined. Housings can be made of flexible pouch foils or rigid metal.

Pouch housings Pouch technology uses multi-layer foil composites with an aluminum core for housings. The housing interior is coated with a thermoplastic layer with a low melting point; polypropylene is preferred. Several adhesive and anti-corrosion layers join the corrosion-resistant sealing compound with the metal core. Robust materials with much higher melting points such as polyamide or polyethylene terephthalate are used for the exterior. Heat sealing is used to fuse both housing parts' sealing compounds to close up the housing. Usually, the housing consists of two deep-drawn half shells with a joined gas pouch. This excess foil facilitates electrolyte filling and is able to absorb gases from the forming and aging processes. It is cut off at the end of the process during the final extraction.

The cell body's shape can easily be deep drawn into the foil. The aluminum core is drawn in the process. While designing the housing, the draw depth, radii, and angles must be planned in such a way that no micro fissures occur in the metal core to preserve its impermeability to water vapor. The deep-draw velocity must be adapted to the metal's drawability. Damage to the thermoplastic layer results in pitting corrosion or cell expansion right after the electrolyte filling.

Electricity feed-throughs are especially critical for all electrochemical cells, not only lithium-ion cells. They are very susceptible to micro leakages or capillaries. Pouch technology uses metal taps (nickel-plated copper for the cathode and aluminum for the anode) with a thickness of up to 0.5 mm as conductors. They are available pre-assembled and are inserted into the sealing compound between both pouch half shells during housing production. They are modified thermoplastics consisting of multi-layer foils which permanently seal the conductor's metal surface with the thermoplastic layer of the sealing compound. Usually, they are already connected to the conductor tabs. Sealing with separate sealing aids is becoming less important. If the seals and the conductor tabs are carried out properly, the housings are virtually impermeable to water vapor. Pouch housings do not have overpressure valves.

The housing yields when gas forms in the cell. The pouch foil is drawn, keeping the overpressure at a minimum.

Rigid metal housings Metal housings are deep-drawn from aluminum or stainless steel panels. The design usually is can and lid, with electricity feed-throughs in the lid as a rule. The feed-throughs are isolated pins, pre-assembled and sealed in frits. The cell body is welded to the pins on the inside. After the cell body is inserted into the cup, the lid is installed. The lid and can are welded with laser welding, roller-seam welding, or ultrasonic welding technology. The metallic weld seam's maximum tightness is also of great importance. The electrolyte filling openings are closed later by laser welding or by forcing in locking balls.

Metal housings have at least one predetermined breaking point or an overpressure valve to prevent cell bursting in the event of malfunction. This safety feature becomes active at medium pressures of around 10 to 15 bar.

17.2.4 Filling

Electrolyte filling is a technologically demanding process step which considerably influences cell functionality. The cell body's pore void volume must be homogeneously wetted with electrolyte. This homogeneity is an important parameter for cell service life and safety. Thin (200 – 300 μm) fissures between two large metal surfaces each (collector foils) must be filled (Fig. 17.5). However, these fissures are not empty capillaries but are filled with electrode material and separator.

Filling is impeded by the fact that the cell housing's cavity is already filled with the cell body. This is why the electrolyte must be dosed very slowly or in several instances. The gases in the pores must be removed before filling to enable the pores' wetting with active materials and separator.

Forming and electrolyte filling are both cell production processes that are time-critical and therefore restrict the throughput. Filling technology strongly depends on the cell design and the materials' and electrolyte's physico-chemical characteristics.

Electrolytes used in lithium-ion technology are highly concentrated lithium-salt solutions in polar organic solvents. They are extremely hygroscopic and therefore corrosive. A good electrolyte has extensive wetting and creep properties. On the one hand, these characteristics facilitate wetting of the electrode structure; on the other hand, they pose a challenge for mechanical engineering. The boiling characteristics of the solvent ingredients also complicate the process because their boiling points are decreased if the pressure is decreased. In practice, this means that each electrolyte can even start boiling at room temperature if the pressure is reduced to a vacuum, causing unwanted electrolyte frothing.

For the reasons mentioned above, the filling process must be developed and verified for every electrochemical system and design.

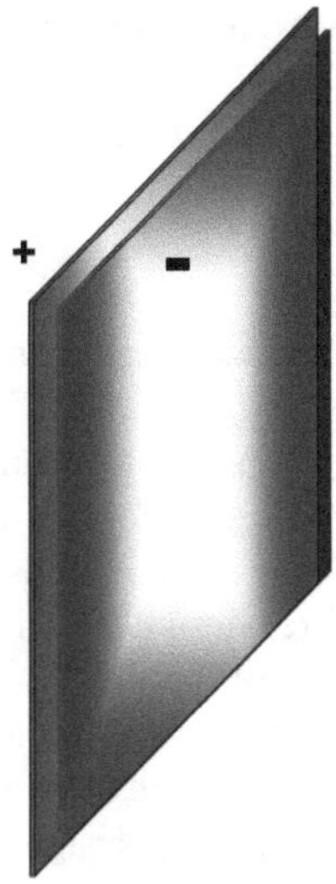

Fig. 17.5 Electrode with a width of 150 to 200 mm and a distance between the collector foils of 200 to 300 μm

17.2.5 Forming

Lithium-ion battery cells are a technology that is categorized as a secondary energy storage system, the cells are uncharged after electrolyte filling. Forming is the process step in which the cell is initially charged and essential layers (solid electrolyte interface, SEI) are produced. It is also a quality management control measure. This is the first instance in which the functional behaviour of the anode, cathode, separator, and electrolyte can be tested. Forming is performed in charging-discharging devices with one individually controlled channel per cell.

The process varies depending on the cell manufacturer. The slowest method, which is also the safest method, comprises two complete charging and discharging cycles; the forming cycle is followed by an additional charging and discharging cycle as a quality management measure. All cells are 100 % tested. The quickest, albeit incomplete, method stops charging when the voltage level is reached at which the SEI is formed. Forming is not completed. Also, there is no subsequent charging cycle for quality management. This method is used for consumer cells. It requires an elaborate overall process because it does not include quality management measures.

Forming can last up to two days. Each manufacturer strives to reduce the time needed for this production bottleneck. The energy required for forming is considerable. To cut process costs and promote "green production", its recuperation as electric power or process heat is recommended.

17.3 Advantages and disadvantages of different cell designs

This section will discuss three basic questions concerning cell manufacturing:

- cylindrical or prismatic cell design?
- foil housing or rigid metal housing?
- cell body in bicell technology or electrodes coated on both sides?

Cylindrical or prismatic cell design? Winding technology is used to produce cylindrical cells. It was developed and tested for the production of condensers, which involves winding thin foils. The winding speeds are high. Electrode compressing is controlled via the tension force during the winding process. A disadvantage of the production of consumer cells of type 18650 is that the electrode bands must be coated intermittently. On the other hand, the high throughput of the cell body winding is an advantage. For collecting high currents, an improved design has lateral conductors along the entire winding. This design enables high currents to be collected in the event of high energy density.

Producing prismatic cell bodies requires more process steps than producing cylindrical cell bodies. Production machines enabling high throughputs are used.

Heat dissipation is the most important characteristic that differentiates a prismatic cell design from a cylindrical cell design. If a prismatic cell is compared with a cylindrical cell with the same cell chemistry and same energy content, heat dissipates more easily from the prismatic cell because it has a larger exterior surface. Prismatic cells are more easily cooled, resulting in less overheating than in cylindrical cells. Cylindrical cells are thicker and the thermal conductivity of the cell body is modest; it therefore takes longer to determine hotspots.

The production technology for both cell body types is similar with the same amount of critical process steps. Cell body production is merely part of the overall process. Anode or cathode cell chemistry do not favor either design. However, the choice of design influences the configuration of the production process, i.e., the production facilities such as workpiece holders, machines, and contact positioning during forming. Once a design has been configured for a production process, modifying it incurs high costs.

Cell body in bicell technology or electrodes coated on both sides? The same electrode layer design can be implemented for the production of three types of cell body. They differ with regard to the resulting energy density and the process yield rate.

Energy density increases from the electrochemical basic cell (single cell) to the bicell (Fig. 17.6, A) to the cell with electrodes coated on both sides (Fig. 17.6, B).

Bicells have 50 % fewer anode collectors than single cells, and cells with electrodes coated on both sides have 50 % fewer cathode collectors than bicells. Gravimetric density increases the most from the single cell technology to the bicell technology, because the anode collectors are made of copper. Cathode collectors are made of aluminum, which only has around one-third of copper's density. This is why the increase

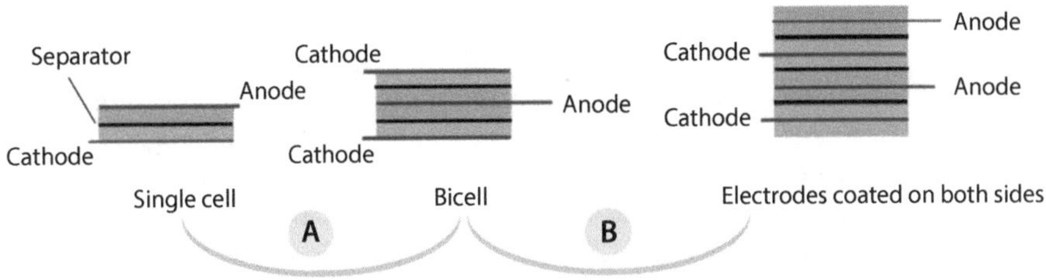

Fig. 17.6 Energy density increase resulting from multiple collector use

in energy density is less pronounced from bicell technology to cells with electrodes coated on both sides (B) than from single cell to bicell technology (A).

The extent to which intermediate products can be verified varies immensely for the different designs. The probability of a cell body error increases proportionately with increasing electrode surface. The stacks of electrodes coated on both sides can only be checked for short circuits after the cell body has been assembled. Bicells, on the other hand, are tested for electrode positioning and short circuits individually before stacking and are removed, if necessary.

A single error renders the entire cell body defective. Therefore, removal of defective parts results in higher yields for bicell-based batteries.

Foil housing or rigid metal housing? The advantage of pouch technology is that the housing parts can be produced easily from foils with deep-drawing or punching tools. Shearing and die cutting are similarly easy procedures for machining the housing material. Hot sealing tools are tried-and-tested devices that are used in the packaging industry. If the seals are manufactured properly (observing the minimum width, soft displacement of the sealing compound), pouches have repeatedly proven that they remain impermeable.

The cell housing is responsible for providing permanent protection against moisture ingress and discharged materials. Both technologies meet this requirement. Alternating temperature tests performed in a condensing atmosphere, i.e., with steep temperature gradients in a saturated water vapor atmosphere, are suitable in this respect and are able to detect even minute leakages. If a leakage occurs, the cells inflate within only a few days.

Pouch technology is a low-cost technology, even for small lot sizes, and it is flexible if slight format changes are necessary during processing. The sealing seams also have proven to be impermeable in the long term if manufactured properly.

In contrast, elaborate, multi-step deep-drawing and cutting tools are needed for producing metal housings. The deep-drawing process is profitable for large lot sizes. Sealing the housing requires complex technologies such as laser or ultrasonic welding. The complexity and cost of deep-drawing tools make it difficult to introduce format changes or produce small batch series.

Both technologies provide high-quality water vapor impermeability. Electricity feed-throughs are surprisingly expensive for both systems, and manufacturers' know-how is required here. The feed-throughs are so costly that they are the cell's most expensive part in some cases. Both technologies differ only slightly in this respect. Cost reductions can only be achieved for medium or large lot sizes.

Both pouch cells and metal housing cells can catch fire in the event of malfunction. Malfunctions are classified according to hazard levels (Table 17.1). A relevant malfunction severity begins at the threshold between hazard levels 4 (blowing-off of a cell) and 5 (blowing-off with fire).

Pouch cells easily even out volume changes between the charged and uncharged states and the increase in thickness that comes with age caused by gas formation. In rigid metal housings, the same malfunction causes a considerable increase in interior pressure or a mechanical load on the cell body.

Pouch-cell housings yield in the event of malfunction and their flexible material enables blowing-off of gases. This occurs when there is an overpressure of only a few mbar. In rigid metal housings, on the other hand, the pressure builds up until the overpressure valve is activated or the predetermined breaking point bursts. As a result, hazard level 4 is often skipped with hazard level 5 taking immediate effect. In the event of malfunction with identical cell chemistry and specific capacity, cell reactions in metal housings are much more severe and result in a fire more often than in pouch cells.

Rigid metal housings are advantageous in terms of their installation and automatic handling. It takes more effort to grasp, position, and process pouch cells to

Hazard Level	Description	Classification Criteria & Effect
0	No effect	No effect. No loss of functionality.
1	Passive protection activated	No defect; no leakage; no venting, fire, or flame; no rupture; no explosion; no exothermic reaction or thermal runaway. Cell reversibly damaged. Repair of protection device needed.
2	Defect/Damage	No leakage; no venting, fire, or flame; no rupture; no explosion; no exothermic reaction or thermal runaway. Cell irreversibly damaged. Repair needed.
3	Leakage Δ mass < 50 %	No venting, fire, or flame; no rupture; no explosion. Weight loss < 50 % of electrolyte weight (electrolyte = solvent + salt).
4	Venting Δ mass ≥ 50 %	No fire or flame; no rupture; no explosion. Weight loss ≥ 50 % of electrolyte weight (electrolyte = solvent + salt).
5	Fire or Flame	No rupture; no explosion (i.e., no flying parts).
6	Rupture	No explosion, but flying parts of the active mass.
7	Explosion	Explosion (i.e., disintegration of the cell).

Table 17.1 EUCAR hazard level classification for batteries

manufacture batteries with them because they are not rigid. The metal housing, which is thicker than a pouch, is more robust and is subject to fewer manufacturing tolerances. For battery assembly, designers, facility designers, and executing engineers prefer rigid metal housings.

17.4 Battery pack assembly

After the finished flat cells have been classified based on their performance, they are combined into a module. It is essential that all cells used in a module have the same performance values. If this is not the case, it has a negative impact on the entire module because the circuitry in the weakest cell in the system determines the module's characteristics (Fig. 17.7).

The cells are preassembled in a frame and their conductors are connected in parallel or in series. The cell conductors are either welded or screwed to the conductor rail. A screw connection has the advantage of ensuring effective handling during assembly and the cells can also be easily exchanged. Its disadvantages are that a screw connection has a low resistance to vibrations and its assembly requires more effort. A sealed material weld connection has better properties in this regard, and it is also advantageous in terms of electrical current flow and transition resistance. Ultrasonic welding, laser welding, and resistance welding are all suitable for conductor welding.

In a next step, the CSC (cell supervisory circuit) board and the cooling plates are installed on the preassembled module. The CSC board monitors the cells and optimizes the way the individual cells work together.

It can also balance out the states of charge of the individual cells. After the frame has been assembled and installed in the housing, the battery module undergoes a final inspection.

The last step involves combining the inspected battery modules with the peripheral electronic components to create the battery pack. The modules are installed in a housing during this process. A contactor rail is then fitted, connecting the individual modules of the battery pack. The battery modules can now be screwed to the housing. The battery management system (BMS) and the power electronics are also installed in the housing. The BMS measures and controls the temperature, state of charge, and voltage of the individual cells. The main task of the BMS is to create a connection between the battery and the other components. An end-of-line (EOL) inspection is performed after the battery pack has been fitted with a high-voltage connection (Fig. 17.8). If the battery pack passes this inspection, it is sealed and charged.

Fig. 17.7 Battery module assembly

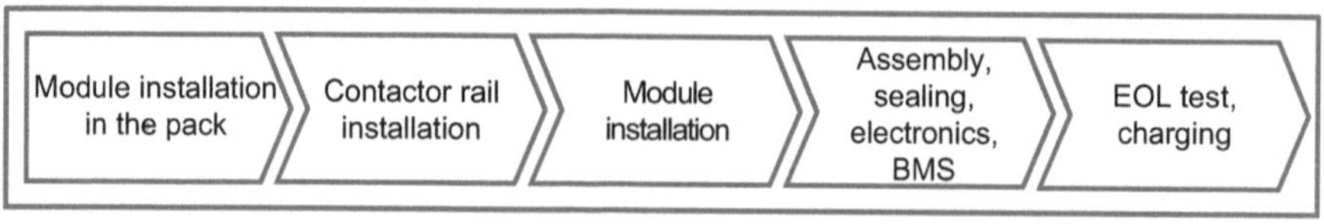

Fig. 17.8 Battery pack assembly

17.5 Technological challenges of the production process

The lithium-ion battery cell production process typically consists of heterogeneous production technologies. These are provided by machinery and plant manufacturers who are usually specialized in individual sub-process steps such as mixing, coating, drying, calendering, and slitting. Each of these sub-process steps is offered by competing machinery manufacturers. They have different technological approaches to the sub-process steps (single-sheet stacking vs. Z-folding, airborne sheet conveyors vs. roll conveyors, slot injectors vs. doctor blade, etc.).

It is a major challenge for cell manufacturers to establish an integrated, technologically balanced production concept with many interfaces and also an extensive variety of technologies. Designing an integrated production structure with coordinated production technologies for cell production requires a systematic approach which involves selecting technologies in cross-company value added chains. In contrast to module and pack assembly, the production of lithium-ion battery cells typically integrates various production technologies and draws on wide-ranging fields of expertise. This is why the machines and plants for cell production are marketed by different companies. There are only a few machinery and plant manufacturers who cover the majority of the entire cell production chain with their own technology (Fig. 17.9). More often than not, the cell production value added chain is served by a large number of specialists who have extended their expertise stemming from a sub-process step to battery production.

Sub-process steps in battery cell production involve a great number of companies that have the know-how for specific production steps and offer various production technologies for these steps. However, these companies have very little know-how regarding the production steps before or after their particular specialism. This means that lithium-ion cell manufacturers face the challenge of identifying suitable machinery and plant manufacturers for every process step. They have to instruct suppliers on which production technology is to be used and ensure that the machinery and systems are combined to form a cost-effective overall process. There is no continuous automation technology, making it difficult for cell manufacturers to transform lithium-ion cell manufacturing into a mass-production process. Overall, the current structures lead to considerable disparities in the quality of the end product. Due to the strongly compartmentalized value added chain, the cell production processes are not cost-effective, neither in terms of production systems or technological design. There is a great deal of potential for improvement.

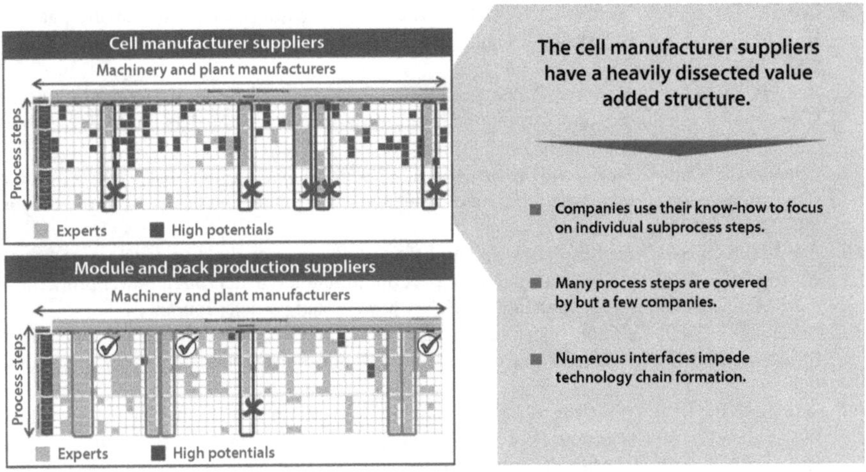

Fig. 17.9 Value added chain structure of machinery and plant manufacturers

Bibliography

1. Haselrieder (2013) Efficient electrode production for lithium-ion batteries
2. Bauer W, Nötzel D (2011) Rheological properties of electrode pastes for lithium iron phosphate and NMC batteries
3. Flynn J-C, Marsh C (2012) Development of continuous coating technology for lithium-ion electrodes
4. Haselrieder (2011) Auslegung und Scale-up des Trocknungsprozesses zur Fertigung von leistungsfähigen Elektroden mit optimierter Struktur und Haftung
5. Zheng Y, Tian L (2012) Calendering effects on the physical and electrochemical properties of Li[Ni1/3Mn1/3Co1/3]O2 cathode
6. Scrosati B (2002) Advances in lithium-ion batteries
7. Spahr M, Goers D, Leone A, Grivei E (2011) Development of carbon conductive additives for advanced lithium ion batteries. J Power Sources 196(7):3404–34138.
8. Sanchez-Gonzalez J, Macias-Garcia A, Alexandre-Franco MF, Gomez-Serrano V (2005) Electrical conductivity of carbon blacks under compression. Carbon 43:741–747
9. Sides CR, Croce F, Young VY, Martin CR, Scrosati B (2005) A high-rate, nanocomposite LiFePO4/Carbon cathode. Electrochem Solid-State Lett 8(9):A484–A487
10. Chen J, Wang JZ, Minett AI, Liu Y, Lynam C, Liu H, Wallace GG (2009) Carbon nanotube network modified carbon fibre paper for Li-ion batteries. Energy Environ Sci 2:393–396
11. Zhamu A, Shi J, Chen G, Fang Q, Jang BZ (2012) Graphene-enhanced anode particulates for lithium ion batteries. US 2012/0064409 A1
12. Buqa H, Holzapfel M, Krummeich F, Veit C, Novak P (2006) Study of styrene butadiene rubber and sodium methyl cellulose as binder for negative electrodes in lithium-ion batteries. J Power Sources 161:617–62
13. Lee J-H, Paik U, Hackley VA, Choi Y-M (2005) Effect of carboxymethyl cellulose on aqueous processing of natural graphite negative electrodes and their electrochemical performance for lithium batteries. J Electrochem Soc 152(9):A1763–A1769
14. Sano A, Kurihara M, Ogawa K, Iijima T, Maruyama S (2009) Decreasing the initial irreversible capacity loss of graphite negative electrode by alkali-addition. J Power Sources 192:703–707

15. Lee JH, Lee S, Paik U, Choi Y-M (2005) Aqueous processing of natural graphite particulates for lithium-ion battery anodes and their electrochemical performance. J Power Sources 147:249–255
16. Zaidi W, Oumellal Y, Bonnet J-P, Zhang J, Cuevas F, Latroche M, Bobet JL, Aymard L (2011) Carboxymethylcellulose and carboxymethycellulose-formate as binders in MgH2-carbon composites for lithium-ion batteries. J Power sources 196:2854–2857
17. Ouatani LE, Dedryvère R, Ledeuil J-B, Biensan P, Desbrieres J, Gonbeau D (2009) Surface film formation on carbonaceous electrode: influence of the binder chemistry. J Power Sources 89:72–80
18. Lee J-H, Kim H-H, Wee SB, Paik U (2009) Effect of additives on the dispersion properties of aqueous based C/LiFePO4 paste and its impact on lithium ion battery high power properties. Hosaka powder technology foundation, KONA powder and particle. Journal 27
19. Lanciotti C (2009) Lithium battery cell manufacturing process. Joint European Commission/EPoSS/ERTRAC workshop 2009, Brussels, Kemet Arcotronics Technologies, Sasso Marconi, Italy
20. Schelisch J (2011) Forschung für die Produktion von Morgen, Portraits der ausgewählten Projekte im BMBF-Programm Forschung für die Produktion von morgen. Projektträger Karlsruhe (PTKA-PFT), Bundesministerium für Bildung und Forschung
21. Freedom CAR: Electrical energy storage system abuse test manual for electric and hybrid vehicle applications; Sandia Report, SAND 2005–3123

Facilities of a lithium-ion battery production plant

18

Rudolf Simon

Contents

18.1 Introduction

This Chapter describes the set-up of a battery production plant. The required manufacturing environment (clean/dry rooms), media supply, utilities, and building facilities are described, using the manufacturing process and equipment as a starting point. The high-level intra-building logistics and the allocation of areas are outlined. Lastly, the Chapter offers an outlook on future challenges and development potential.

18.2 Manufacturing process and requirements

Lithium-ion cell production can be divided into three main stages: electrode production, cell assembly, and electrical forming. Fig. 18.1 shows a design concept for a pilot production site with the main manufacturing areas placed according to their position in the process sequence.

R. Simon (✉)
M+W Group GmbH, Löwentorbogen 9b, 70376 Stuttgart, Germany
e-mail: rudolf.simon@mwgroup.net

R. Korthauer (ed.), *Lithium-Ion Batteries: Basics and Applications*,
https://doi.org/10.1007/978-3-662-53071-9_18

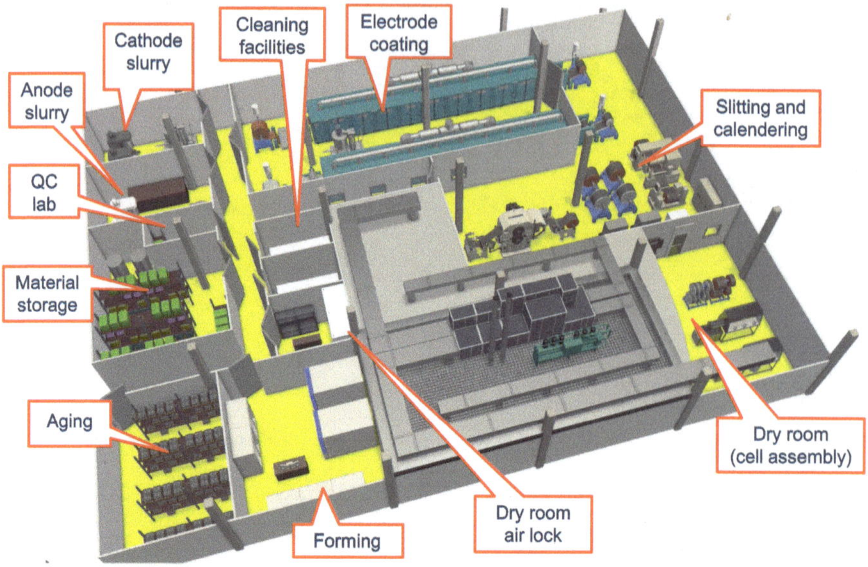

Fig. 18.1 Design concept for a pilot production line

During the slurry mixing process, it is particularly important to extract particulate matter (dust), control solvent quantities, and protect materials from cross-contamination. In the subsequent cathode and anode coating stage, including various drying and calendering processes, the most important factors are protection from cross-contamination, treatment of solvent exhaust, including solvent recovery, and humidity control.

Cell assembly takes place in a special dry and clean environment with a dew point of −40 °C or even less. Dry rooms must be specifically designed to prevent any electrostatic charge since such a design dramatically reduces dissipation through the air's residual conductivity. The electrical forming process takes up to several days, causes high power consumption, and requires smart energy management, process gas extraction, and fire and explosion protection.

18.3 Environmental conditions in the production area

There are a variety of specific requirements for lithium-ion cell production, in particular strict control of the indoor climate and cross contamination. These factors have a significant impact on the quality, safety, performance, and service life of cells.

The recommended ambient conditions for temperature and humidity for each of the major production stages are divided into groups with similar requirements (Table 18.1).

Production plant planning seeks to minimize the different climatic environments within the production plant for reasons of cost. ISO 7 or ISO 8 classified clean

Table 18.1 Environmental conditions within different production areas

Production step	Relative humidity	Dew point	Temperature	Clean room/ environment
Slurry mixing anode	45 % ± 15 %		22 °C ± 2 K	Controlled
Slurry mixing cathode	45 % ± 15 %		22 °C ± 2 K	Controlled
Coating anode	< 15 %	−8 °C	22 °C ± 2 K	ISO 7
Coating cathode	< 15 %	−8 °C	22 °C ± 2 K	ISO 7
Calendering anode	< 15 %	−8 °C	22 °C ± 2 K	ISO 7
Calendering cathode	< 15 %	−8 °C	22 °C ± 2 K	ISO 7
Cell assembly	< 1 %	−40 °C	22 °C ± 2 K	ISO 7
Filling with electrolyte*	< 1 %	−40 °C	22 °C ± 2 K	ISO 7
Cell forming	45 % ± 15 %		22 °C ± 2 K	Controlled
Cell aging	45 % ± 15 %		22 °C ± 2 K	Controlled
Testing and classification	45 % ± 15 %		22 °C ± 2 K	Controlled
Battery pack assembly	45 % ± 15 %		22 °C ± 2 K	Controlled

* encapsulated machine parts

rooms are recommended for the electrode production and cell assembly areas. Fig. 18.2 shows the different environmental zones in a manufacturing area layout.

The anode and cathode coating and drying processes require controlled pure air and relative humidity below 15 %. After that, humidity decreases gradually to levels under 1 % during electrolyte filling.

18.4 Dry room technology

The extremely low humidity requirements during cell assembly and, particularly, for the electrolyte filling step, are a challenge in lithium-ion battery manufacture. Depending on the product quality requirements, a dew-point down to −60 °C is necessary, which corresponds to a relative humidity of less than 0.1 % in the temperature range of 21 °C ± 1 K.

The low humidity requirements are due to the chemical processes taking place inside the cell. Residual water content during lithium-ion cell production leads to undesirable side effects that reduce cell safety. Furthermore, water in the anode graphite layers also blocks access for lithium ions, thus lowering the cell's capacity [1], [2].

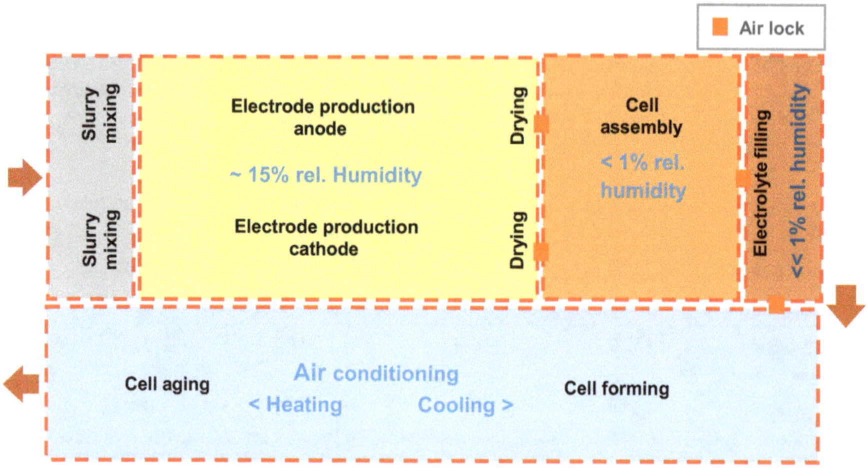

Fig. 18.2 Environmental zones in lithium-ion battery production

Such low humidity values cannot be reached with conventional air conditioning units. Therefore, condensation dehumidifiers are installed with a downstream adsorption dryer. To reduce operating costs, all walls, ceilings, floors, and air circulation systems, including connection joints, must be vapor tight. This minimizes both external humidity ingression and leakage of expensive conditioned dry air. This system is thus significantly different from the conventional clean room technology used in electronic industries, which is permeable to air. Ingression of particles can be prevented by overpressure, but a dry room must minimize water diffusion to ensure cost-efficient operation and high production quality. This minimization also requires reducing the number of operators to a minimum. A person can emit up to 150 g moisture per hour when physically active in a dry room. All materials also need to be pre-dried and enter and exit through interlocks.

Fig. 18.3 shows a typical dry room system. The installation consists of the sealed dry room, an air dehumidifier, diffusion-resistant duct work, customized personal and material interlocks, and filter-fan units, which create a clean environment by circulating air within the dry room.

The dry air is treated in several process steps. The recirculation air is filtered, heated, and then routed to a cooling grid, where the dew point falls and water condensates. Losses due to exhaust and leakage are compensated with outside air. The two airflows are mixed before passing through the adsorption wheel, which conditions the air to the required dew point. The dry air is then cooled if necessary.

The key component in the dehumidifying system is the adsorption wheel, which must be reconditioned after drying supply air. Part of the dry room exhaust air is used for this purpose. Sorption rotor reconditioning, also called desorption, takes place at 150 °C. A combined plate heat exchanger and electric heater supplies the required thermal energy to heat the exhaust air to this temperature. Process waste heat, steam, or gas can also be used here. Fig. 18.4 shows an example of a dry room

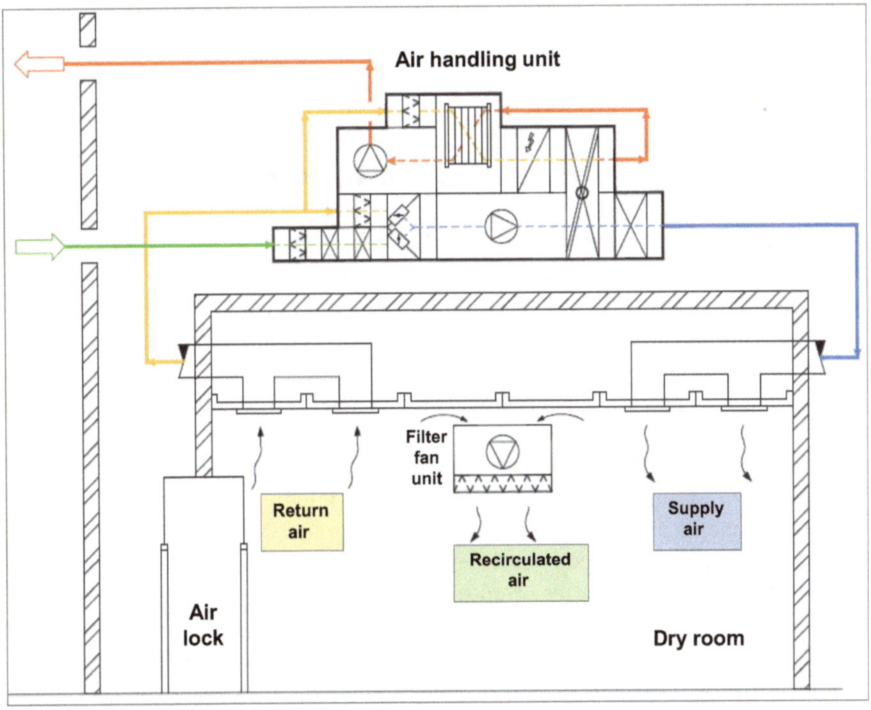

Fig. 18.3 Dry room schematic diagram

Fig. 18.4 Dry room for mass production under construction (*Source* M+W Group)

for mass lithium-ion cell production. The photo highlights ongoing complex sealing work, the conductive, diffusion-resistant floors, and preparation work for filter fan units.

18.5 Media supply and energy management

Media supply for a battery production plant (Fig. 18.5) can be divided into two categories. On the one hand, there are process media, which are required for the actual manufacturing process itself. This part includes DI water and/or the organic solvent for the slurry paste, process exhaust, process cooling water, and compressed dry air. On the other hand, there are the building facility systems, which provide the required manufacturing environment and the related media. These comprise the HVAC and electrical systems.

Intelligent energy management systems are required to conserve resources. The following two examples show the possibilities of intelligent energy use, which is one of the benefits provided by a professional integrated design.

Drying the coated electrode generates a high volume of heated air. The temperature of this exhaust air can reach up to 160 °C, depending on the manufacturing technology used. Once any solvent or water is separated, this hot air can be fed back into the drying tunnel to save energy. Another example of intelligent energy management is using the electrical energy from discharging to charge the next cell load during the forming process. Some vendors already offer such functions integrated into their process equipment.

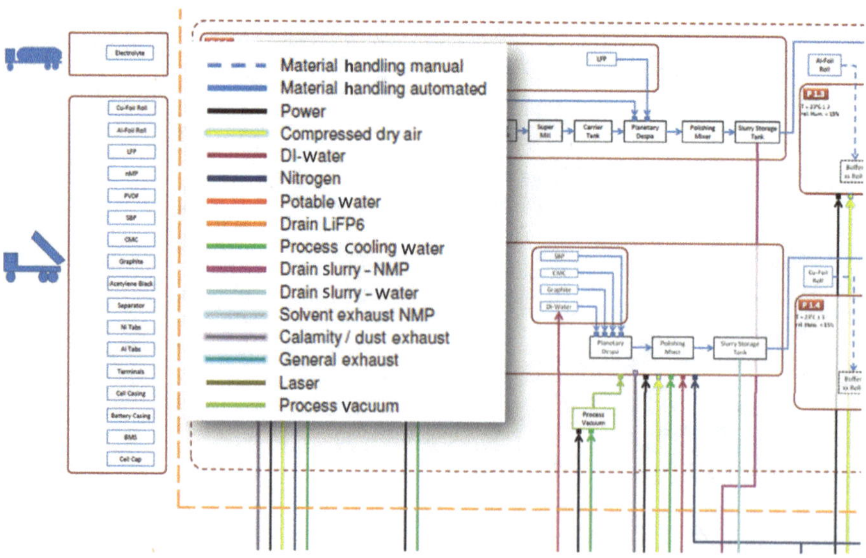

Fig. 18.5 Media supply schematic diagram

18.6 Area planning and building logistics

Besides the manufacturing floor, other areas are needed for other functions to operate a battery production plant. They meet production, material supply logistics, security, and personnel requirements and protect against external conditions such as the weather (Figs. 18.6, 18.7)

Manufacturing support All functions that assist actual production itself, such as material and personal interlocks, electrolyte dosing units, and quality management.

General support In addition to manufacturing support areas, support functions not directly related to the production fall into this category. These include maintenance workshops, locker and gowning rooms.

Logistics Storage areas are required for receiving/shipping and for storing raw material and finished products. The sizes needed depend on the operating concept, the variety of products, and the required storage period.

Utilities All systems for process media generation and building technology, including their distribution networks, are usually installed in a centralized location to achieve significant savings and synergy effects.

Administration Central administration areas are necessary. This includes offices for administrative functions such as management, purchasing, and human resources plus cafeteria and recreational areas.

The following aspects must be taken into account when the area concept is produced:

Personnel flow People enter the production plant through staff entrances and then proceed to the work areas. The production staff reach their workstation through the dressing rooms. Depending on the location, this path may be complex (dry room) or less complex (warehouse).

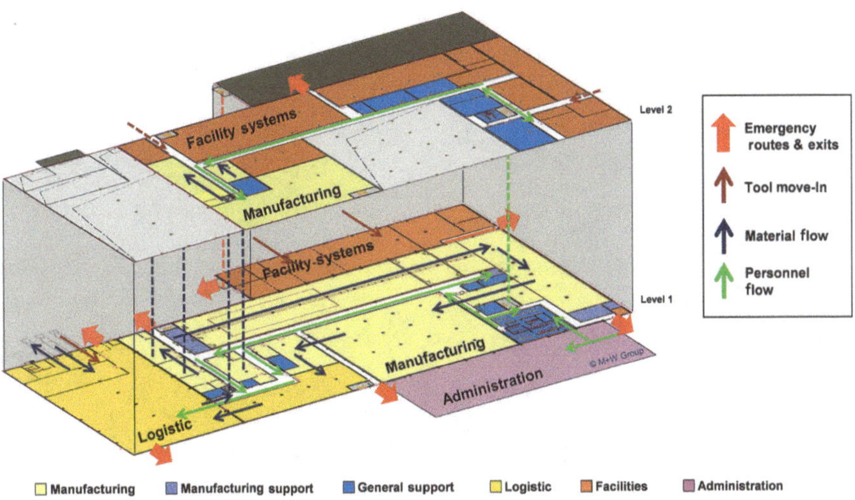

Fig. 18.6 Area plan and building logistic

Material flow Raw material and semi-finished parts are delivered to the production plant logistics area. After incoming inspection, the parts are (manually) transported to the individual areas. To compensate varying cycle times for the individual process steps, in-line buffers are required, which should be kept as low as possible. Material is generally carried by hand within an electrode manufacturing area. The degree of automation is significantly higher for cell assembly (in dry room). The cut electrode rolls and later the battery cells are combined to batches and transported on work piece carriers or conveyors before returning, as finished products, to the production plant logistics area. There, the cells are packaged in batches for transport, or they go to an adjacent battery pack assembly line.

Equipment move-in It must be ensured that any new equipment can be moved in and existing process equipment replaced during operation. New machines are usually delivered to the logistics area and small machinery can be transported from there directly to their place of operation via corridors and elevators. A movable platform might be needed on the outer facade to move in machines that will not fit through corridors or into elevators. In extreme cases, heavy-duty lifting equipment is needed to raise machines to the required height.

Escape and emergency routes Another important issue is the escape and emergency routes concept. Regulations are country-specific, but, as a fundamental rule, it must be ensured that all people in the building can reach an emergency exit within a specified maximum distance. The building concept (length, width, height, and number of floors) is often determined by such a maximum distance to an exit to the outside, stairwells, or a separate fire lobby.

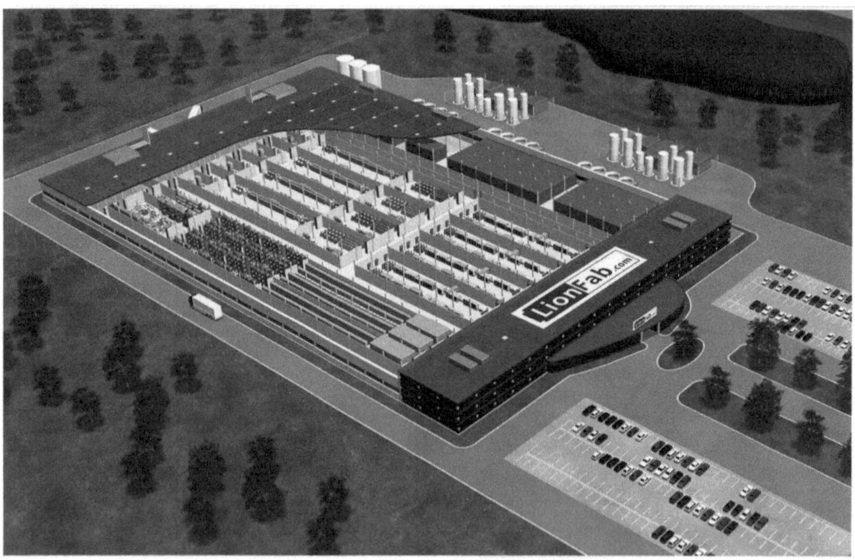

Fig. 18.7 Overview of a battery production plant for large-scale production

18.7 Outlook and challenges

- Over the coming years, there will be a large market for high-performance lithium cells. This will require significant progress in quality, safety, energy and power density, durability, and costs. These goals call for superior quality standards, a high degree of automation, first-class logistics, high-grade manufacturing technologies, and maximum resource efficiency.
- As a source of moisture, humans have a considerable impact on the investment and operating costs for a dry room. Theoretically, savings can be achieved by reducing the number of staff and using automated processes instead, or by encapsulating different stages in the process.
- A precondition for successful electric mobility is a reduction of production costs in the short term and achievement of a positive environmental balance for the entire production chain.
- Dynamic development of battery production technology requires production plant planning to be flexible. Production plants must be able to adapt to dynamic external and internal influences and respond flexibly.
- Similarly, a modular production plant concept (with exchangeable manufacturing modules) is recommended for easy adaptation to technological advances and reaching greater availability by providing redundancy in parallel production lines in mass production.
- With currently available manufacturing concepts (2014), economic operation of a pouch cell production can only be achieved at a capacity of above 500 MWh per year.
- The high-tech strategy of the German government, as in the "Industry 4.0" project, can contribute significantly to the development of a globally competitive battery production plant [3].
- To achieve these goals, battery producers will be increasingly required to show expertise in building technologies, production plant planning, and automation.

Bibliography

1. Y.P. Wu et al. / Journal of Power Sources 112 (2002) 255
2. I. Profatilova et al. / Electrochemica Acta 54 (2009) 4445
3. Industrie 4.0 : http://www.hightech-strategie.de/de/59.php

Production test procedures

19

Karl-Heinz Pettinger

Contents

19.1 Introduction

Lithium-ion cell production processes, starting with the active materials, always have high vertical integration rates in spite of production technology developments [1] [2] [3].

Table 19.1 shows an example of the production process, in this case that of a bicell process. The process comprises 15 production stages from paste preparation to final sealing. It is geared toward high throughput, and each of the stages is based on the preceding production stage.

Sequencing 15 production stages places extensive demands with regard to the quality of the product from the previous production stage. The individual stage yields do not merely add up to generate an overall yield; there is a multiplier effect! In reality, yields of 100 % rarely exist. This means that lower yields of 95 to 99 % have the following effect on the overall yield:

K.-H. Pettinger (✉)
Technologiezentrum Energie, Hochschule Landshut, Am Lurzenhof 1,
84036 Landshut, Germany
e-mail: karl-heinz.pettinger@haw-landshut.de

R. Korthauer (ed.), *Lithium-Ion Batteries: Basics and Applications*,
https://doi.org/10.1007/978-3-662-53071-9_19

Yield per stage **99 %:** Overall yield = $(0.99)^{15}$ = 0.86 = **86 %**
Yield per stage **98 %:** Overall yield = $(0.98)^{15}$ = 0.74 = **74 %**
Yield per stage **97 %:** Overall yield = $(0.97)^{15}$ = 0.63 = **63 %**
Yield per stage **96 %:** Overall yield = $(0.96)^{15}$ = 0.54 = **54 %**
Yield per stage **95 %:** Overall yield = $(0.95)^{15}$ = 0.46 = **46 %**

This example shows how essential testing and sorting stages are for the production process. Product material costs are more important than production costs in high-yield plants. Optimizing yield is therefore paramount for a plant to be operated economically.

This does not only apply to the above-mentioned battery production process with bicells, but to all processes. To ensure low failure rates during the battery warranty period, a great deal of control stages are required; they detect faults during the production process. Thin areas in separators can cause long-term effects such as increased self-discharging. It is possible that these are not detected by process control measures. Hence, reference samples are monitored to detect long-term effects.

Failure rates of < 1 ppm are called for internationally; this means that only every millionth battery should fail. The respective control measures control and secure the process. The lithium-ion cell production process can last up to two weeks. Detecting a manufacturing error as late as at the end of the process chain during the final inspection is the worst case scenario. However, the process still continues, and the

Table. 19.1 Lithium-ion cell production stages (bicells, pouch housings)

Process step	Production stage
1	Paste preparation
2	Foil casting
3	Anode band production
4	Separator application
5	Cathode application
6	Bicell lamination
7	Separation
8	Stacking
9	Conductor welding
10	Housing in foil pouch
11	Drying
12	Electrolyte dosing
13	Tempering
14	Forming
15	Evacuation and final sealing

cells produced during this time span need to be blocked. This can entail immense costs; therefore, manufacturers strive to implement extensive control stages in the process. The most important stages are described below.

19.2 Test procedures during coating

The essential foundation for continuously high cell quality is homogeneous electrode band coating.

Because the throughputs are very high, this is a very challenging task. For example, a small production line with a throughput of two million 20-Ah cells per year processes 26,000 km of anode and cathode band, respectively; this is equal to more than 14,000 m^2 flawless surface per day.

The coating parameters that are examined are layer thickness and inline surface quality. Measuring systems using radioactive radiation absorption or, of late, high-resolution laser systems provide reliable layer thickness measurements. Both procedures ensure that the desired control values are achieved. Measuring point positioning and density are key. As it is almost impossible to test the full cross-section, either measuring points close to each other are used or traversing measuring heads. For the traversing systems, it is important that their horizontal speed is concurrent with the plant's coating speed, which can reach up to 40 m/min. In an adequately stable process, these measurements are sufficient for testing the profile. The most common deviations are a tapered gradient in the profile's coating thickness or so-called "chatter marks", i.e., a rippled gradient, in machine direction.

The quality of the entire surface should be tested with camera systems after the dryer and before the winder. This enables detection of defects, contaminations, and cavities.

The most important part is handling the detected defects. Coating is a continuous process that is best not stopped. Coater stabilization can last up to one hour; it is not viable to stop the machines. Thus, the detected defects must be flagged and removed later in the process. Flagging can be performed by marking the defects on the non-coated edge with a color. Later, during electrode separation and before cell assembly, they can be removed. This requires complex production data control and linking of the individual process stages.

19.3 Test procedures during cell assembly

Cell body production The electrode body can be manufactured with different technologies such as winding (round or prismatic) or stacking (bicell or electrodes coated on both sides). Irrespective of the production technology, it is important to provide full electrical isolation between the anode and the cathode. Also, each cathode part must face an anode part, isolated only by the separator. To ensure these requirements are met, it is necessary to test the finished electrode stack to verify that the anodes and the cathodes, including the conductor, are electrically separated and to check that the individual electrodes are correctly positioned.

Testing the cell body for short circuits In the most harmless of cases, defective electrical anode-cathode separation causes increased self-discharging, while the worst case scenario is a direct short circuit of the electrode body. Then, not even cell forming is possible.

The cell body must be tested for short circuits at the end of the assembly at the latest. Short circuits can be caused by defective separators, uneven electrode cutting edges, wear and tear of the production machines, incorrect handling, or process stages such as ultrasonic conductor welding. Ultrasonic welding has the advantage that it results in very low weld transition resistance; however, it introduces mechanical energy into the electrode body. This can damage the separator in vulnerable locations, or the conductive electrode mass can be rattled loose from the collector and deposited between the electrodes, causing soft short circuits.

The test comprises a resistance measurement, either as an AC or DC measurement.

Standard resistance measuring bridges are used to measure AC resistance. At the beginning of the measurement, however, these measuring bridges do not provide constant measured values because the cell body has a high capacitive share in the overall resistance. Usually, measured values increase continuously at a decreasing speed until they reach a certain limit. If this level is very high, i.e., R > 1MOhm, it is assumed that the cell body does not have any detrimental short circuits. If the measured value is in the kOhm range or even below 1 kOhm, the cell body will not be able to take on more charge during forming. The flowing charge is not stored but transformed into heat by means of ohmic resistance.

Insufficient reproducibility is another disadvantage of AC measurement technology. It is negatively influenced by the measured sample's moisture content. Water has a relatively high dielectric constant, which influences the capacitive share of the overall resistance. Electrode and separator moisture content therefore influences the absolute measurement value, so that measured values cannot be replicated if, for example, humidity levels change. AC resistance measurement is a simple and very quick procedure for detecting hard short circuits. The operator must be aware that the measured overall resistance has an ohmic and a capacitive share. This method is used for process control under consideration of the above-mentioned issues.

DC resistance measurement is another measurement method. Direct voltage is applied to the test specimen and the flowing current is measured. The condenser, consisting of the electrodes separated by the separator, must be charged first, similar to the AC resistance measurement method. The resulting limit current is independent of the capacity; it is almost the same as the sheer leakage current. Measured values are adjusted more quickly than with the AC method. Low voltages of < 20 V are sufficient for detecting direct short circuits, while higher DC voltages, up to 200 V, are used to detect soft short circuits. The latter cause the separator to "burn through" in thin areas. Handling higher voltages such as these requires special equipment.

The short circuit tests must have a clock speed that is compatible with that of the process. There is only a window of a few hundred milliseconds for measuring individual bicells during cell body assembly. If bicells are used in cell body design, there is the advantage that small cell body units can be tested for functionality during assembly and removed, if necessary. Testing and removing small individual parts decreases the overall process reject rate. After inserting the cell body into the housing and connecting the conductors, another short-circuit test during the production process must be carried out. Both the isolation of the electrodes from each other and contact to the housing are tested. It is not possible to forgo this test stage because changes to the electrode stack and the conductors can occur during cell body installation in the housing. Every time the cell body is handled, it can be damaged.

Testing cell body thickness Testing electrode body thickness is necessary for two reasons: Before installation, it guarantees that the electrode body physically fits into the housing. The desired finished cell thickness must be met; this thickness must also comply with the tolerances specified in the data sheet. The measured value once more enables monitoring of coating on the electrode mass and, therefore, also of cell balancing. Coating thickness variations are multiplied in the electrode body because the coating comprises several layers. Coating irregularities of the cell body such as tapered coating profiles can thus be tested quickly. Technologies with a sufficient resolution such as mechanical, optical, or laser systems are suitable for these tests.

Electrode positioning Out-of-position electrodes not only influence cell functionality but also pose a safety hazard. As a rule, each cathode part must face a correctly paired anode part, as unpaired cathode parts are preferred sources for surplus ions during metallic lithium separation in the event of overcharging (Fig. 19.1).

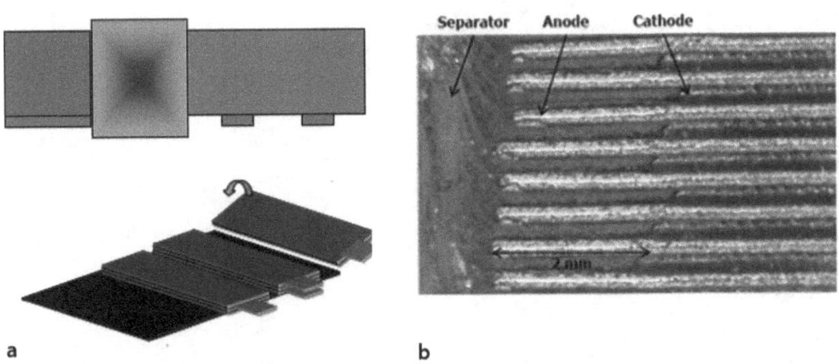

Fig. 19.1 **(a)** Assembly in theory: conductor notching from the electrode band and prismatic electrode winding [4], **(b)** Assembly in practice: cell body cross section with electrodes coated on both sides [4]

Production tolerances are taken into account when designing the cell by implementing an anode that is always slightly larger than the cathode. This surplus anode area is inactive during charge storage. Its size must strike a balance between providing optimum energy density and handling process tolerances.

The electrodes are pre-assembled from the coated bands or punched out as a whole. Before cell assembly, the correct electrode matrix position in the punched-out bands must be tested. Electrode position is tested inline to 100 % using camera systems.

Fig. 19.2 shows a camera system for controlling notching positions in an electrode band. The camera and light source are under a protective cover to prevent incidences of extraneous light. The electrode band feed reaches up to 10 m/min. Ten electrodes per second can be measured with such systems. Process data analysis provides information on process stability and statistical deviations.

Electrode positioning is therefore essential. Fig. 19.3 shows a cell stack cross-section on the right. Here, the anodes are 2 mm larger than the cathodes. The position

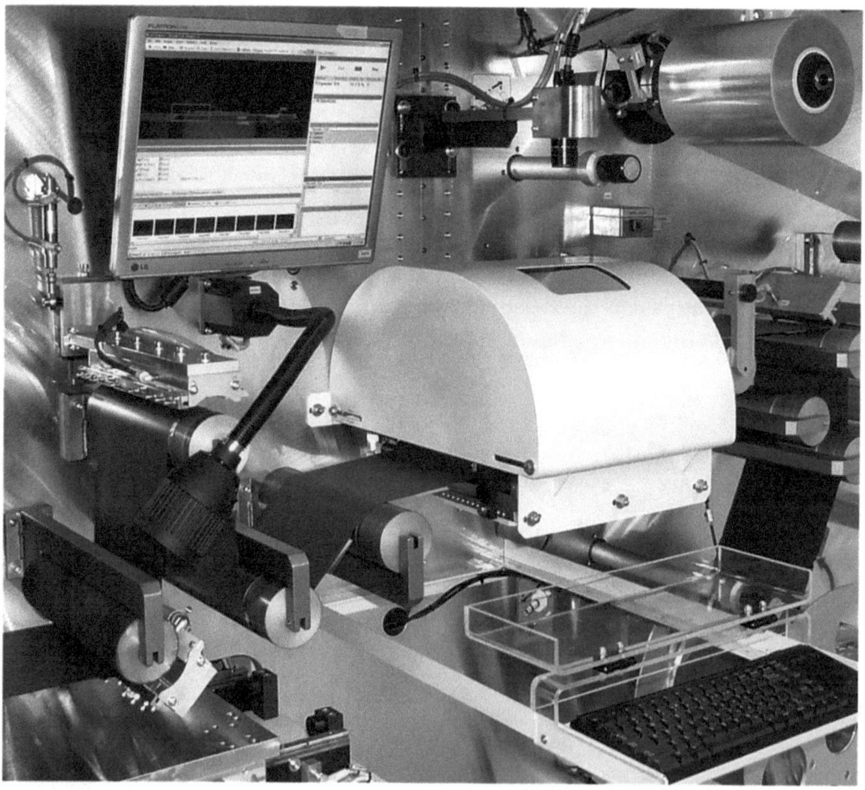

Fig. 19.2 Inline electrode geometry measurement. The anode band being measured is moved from left to right [4]

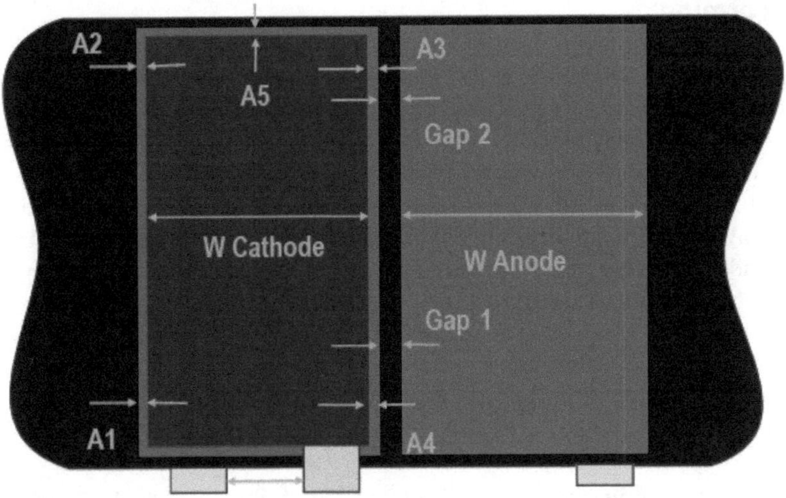

Fig. 19.3 Test dimension examples for correct electrode positioning [4]

tolerances are visible near the electrode edges. They can result from electrode positioning and the coating's position on the collector. Misaligned electrodes must either be removed at once or marked for later removal. This can be carried out, for example, by marking the electrode conductors with a color.

Short circuit measurements and visual inspection that the electrodes are correctly positioned are essential test stages in the cell production process. Currently, surface contamination detection can be performed for a particle size of > 30 µm. Efforts are currently underway to improve small particle optical resolution during continuous operations. Electrode and band contaminations can be removed by cleaning. Very soft rotating brushes, connected to a suction system, are used to clean the surfaces.

19.4 Electrolyte dosing

During so-called activation or electrolyte filling, the metered electrolyte amount must be checked. Both electrolyte overdosage and underdosage must be prevented. Underdosage means that the cell does not have sufficient electrolyte to achieve the specified performance characteristics. Purely volumetric control of the metered electrolyte amount is not recommended because gas cavities are also captured during the measurement. This is why gravimetric control is the preferred procedure, i.e., weighing the cell before and after metering.

19.5 Forming

Forming is, on the one hand, the cell's "electric birth" and, on the other hand, the first extensive electrical test step. It comprises the first charging of the cell. Also, the required protective layers such as the solid electrolyte interface are formed and unwanted contaminations are removed electrochemically.

Initial charging of the cell uses a considerable amount of energy to form the protective layers and for nonrecurring secondary reactions. This means that the first charging and discharging cycle introduces much more energy into the cell than it takes out. This overcapacity quickly decreases with the number of charging cycles. The charging capacity is higher in the first cycle than in the second cycle. The charge is irreversibly consumed during these forming processes; it is used to form protective layers (solid electrolyte interface, SEI), to silently burn residual water, and to oxidize contaminations. The discharging capacities of the first and second cycle, on the other hand, are very similar.

The charge factor, also called the cycling efficiency, is a very sensitive parameter that mirrors the process and battery quality. It is the ratio of the removed capacity in relation to each cycle's introduced capacity:

$$\text{Charge factor}[\%] = \frac{\text{Discharging capacity}[\text{mAh}]}{\text{Charging capacity}[\text{mAh}]}$$

The charge factor approximates 100 % after several charging and discharging cycles. High-quality lithium polymer batteries have charge factors of 99.5 % or more. If the charge factor changes, this is a sign of slowly creeping-in process changes. There is no other measurement parameter that gives a better indication of the overall process status. It is a sum parameter that reflects the following qualities: dispersion, coating, cell body assembly, separator, electrolyte filling, electrolyte distribution, and active material utilization.

During forming, a 100 % test is conducted for:

1. discharging capacity,
2. charging efficiency,
3. internal resistance.

These three parameters are tested in one test step; the cell only needs to be handled and connected one time. The parameter for discharging capacity in the data sheet is checked. Charging efficiency provides information about the individual cell's electrochemical system status (such as wetting, purity, process stability) and, if need be, about short circuits.

It makes sense to also check internal resistance in the testing device. Internal resistance is determined by the voltage drop during the short discharging current pulse (< 1 s) in accordance with Ohm's law. The pulse duration must be so short that the diffusion effects do not influence the measurement and only pure internal resistance is measured.

19.6 Final inspection after ripening

Forming is frequently followed by ripening, also known as test storage. During test storage, the charged batteries are matured for several days at increased temperatures. During maturing, the final distribution of electrolytes occurs as well as the conversion of residual byproducts and contaminations. The battery stabilizes and possible self-discharging is forced.

After test storage, the next step in the process is the final extraction and sealing, if required. These steps are the last mechanical stages of cell production. An OCV (open circuit voltage) measurement after test storage determines whether the battery has soft short circuits or has been damaged during final sealing. The OCV values measured after final sealing are compared to the OCV values measured after forming. Internal resistance is measured again and the final mechanical dimensions are checked.

Cell grading, i.e., classification, can now take place based on measurement data acquired during the process. It is performed in compliance with customer's specifications, e.g., cell grading in accordance with internal resistances.

19.7 Reference sample monitoring

Aside from quality control, reference samples are monitored to obtain aging data and to verify long-term series quality.

This means that during production, batteries are taken out of the production line at certain time intervals and for each product change. The removed batteries are stored semi-charged and permanently inspected at increasing time intervals, e.g., at the time of removal, after one month, after three, six, nine, 12 months, and then every six months.

For the test procedure, the cell is subjected to two entire discharging and charging cycles with the following sequence:

1. discharge
2. full charge
3. discharge
4. full charge
5. semi-charge

Each cell's state of charge at the time of storage is known from the previous reference sample measurements. The first discharging determines the still existing charge and, therefore, the self-discharging rate. By subsequently fully charging and discharging the cell, changes in storage capacity caused by age can be identified. The second discharging reflects irreversible aging. At the end of the testing program, the cell is charged to a defined state and stored again. Simultaneously, internal cell resistance is measured and recorded. This reference sample monitoring generates real-time data on aging and self-discharging. Because reference sample storage quickly increases during the course of production, the number of reference samples must be determined realistically.

Bibliography

1. Lithium Ion Cell Production Processes, Ralph Brodd, Kazuo Tagawa, Advances in Lithium-Ion Batteries, 2002, pp. 267–288
2. Production Processes for Fabrication of Lithium-Ion Batteries, Kazou Tagawa, Ralph J. Brodd, Lithium-Ion Batteries, 2009, pp. 181–194
3. Lithium ion battery production, Antti Väyrynen, Justin Salminen, The Journal of Chemical Thermodynamics, Vol. 46, March 2012, pp. 80–85
4. Fig. 19.1, 19.2, 19.3: Courtesy of Kemet Electronics Italia S.r.l (Arcotronics), I-40037 Sasso Marconi, Italy

Part IV

Interdisciplinary Subjects – From Safety to Recycling

20 Areas of activity on the fringe of lithium-ion battery development, production, and recycling

Reiner Korthauer

Producing lithium-ion batteries is a complex technical process consisting of multiple steps that have to be seamlessly integrated. A perfect cell can – but does not have to – result in the most suitable battery for the respective application. The specific application has to be taken into account in the cell design stage. All steps in the production process require a high standard of diligence, which can only be achieved with implementing state-of-the-art, safe production methods.

Occupational safety and health The individual stages in the production of lithium-ion batteries are not yet all fully automated; human intervention is still necessary. As the cells partially exhibit a high chemical energy potential already after only a few steps in the production process, they have to be handled with particular care. In addition, producing cells involves using chemical substances with varying characteristics and, in some case, hazardous substances. Manufacturers have to comply with all current regulations regarding the handling of such substances. Their production premises must therefore be appropriately equipped.

Chemical safety During their day-to-day use, lithium-ion batteries are exposed to a variety of environmental influences. The range of possible incidents can, at worst, result in battery destruction. Moreover, batteries can also be subjected to electric, thermal, and mechanical abuse without any direct external influence. In almost all cases, this results in a significant increase in the cell temperature, which can result in thermal cell degradation. Thus, the thermal behavior of the most important components plays a central role.

Electrical safety Against the backdrop of energy policies promoting a reduction of reliance on fossil fuels, traction batteries for electric mobility and batteries in

R. Korthauer (✉)
LIS-TEC Gmbh, Kriftel, Germany
e-mail: korthauer@lis-tec.de

R. Korthauer (ed.), *Lithium-Ion Batteries: Basics and Applications*,
https://doi.org/10.1007/978-3-662-53071-9_20

stationary applications are operated at high voltages; this requires the personnel operating or servicing the systems to exercise utmost care. To avoid accidents resulting in injuries or material damage, operators have to pay particular attention to electrical safety. In this respect, every possible effort should be made to ensure safety, for both current and future operations.

Functional safety In recent years, there has been a considerable increase in the number of electrical/electronic systems used in vehicles. They offer the driver many benefits – as long as they perform to specification. Indeed, the risk is not only that a failure results in a function breaking down, but in some cases it can also result in a critical driving situation. To reduce malfunctions of this kind to a tolerable level, the ISO 26262 standard has been introduced. It outlines the requirements for developing safety-oriented electrotechnical/electric systems.

Functional and safety test The functional and safety test covers all components (from the cell to the overall arrangement) in lithium battery development, testing, and series production. Comprehensive safety technology is essential for all of the necessary tests, because the majority of these are implemented on charged batteries with the highest hazard level, class 7, which is the level at which a battery can explode. Battery testing systems consist of many different modules – ranging from a temperature chamber to a data logger. Nowadays, not only battery manufacturers implement functional and test systems, but also vehicle manufacturers use them for tests in the area of electric mobility.

Transport The finished lithium-ion battery has to be transported to the site where it will be deployed or to the customer. This can be by road, water, or air. Lithium-ion batteries are hazardous goods, which requires compliance with extensive safety regulations for all of the above-mentioned means of transport. In recent years, diverse public authorities and organizations responsible for transport have been involved in developing a comprehensive set of provisions for this purpose.

Recycling The service life of a lithium-ion battery is finite, like for all economic goods. It depends essentially on the environmental conditions to which the battery is exposed as well as on the manner in which the battery is handled. The more that is known about the design of the battery and its built-in cells, the easier the recycling process. This greatly influences the chemical process steps applied and the battery materials that have to be extracted in this process. In the future, production will be performed in closed loops for many applications; this must also be the ultimate aim for lithium-ion batteries.

Vocational education and training A well-trained workforce is the foundation of zero-defect production. Employee skills are extremely important, in particular for a high-tech product such as lithium-ion batteries. Industry has become aware of this challenge and reacted to demands to familiarize employees with company processes early on. Today all companies offer their own process-integrated vocational training and continuing education. Today's electrical engineering professions also cater to the new business segment of electric mobility, covering all of its facets. The vocational education ranges from electronics technician and mechatronics fitter through to production technologist.

Standardization Standards, together with their technical specifications, also form the basis for developing and producing a state-of-the-art, new technology such as lithium-ion batteries. Standards regulate battery tests, constitute a basis for transporting batteries safely worldwide and provide users with security when deploying battery systems in almost all fields of application. Standards are not static; i.e., they are revised on a regular basis and adjusted to changing circumstances. Standardization is also essential for lithium-ion battery technology and its diverse fields of application.

Occupational health and safety during development and usage of lithium-ion batteries

21

Frank Edler

Contents

21.1 Introduction

It has been repeatedly shown that the usage of lithium-ion batteries can harbor dangerous surprises, for example, in an incident in which a car battery caught fire several weeks after testing [1]. In addition to ensuring the safety of end users, this new storage technology poses new challenges in regard to occupational health and safety in industrial applications.

Everybody who handles electrochemical energy storage systems should be aware that, independent of whether a battery is "full" or "empty",

- there is no or only a limited possibility of switching off an individual battery cell's voltage.
- it contains a considerable amount of chemical energy.
- its battery cells' contents and reaction products are hazardous to health.

F. Edler (✉)
elbon.de, Freibadstrasse 30, 81543 Muenchen, Germany
e-mail: frank.edler@elbon.de

R. Korthauer (ed.), *Lithium-Ion Batteries: Basics and Applications*,
https://doi.org/10.1007/978-3-662-53071-9_21

Labor legislation's safety objectives have been established on a European Union level and are anchored in the German Occupational Health and Safety Act [2]:

> The employer has the duty to take the necessary occupational health and safety measures [...] to influence the health and safety of workers.
>
> He has to examine the effectiveness of the measures and, where necessary, adapt them to changing circumstances [...].

> When planning and implementing the measures [...] the employer shall, in the light of the nature of the activities, [...]
>
> 1. guarantee that they are appropriately organized and provide the necessary means, and
> 2. take precautions so that the measures are, if required, observed when performing all activities and incorporated into the management structures, and workers are able to fulfill their duties to cooperate.

The specific safety requirements from Table 21.1 have been developed on the basis of these regulations.

The requirements and safety measures depicted in Table 21.1 apply to every type and size of battery. However, the following factors specifically apply to industrial applications of lithium-ion batteries:

- Usage of batteries larger than consumer batteries means that there is more chemical energy that might be released in the event of an incident.
- Series connection of hundreds of individual cells results in higher voltages, which are perilous in the event of an accident.
- Operating conditions can change owing to circumstances, e.g., in road vehicles or depending on the load profiles.

Table 21.1 Battery-specific safety requirements and measures

Battery safety requirements	Safety measures, examples
1. Protection of electrical potential against inadvertent conductance	Galvanic isolation, sufficient clearance and creepage distances, electric shock protection
2. Securing controlled electrochemical processes	Safety-tested cells, electronic battery management, suitable storage temperatures, cooling
3. Verification of the battery cells' mechanical integrity	Casing, protected installation location, robust transport packaging

Moreover, the industry developing these new applications is to a certain extent navigating in uncharted waters. Newcomers and companies from other areas are joining the emergent battery production market. They do not have any experience regarding specific occupational health and safety.

21.2 Occupational health and safety during the battery life cycle

Table 21.2 displays the battery life of industrial applications. The core life cycle and all of its activities are shown side by side. The activities normally apply to each of the individual phases. On the left, logistics activities such as storage and transport are depicted; on the right, the accompanying activities development and incident handling are depicted.

Each of the activities has specific occupational health and safety requirements:

Cell production As cell production includes handling hazardous chemicals, the occupational health and safety requirements are similar to those of the chemical

Table 21.2 Battery life cycle

<table>
<tr><th colspan="6">Industrial battery life cycle phases</th></tr>
<tr><td rowspan="10">Storage</td><td rowspan="10">Transport</td><td>Cell production</td><td>Processing materials and substances into completed and functioning battery cells</td><td rowspan="10">Development/testing</td><td rowspan="10">Incident handling</td></tr>
<tr><td>Battery system cell integration</td><td>Electrical and mechanical cell integration. Integration of electronics and further interfaces (cooling, housing, etc.)</td></tr>
<tr><td>Battery integration in application system</td><td>Battery integration with high-voltage electrical system, electronic control, mounting, cooling system, etc. in an overall system, e.g., an e-drive.</td></tr>
<tr><td>Application system provision</td><td>Commissioning, testing, customer sale (if applicable)</td></tr>
<tr><td>Normal operation</td><td>Usage under intended operating conditions (vehicle, industrial application, etc.)</td></tr>
<tr><td>Maintenance and service</td><td>Periodical testing, troubleshooting, etc.</td></tr>
<tr><td>Repair</td><td>Dismounting and replacement of batteries</td></tr>
<tr><td>Decommissioning</td><td>Application system close-down, battery decommissioning</td></tr>
<tr><td>Disassembly</td><td>Disassembly of battery systems into modules, cells, other parts</td></tr>
<tr><td>Disposal/recycling</td><td>Suitable scrapping of cells, reclaiming of materials</td></tr>
</table>

industry. Correct cell production also serves as an important foundation for occupational health and safety in the subsequent phases, because manufacturing errors, which possibly lead to cell fires at a later stage, pose a potential danger for processing and usage.

Battery system cell integration During high-voltage battery production, the overall voltage increases with each cell connected in series. Electrical safety issues must be considered from 75 V DC upwards, because then the Low Voltage Directive applies [3].

The most important occupational health and safety measures during cell integration are:

- battery assembly only by qualified personnel (qualified electricians) with a lithium-ion battery safety briefing,
- employment of special tools and working clothing when working on live components,
- installation of warning and access control signs for non-qualified personnel.

Normally, the battery assembly results in a closed system, which not only integrates the mechanically and electrically connected cells, but also the following components:

- battery isolator and fuse
- electrical connectors (plugs) for high and low voltages as well as for communication purposes
- battery management electronics and sensor technology
- cooling system connections, if applicable
- housing

A system completed in such a fashion (Fig. 21.1) is not considered a direct hazard when used appropriately under the following conditions:

- The battery poles' electrical potential has been disconnected from the mains.
- The isolation resistance of high-voltage circuits to tangible parts is sufficiently high.
- During assembly and testing no errors have occurred that could cause cell damage.

As with cell production, appropriate quality assurance measures are the foundation for occupational health and safety at a later stage. Functional safety, i.e., the effectiveness of the battery management's electronic safety functions that protect the battery system against dangerous overloads during operation, is very important. The effectiveness of such protective functions is essential for safety during system integration tests and operating mode.

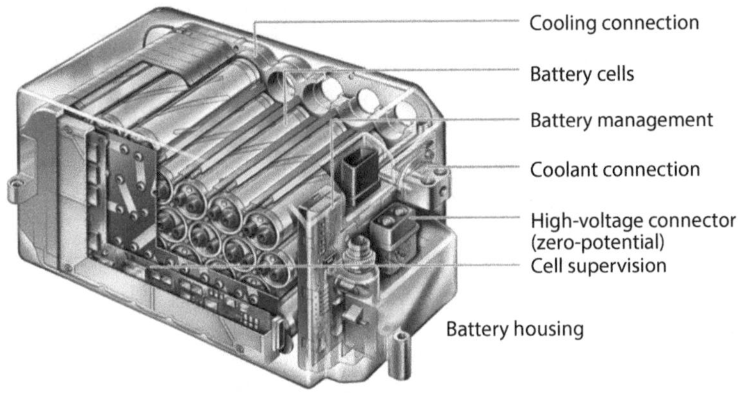

Fig. 21.1 Closed battery system

Battery integration in application systems Appropriate usage (prevention of mechanical or thermal overloads) of batteries, which are usually encased, is a requirement.

Based on these conditions, the occupational health and safety measures focus on electrical safety.

The foundation for occupational health and safety consists of qualified personnel (qualified electricians aware of battery-specific risks), suitable specialized tools for working on live components, and warning and access control signs for non-qualified personnel.

Application system provision and appropriate operation The occupational health and safety of the operators must be ensured if an application system with an integrated battery is used.

As a battery system has a design that enables it to be operated safely in normal operating mode, occupational health and safety measures generally focus on safety briefings and determining organizational procedures:

- briefings on identification of hazards (lightning sign, etc.)
- briefings on installation positions of batteries
- protection of batteries against unauthorized manipulation (insecure areas, if applicable)

Maintenance and service There is a distinction between the maintenance and service of the entire application and of the battery system.

In terms of occupational health and safety, it is most important that

- the battery cannot be damaged dangerously by modifications to the overall system (e.g., by introducing more heat).
- batteries cannot be damaged unknowingly (e.g., repair welding).

Furthermore, batteries that are out of order must be identified (e.g., via battery diagnostic interfaces), and battery repairs should only be performed by approved specialists. These last aspects require extensive organizational measures, which usually have to be settled in a timely manner between the OEM and the battery manufacturer (e.g., in the form of maintenance contracts). All diagnoses and, if applicable, alarm functions must be designed so that safety-relevant information can be assessed and processed in a target-oriented fashion.

Storage Appropriate storage of cells or batteries is a primary requirement for safe handling. This applies to

- compliance with specified storage temperatures and permissible climatic conditions. In particular, high temperatures damage the cells; in extreme cases this can result in so-called thermal runaway.
- storage with suitable states of charge. An extreme deep discharge must especially be prevented, because it might lead to irreversible cell damage.
- protection from mechanical damage (e.g., by means of adequate packaging).

Also, the storage location must be suitable to facilitate fire protection ensuring that

- batteries are protected from other fire hazards (e.g., in containers with fire protection classification F90).
- battery fires do not spread to other parts of the storage location (e.g., by means of storage in separate buildings).

Transport Aside from the extensive legal requirements concerning transport safety on public roads [6], the following points should be observed for all transport:

- there should be no mechanical overload (vibration, shock, etc.).
- the temperature limits must be heeded.
- there is appropriate protection against climatic influences.

Caution must also be exercised during packing and unpacking. Especially the elastic surface of pouch cells could be critically damaged by sharp or pointed objects. It has been reported that severely scratched cells are frequently rejected because their boxes are opened with a knife [4].

In terms of occupational health and safety, the quality assurance measures enable safe handling of cells and batteries. This includes performing appropriate visual and functional tests on cells (if applicable) before they are installed in a battery system.

Development/testing The battery life cycle depicted at the very beginning is of course an ideal; it is based on harmonized procedures and sophisticated technology. However, the innovative field of new battery applications requires further development and optimization of both the products and the processes. This is why it is advisable to take into consideration the respective level of development to add occupational health and safety measures, if applicable, especially in the following situations:

- testing of cells without safety certification
- testing of systems without validated safety functions
- relevant modifications of cells, battery, and application system design.

Developing cells, batteries, and systems with a focus on safety will remain an essential requirement for occupational health and safety in regard to their handling.

The slogan for industrial battery application development must be "Design for Safety". Chemical, mechanical, electrical, and electronic cell design requires utmost diligence.

To fight electrical hazards, it should be rendered impossible or at least very difficult for assembly staff to touch dangerous circuitry during the construction of modules due to simple handling mistakes. This can be achieved by employing suitable electrical connector geometric designs. In addition, fully integrating the battery should factor in risk-reducing elements in the event of a battery fire: Potential fire effluents should be diverted into inaccessible areas, preferably into the open air.

Destructive battery tests and tests with an increased cell damage risk should be performed in special laboratories with the facilities to competently handle reaction products that pose an explosion or health hazard.

Fig. 21.2 Electric vehicle after a battery fire following a crash test. The subsequent fire spread to four nearby vehicles [1]

The incident in which a battery fire started three weeks after a vehicle crash test and spread to other electric vehicles shows that peculiar risks inherent to batteries can surprise even experienced institutions (Fig. 21.2).

Incident handling As long as shortcomings in the quality of integration and supply cannot be completely ruled out, each involved organization needs to incorporate the eventuality of a cell or battery fire into their emergency regulations. The more cells involved in a fire, the higher the release of

- heat,
- toxic and very acidic substances,
- and possible build-up of explosive compounds.

Firefighting can become problematic because lithium-ion battery fires are "self-sustaining", so damping down the fire is very difficult. Other firefighting issues are the battery casing, which is necessary to mechanically protect the batteries, and the almost inaccessible and heavily protected locations where batteries are installed (e.g., in the vehicle undercarriage).

It is therefore important to prevent the fire from spreading and to ensure that personnel or bystanders seek shelter. Protective breathing apparatuses, flame-resistant clothing, and shelters, if applicable, must be available for personnel in an emergency situation.

A specific disaster recovery plan for batteries should be agreed on with the (plant) fire brigade and practiced regularly. When defining disaster recovery measures for electrical accidents, it needs to be taken into consideration that direct current has additional physiological effects on organisms compared to alternating current in the same voltage range. Chemical disintegration within the human body can be more pronounced and, in some cases, requires special treatment in a hospital.

21.3 Company-specific occupational health and safety

Only a competent target-performance analysis can determine whether and to what extent additional occupational health and safety measures need to be employed in companies that develop, test, and process lithium-ion batteries. Practice has shown that this requirements analysis needs to be performed in a structured manner.

A systematic record should be kept of premises in which batteries or battery cells are transported, stored, processed, tested, or used (Fig. 21.3). The activities there should also be recorded, the risk hazards determined, and safety measures defined. Table 21.3 shows a simplified example. For more detailed planning and implementation of safety measures, the know-how of several areas of expertise are needed: technology, occupational health and safety, fire protection, and organization.

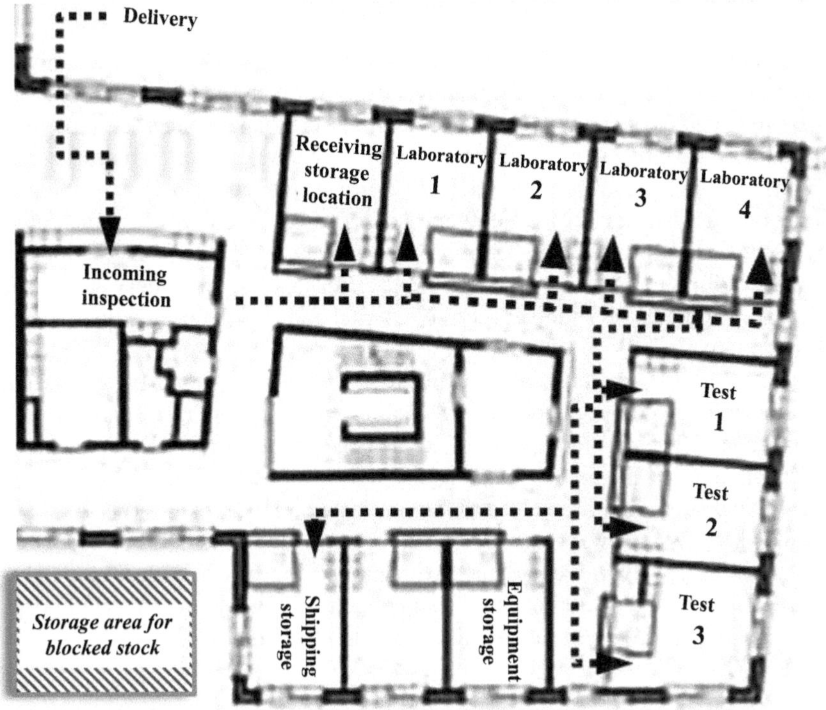

Fig. 21.3 Fictitious diagram of battery prototype production (example)

Table 21.3 Determination of specific safety measures (example)

Location/path	Activities	Risk hazards	Safety measures
Delivery	Incoming goods (cells); incoming goods (rejects)	Falls/shocks; contact with water; contact with heat; pre-damaged rejects	Level paths; roofing; immediate incoming goods (rejects) in quarantine (store)
...	...	...	...
Laboratory 1	Electrical module assembly	Exposed live parts; electric arc; mechanical cell damage	Isolated tools; personal protective equipment; constructional design of connectors; ...
...	...	...	...

21.4 Outlook

Before introducing or extending company activities to the industrial production of lithium-ion batteries, it is essential that awareness for battery-specific safety issues is acquired. This also applies to management. All personnel concerned must have a basic knowledge of the potential electrical and chemical hazards, because small product defects or carelessness can have a grave outcome. Systematic battery-specific occupational health and safety measures can be constructively implemented and tailored to company needs. Currently, no official standardized regulations for the safe handling of batteries are available [5]. For this reason, battery safety specialists need to be involved in addition to external and internal occupational health and safety experts.

All companies, from those responsible for cell production and battery assembly to those involved in rejection and recycling, need to employ special activities because the risk hazards and operators' knowledge about the technological characteristics of batteries and battery systems differ.

Health and safety requirements for the implementation or extension of battery-related operations are always tailored to a company's specific needs. For this reason and because of the limited experience with lithium-ion battery industrial production, this topic should be approached from an interdisciplinary perspective in order to implement constructive measures adapted to the technology.

Bibliography

1. Smith B (2012) Chevrolet volt battery incident report no DOT HS 811 573, NHTSA
2. Gesetz über die Durchführung von Maßnahmen des Arbeitsschutzes zur Verbesserung der Sicherheit und des Gesundheitsschutzes der Beschäftigten bei der Arbeit (Arbeitsschutzgesetz – ArbSchG) Bundesgesetz, 07. Aug. 1996
3. DIRECTIVE 2006/95/EC OF THE EUROPEAN PARLIAMENT AND OF THE COUNCIL of 12 December 2006 on the harmonisation of the laws of Member States relating to electrical equipment designed for use within certain voltage limits
4. Spek E (2011) Safe handling of high voltage battery systems. SAE seminar, Birmingham, 21. Sept. 2011
5. Edler F (2010) Elektromobilität – aber sicher! AUTOMOBIL-ELEKTRONIK 2010–06
6. UN Manual of tests and criteria, subsection 38.3

Chemical safety 22

Meike Fleischhammer and Harry Doering

Contents

22.1 Introduction

Unlike other battery types such as lead-acid, nickel-cadmium, and nickel metal hydride, today's standard lithium batteries, by definition, must not trigger secondary reactions apart from those during charging and discharging. Secondary reactions in standard batteries allow overcharging in the form of gassing (electrolyte degradation), for instance. In closed systems, this causes a temperature increase due to the oxygen cycle, but not a system failure. In lithium batteries, overcharging also results in electrolyte degradation, but the process is irreversible and leads to system failure.

Lithium-ion batteries can be damaged in different ways, causing performance and capacity loss and even destroy the cell/battery, possibly resulting in "catastrophic failure". Uncontrolled, quick reaction processes in components can cause gassing, venting, a high temperature increase, or even fire. Gas mixtures may form an explosive atmosphere when combined with air in unfavorable conditions.

M. Fleischhammer (✉)
ZSW, Clemensstrasse 18, 80803 Muenchen, Germany
e-mail: mfleischhammer@gmx.de

H. Doering
ZSW, Liese-Meitner-Strasse 24, 89081 Ulm, Germany
e-mail: harry.doering@zsw-bw.de

R. Korthauer (ed.), *Lithium-Ion Batteries: Basics and Applications*,
https://doi.org/10.1007/978-3-662-53071-9_22

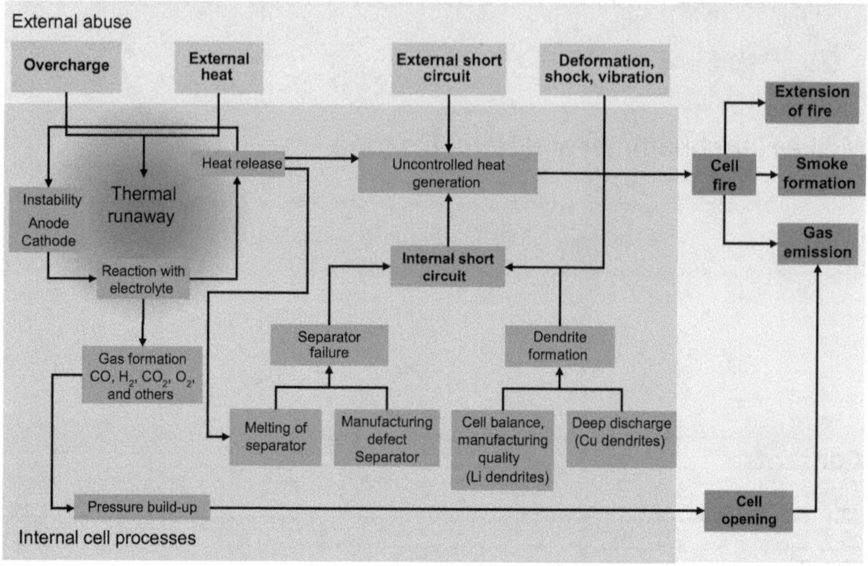

Fig. 22.1 Trigger and thermal degradation process in a lithium-ion battery (*Source* ZSW)

Fig. 22.1 shows an overview of different triggers and the thermal degradation process. External triggers are electrical, thermal, and mechanical abuse. Internal triggers are contamination with metals, defective separators, and lithium plating on the anode. These triggers cause cell degradation.

Irrespective of the trigger type, the result is generally an increase in cell temperature, causing thermal degradation in other cell components and additional heat generation, which, in turn, may trigger further degradation processes. This self-propelling process is called thermal runaway. It often causes unchecked cell heating and cell fire.

Chemical safety in lithium-ion batteries is predominantly based on cell component thermal stability and degradation product hazard potential. Typical degradation products are flammable gases such as H_2, CO, and CH_4, toxic compounds such as CO, HF, and PH_3, substances hazardous to health such as aldehydes and carcinogenic cathode oxide dusts of cobalt and nickel. The main source of gas formation is the electrolyte; the gas formation is caused by reduction on the anode or oxidation on the cathode. The following sections thus focus on the thermal behavior in the electrolyte, cathode, and anode.

22.2 Electrolyte

Commercial electrolytes are based on the conducting salt $LiPF_6$, dissolved in a mixture of different organic carbonates such as ethylene carbonate (EC), dimethyl carbonate (DMC), and diethyl carbonate (DEC). Thermal degradation of standard electrolyte mixtures occurs at around 250 to 300 °C. The hazard potential from lithium batteries is due to the sometimes low boiling points of solvents (Table 22.1) and their high flammability.

Thermal degradation reactions between $LiPF_6$ and solvents overall increase the safety hazard in the cell. The reactions are dependent on the exact electrolyte composition, but, first and foremost, are based on $LiPF_6$ degradation while forming PF_5 and on further reactions between PF_5 and the solvents. Fig. 22.2 shows the main electrolyte degradation reactions.

Water content is a key parameter in terms of the electrolyte's thermal stability. First of all, even low quantities of water can shift degradation to lower temperatures [1, 2]. Water also reacts directly with the salt, forming POF_3:

$$LiPF_6 + H_2O \rightarrow POF_3 + LiF + 2HF$$

Research has shown that 300 ppm of water causes formation of POF_3 and HF [2]. However, it was only possible to demonstrate minor reactions between POF_3 and the solvents EC, DMC, and EMC. This is why additional catalysts such as alcohols or cathode active materials are probably required for the reactions between salt and solvents.

Thermal electrolyte degradation yields the following degradation products: $CO_2, H_2O, CH_3F, (CH_3)CHF, CH_3CH_2F, FCH_2CH_2Y\,e(Y = OH, F, \ldots)$, $CH_3CH = CH_2, (CH_3)_2CHOH$.

Isolated reactions between the conducting salt and the electrolyte are just one type of reaction. There are other reactions between the electrolyte or electrolyte degradation products and the anode and cathode.

Although electrolytes vary in their chemical composition, the hazard posed by each electrolyte mixture is similar. This is mainly due to the low solvent flash points and the multitude of reactions.

Increasing electrolyte safety is currently a very important focus of research. One approach is to use additives which reduce flammability in the electrolyte or increase its thermal stability. Another possibility is the development of new electrolytes, such as solid polymer electrolytes or ionic liquids.

Table 22.1 Physicochemical characteristics of solvents for lithium batteries as compared to hydrogen

Substance	Abbreviation	Boiling point [°C]	Flash point [°C]	Ignition temperature [°C]	Explosion limit [%]
Dimethyl carbonate	DMC	90	16	465	10–25
Ethylene carbonate	EC	250	150	465	3–16
Propylene carbonate	PC	240	135	510	>2
Hydrogen	H2	−253		560	4–75.6

$LiPF_6$ degradation:

$$LiPF_6 \rightarrow LiF + PF_5$$
$$PF_5 + H_2O \rightarrow PF_3O + 2HF$$
$$PF_5 + HOR \rightarrow PF_3O + HF + RF$$
$$PF_3O + H_2O \rightarrow POF_2(OH) + HF$$

$LiPF_6$ + DMC:

PF_3O + DMC:

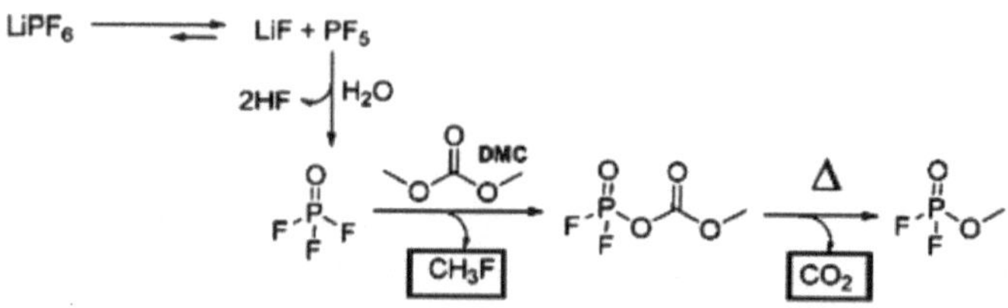

PF_3O + EC:

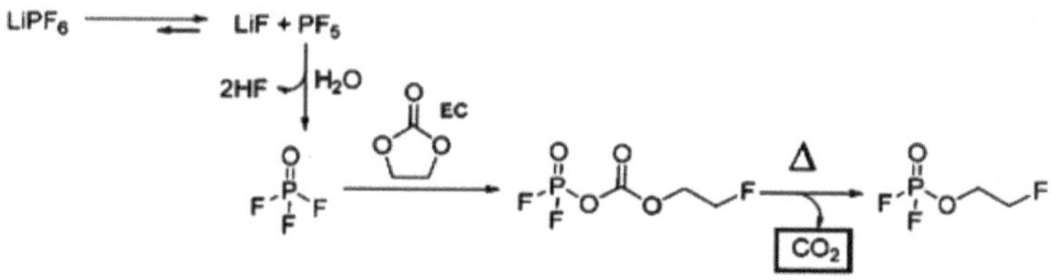

HF + EC:

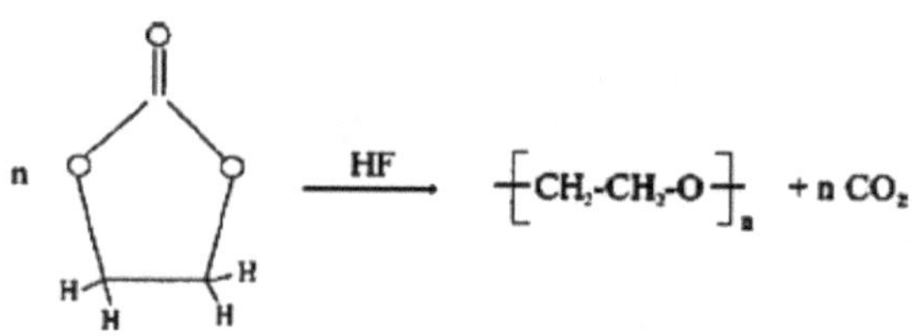

PF_5 + DEC: $(C_2H_5O)2CO + PF_5 \rightarrow C_2H_5OCOOPF_4 + HF + CH_2{=}CH_2$

$$(C_2H_5O)_2CO + PF_5 \rightarrow C_2H_5OCOOPF_4 + C_2H_5F$$
$$C_2H_5OCOOPF_4 \rightarrow PF_3O + CO_2 + C_2H_4 + HF$$
$$C_2H_5OCOOPF_4 \rightarrow PF_3O + CO_2 + C_2H_5F$$
$$C_2H_5OCOOPF_4 + HF \rightarrow PF_3O + CO_2 + C_2H_5F$$

Fig. 22.2 Potential degradation reactions in $LiPF_6$-based electrolytes [1, 3–7]

22.3 Anode

At present, the most widely used materials for anodes are graphite and carbon. Another material which looks very promising is the titanate $Li_4Ti_5O_{12}$. Graphite and carbon have a high thermal stability. However, this decreases considerably when lithium intercalates into the graphite or carbon structure. Table 22.2 shows research results from different work groups on thermal stability in graphite-based lithiated anodes. The following exothermal reactions have been demonstrated:

- Primary SEI degradation and secondary SEI formation (A)
- Secondary SEI degradation and direct reaction between intercalated lithium and the electrolyte (B)
- Reaction between intercalated lithium and the binding agent (C)
- Reactions between binding agents and degradation products (D)

The so-called primary SEI (solid electrolyte interphase) is a protective layer formed on the anode during the first charging cycle. It is caused by electrolyte reduction on the anode surface and it consists of various inorganic components such as LiF, Li_2CO_3, and Li_2OH and metastable organic components such as $(CH_2OCO_2Li)_2$, $ROCO_2Li$, or ROLi. The first reactions in metastable organic components start at a cell temperature as low as 80 °C, forming the secondary SEI. After the SEI is destroyed, the electrolyte is reduced on the anode while forming CO, CH_4, C_2H_4, C_3H_6, etc. (Table 22.3). Higher temperatures (> 250 °C) can trigger reactions between the different degradation products and the PVDF binding agent.

Therefore, SEI stability plays an important role in graphite anode safety. It can be controlled by means of materials science (morphology, particle surface and size), but such correlations are not always clear-cut. SEI stability can also be increased with electrolyte additives such as LiBOB [13].

The so-called lithium titanate $Li_4Ti_5O_{12}$ is another standard anode material. In lithiated state, titanate reacts exothermally. However, the generated heat is much lower than for graphite [14, 15]. This why the material is considered more thermally stable and therefore safer.

Heat generation is generally lower on anodes than on cathodes. Reactions on the anode, however, start at relatively low temperatures (SEI degradation at > 80 °C) and can trigger subsequent degradation reactions and even thermal runaway if heat dissipation is insufficient.

22.4 Cathode

During charging, the cathode active material releases lithium ions (delithiation). These ions are reversibly incorporated into the crystal structure during discharging (lithiation). As a general rule, a delithiated structure's stability is less than that of a lithiated structure. Cathode materials such as $LiCoO_2$ and $LiNi_{0.8}Co_{0.15}Al_{0.05}O_2$ exothermally degrade at 240 °C and 250 °C, respectively, releasing oxygen [21–23].

Table 22.2 Reaction temperatures of different graphites in lithiated state [8–12]

Material			Exothermal reactions			
			(A) T_{max} [°C]	(B) T_{max} [°C]	(C) T_{max} [°C]	(D) T_{max} [°C]
Yamaki et al.	Natural graphite	Fully charged	140	280		
	1 M $LiPF_6$ in EC/DMC (1:1)					
	Half coin cell					
Wang et al.	Graphite	Fully charged	101	217	234	249
	1 M $LiPF_6$ in EC/DEC (1:1)					
	Half coin cell					
Yang et al.	Natural graphite	Fully charged	100	283	336.	
	1 M $LiPF_6$ in EC/EMC (3:7)					
	Half coin cell					
Watanabe et al.	Graphite	Fully charged	105	260		
	1 M $LiPF_6$ in EC/DMC (1:1)					
	Half coin cell					
Shu et al.	Hard carbon spherule	Fully charged	104	213	274	292
	Artificial graphite		92	215		
	Natural graphite					
	1 M $LiPF_6$ in EC/DMC (1:1)		119	207		
	Swagelok, vs. Li					

The oxygen, in turn, causes a strongly exothermal electrolyte oxidation that could cause a cell fire.

In summary, the cathode safety hazard is largely dependent on instability in the host structure and on oxygen release. Ceder [24] classifies cathode material safety according to decreasing chemical oxygen potential as follows:

Layered oxides ($LiMO_2$, M = Al, Co, Mn, Ni,) < Li-Mn spinel ($LiMn_2O_4$) < phospho-olivines ($LiMPO_4$, M = Co, Fe, Mn) (Fig. 22.3).

Oxygen formation on the cathode is a decisive factor for thermal runaway and, consequently, for lithium-ion cell safety. Oxygen release from oxides even enables or promotes oxidation and, therefore, also energy release.

Oxygen release also impedes extinguishing of battery fires, because merely excluding oxygen is not sufficient to extinguish such fires.

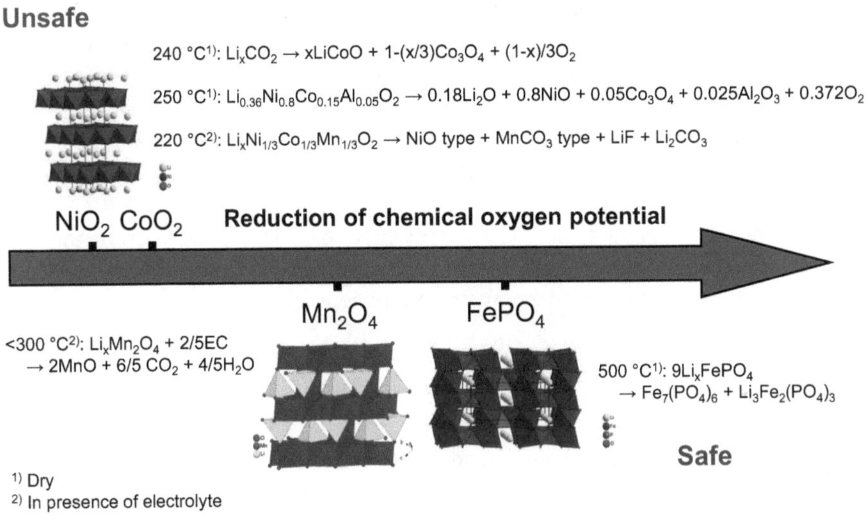

Fig. 22.3 Safety classification of cathode material according to chemical oxygen potential [21, 23–28]

Dynamic scanning calorimetry (DSC) also documents cathode material stability differences (Fig. 22.4). The peaks show the temperature at which energy is released. Peak height and peak surface correlate with the amount of energy released.

Current calculations [29], however, suggest that cathode materials crystallizing in the phospho-olivine structure $LiMPO_4$ (M = Co, Fe, Mn) are basically not safe. Such materials containing Co or Mn are an interesting option for so-called high-voltage

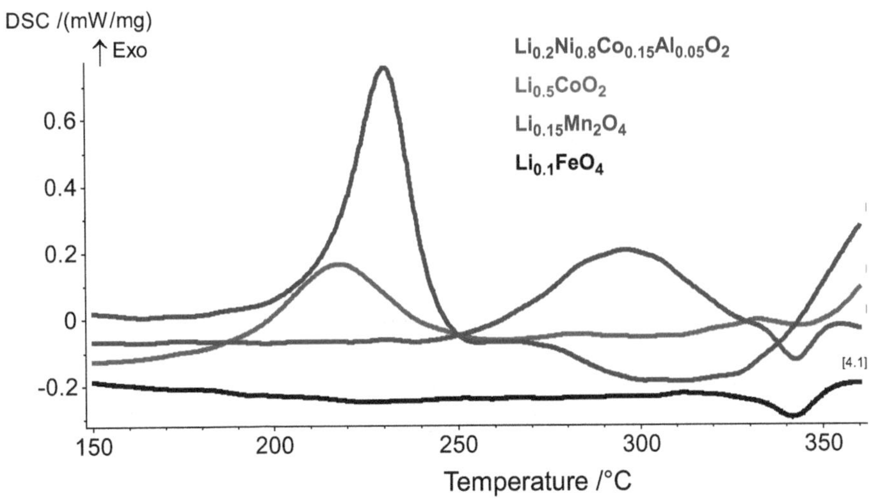

Fig. 22.4 DSC measurements of different cathode materials *(Source* ZSW)

or 5-V materials. Thermal stability in this group of materials decreases as follows: Fe > Mn > Co. The calculations concur with experiments [30] that indicate a high thermal stability for $LiFePO_4$, as opposed to $LiCoPO_4$ with a low thermal stability.

Gas release during thermal cathode degradation is mainly due to electrolyte oxidation while CO_2 is formed (Table 22.4). CO_2 formation is less dependent on the cathode type but more on the electrolyte composition instead. Aside from CO_2, formation of C_2H_2 and C_2H_5F is also possible [31].

These materials or their degradation products can be propelled from the cell during rapid cathode material degradation. The fine dusts or nanoparticular oxides such as NiO, MnO, and Co_3O_4 are again carcinogenic.

22.5 Other components

In addition to electrolyte and anode and cathode active materials, a cell consists of other components such as a separator, binding agent, and conducting additives. The conducting additives are mainly carbon black, carbon, and graphite. They feature high thermal stability and therefore can be disregarded. The binding agent used the most in commercial applications is polyvinylidene fluoride (PVDF). PVDF is thermally stable (> 450 °C); however, reactions between the binding agent and degradation products may occur in the anode at temperatures of 280 to 350 °C [20] (Table 22.3).

The separator prevents direct contact between the cathode and anode. When this is no longer possible, an internal short circuit occurs that usually triggers a thermal runaway. Commercial shut-down separators melt at 135/165 °C and prevent further electrical current flow. Separators with a ceramic coating are manufactured to provide greater temperature stability. Another option is an anode coating with a heat resist layer (HRL). Fig. 22.5 summarizes different potential cell component reactions as a function of temperature.

The safety hazard posed by a lithium-ion battery is determined by the individual cell component thermal instabilities and the degradation reactions described above. However, reactions between components also need to be taken into account, such as electrolyte oxidation on the cathode. The whole cell therefore needs to be considered. The behavior of two lithium cells (pouch, NCM) during 1C-overcharging is discussed below. In one instance, overcharging causes electrolyte degradation and, subsequently, cell opening without thermal runaway. In the other, overcharging causes thermal runaway. These experiments were conducted in a closed, pressure-resistant vessel (autoclave) to record the gas volumes and compositions formed. Fig. 22.6 shows the overcharging process in a cell without thermal runaway. A more pronounced temperature increase is recorded above around 40 % overcharging and the autoclave pressure slowly increases due to gas formation in the cell and its inflation.

The sudden increase in pressure, and small voltage drop, after around 53 minutes of overcharging means that the cell has opened. Further overcharging causes a further temperature increase and further gas emission. Overcharging was stopped

Table 22.3 Degradation reactions on the anode [10, 16–20]

Primary SEI formation	
EC	$2(CH_2O)_2CO + 2Li^+ + 2e^- \rightarrow (CH_2OCOOLi)_2\downarrow + C_2H_4\uparrow$
	$(CH_2O)_2CO + 2Li^+ + 2e^- \rightarrow Li_2CO_3 + C_2H_4\uparrow$
DEC	$(C_2H_5O)_2CO + Li^+ + e^- \rightarrow C_2H_5OCOOLi + C_2H^{\cdot}_5$
DMC	$(CH_3O)_2CO + Li^+ + e^- \rightarrow CH_3OCOOLi\downarrow + CH^{\cdot}_3$
$LiPF_6$	$LiPF_6 \rightarrow LiF + PF_5$
	$PF_5 + H_2O \rightarrow 2HF + PF_3O$
	$PF_5 + nLi^+ + ne^- \rightarrow LiF + Li_xPF_y$
	$PF_3O + nLi^+ + ne^- \rightarrow LiF + Li_xPOF_y$
	$H_2O + Li^+ + e^- \rightarrow LiOH\downarrow + 0{,}5H_2$
Primary SEI degradation and secondary SEI formation	
	$(CH_2OCO_2Li)_2 \rightarrow Li_2CO_3 + C_2H_4 + CO_2 + 0{,}5O_2$
	$2Li + (CH_2OCO_2Li)_2 \rightarrow Li_2CO_3 + C_2H_4$
	$(CH_2O)_2CO + 2Li^+ + 2e^- \rightarrow Li_2CO_3 + CH_2 = CH_2$
	$LiPF_6 + H_2O \rightarrow 2HF + LiF + POF_3$
	$2HF + Li_2CO_3 \rightarrow 2LiF + H_2CO_3$
	$PF_5^- + nLi^+ + ne^- \rightarrow LiF + Li_xPF_y$
	$PF_3O + nLi^+ + ne^- \rightarrow LiF + Li_xPOF_y$
	$PF_3O + Li^+ \rightarrow LiF + Li_xPOF_{3-x}$
Electrolyte reduction on the anode	
DMC	$(CH_3O)_2CO + 2Li^+ + 2e^- + H_2 \rightarrow Li_2CO_3\downarrow + 2CH_4\uparrow$
	$2(CH_3O)_2CO + 2Li^+ + 2e^- + H_2 \rightarrow 2CH_3OLi\downarrow + CH_4\uparrow$
	$(CH_3O)_2CO + 2Li^+ + 2e^- \rightarrow 2CH_3OLi\downarrow + CO\uparrow$
EC	$(CH_2O)_2CO + 2Li^+ + 2e^- \rightarrow 2CH_2OLi\downarrow + CO\uparrow$
	$2(CH_2O)_2CO + 2Li^+ + 2e^- \rightarrow (CH_2OCO_2Li)_2\downarrow + C_2H_4\uparrow$
	$2CH_3(CH_2O)_2CO + 2Li^+ + 2e^- \rightarrow CH_3(CH_2OCO_2Li)_2\downarrow + C_3H_6\uparrow$
PC	$2C_4H_6O_3 + 2Li^+ + 2e^- \rightarrow CH_2(CH_2OCO_2Li)_2\downarrow + C_3H_6\uparrow$
	$PC + 2e^- + 2Li^+ \rightarrow Li_2CO_3\downarrow + CH_3CH = CH_2\uparrow$
DEC	$(C_2H_5O)_2CO + 2Li^+ + 2e^- \rightarrow CH_3CH_2OLi\downarrow + CO\uparrow$
	$(C_2H_5O)_2CO + 2Li^+ + 2e^- + H_2 \rightarrow Li_2CO_3\downarrow + 2C_2H_6\uparrow$
	$(C_2H_5O)_2CO + 2Li^+ + 2e^- + H_2 \rightarrow CH_3CH_2CO_2OLi\downarrow + 2C_2H_6\uparrow$
	$H_2O + Li^+ + e^- \rightarrow LiOH + H_2$
Binder reactions	
	$(CH_2C_2F_2) + Li \rightarrow LiF + -CH = CF- + 0{,}5H_2$
	$(CH_2CF_2) + Li_2O \rightarrow 2\ LiF + H_2O + 2C$

Table 22.4 Electrolyte degradation reactions on the cathode [31]

CO_2 formation:	Electrolyte oxidation
DMC:	$(CH_3O)_2CO + 3O_2 \rightarrow 3CO_2\uparrow + H_2O$
DEC:	$H_2O(C_2H_5O)_2CO + 6O_2 \rightarrow 5CO_2 + 5H_2O$
EC:	$(CH_2O)_2CO + 2{,}5O_2 \rightarrow 3CO_2\uparrow + 2H_2O$
C_2H_2 formation:	C_2H_4 formation
	$C_2H_4 \rightarrow C_2H_2 + 2H+ + 2e$
	$C_2H_2 + 3O_2 + 2H^+ \rightarrow 2CO_2\uparrow + 2H_2O$
C_2H_5F formation:	Surface reaction
	$C_2H_6 \rightarrow C_2H_5^+ + H^+ + 2e^-$
	$C_2H_5^+ + F^- \rightarrow C_2H_5F$

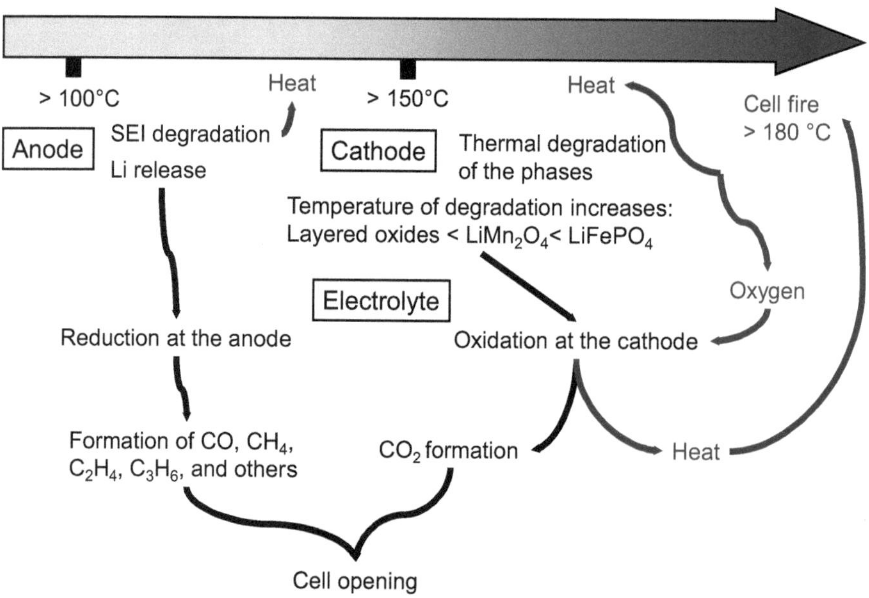

Fig. 22.5 Degradation reaction process in lithium batteries as a function of temperature (*Source* ZSW)

on reaching a state of charge of 200 % according to test specifications. Final pressure of around 25 kPa is equal to a gas volume of around 11 liters.

Fig. 22.7 shows the time period during cell overcharging within which thermal runaway takes place. Voltage breakdown occurs at an SOC of around 167 % and a sudden pressure increase is observed. Pressure then increases to its maximum of 1.23 MPa while the measured cell temperature reaches about 580 °C. Pressure

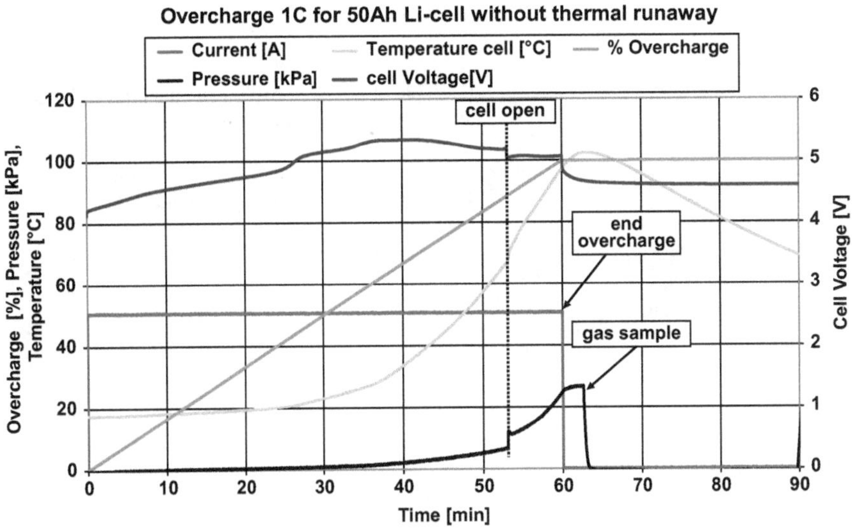

Fig. 22.6 Progression of current, voltage, temperature, and pressure during lithium cell overcharging up to 200 % SOC without thermal runaway *(Source* ZSW)

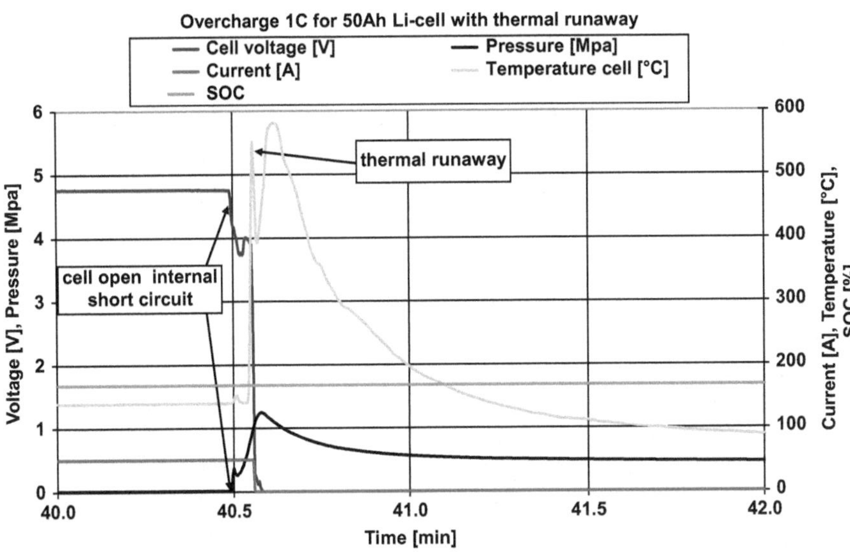

Fig. 22.7 Progression of current, voltage, temperature, and pressure during lithium cell overcharging with thermal runaway (*Source* ZSW)

Table 22.5 Gas analysis results for lithium cell overcharging experiments with and without thermal runaway *(Source* ZSW)

		Example of gas composition and material emission after a 1C-overcharging "event" in a 50-Ah cell	
	Unit	Cell opening at around 100 °C without thermal runaway	Thermal runaway at around 600 °C
Gas volume emitted	Liter	11	140
Fluoride obtained from HF	mg	10.1	500.6
O_2	Liter	9.2	0.7
N_2	Liter	36.0	28.9
H_2	Liter	< 0.05	52.4
CO_2	Liter	3.28	51.5
CO	Liter	< 0.05	14.4
Phosphine	µg	1.3	530
Formaldehyde	µg	90	< 20
Acetaldehyde	µg	2,000	15,000
Propionaldehyde	µg	170	6,000
Butyraldehyde	µg	20	3,800
Valeraldehyde	µg	< 10	400
Methane	ml	230	2
Ethane	ml		1
Ethene	ml	150	8,400
Propane	ml	10	1,200
Propene	ml	< 1	2,700

then decreases to a steady 0.41 MPa following the reactions, cooling down, and condensation of reaction products. This is equal to a gas volume of about 140 liters, signifying thermal runaway.

Table 22.5 summarizes the gas analysis results. While a cell opening without thermal runaway generates electrochemical oxidation products, overcharging with thermal runaway produces hydrogen, ethene, CO, and CO_2 in large amounts. The high proportion of non-oxidized products can be explained by the limited amount of oxygen in the autoclave. Fluoride emission is very pronounced with thermal runaway. It can be assumed that it was present as HF in the gas during an intermediate stage. The toxic substance phosphine (PH_3) is also detected in the low-oxygen autoclave atmosphere.

Bibliography

1. Kawamura T, Kimura A, Egashira M, Okada S, Yamaki J-I (2002) J Power Sources 104:260–264
2. Yang H, Zhuang GV, Ross PN Jr (2006) J Power Sources 161:573–579
3. Li W, Lucht BL (2007) Electrochem Solid-State Lett 10:A115–A117
4. Kawamura T, Okada S (2006) J.-i. Yamaki. J Power Sources 156:547–554
5. Gachot Gg, Ribière P, Mathiron D, Grugeon S, Armand M, Leriche J-B, Pilard S, Laruelle Sp (2011) Analytical chemistry 83:478–485
6. Campion CL, Li Wentao LBL (2005) J Electrochem Soc 152:A2327–A2334
7. MacNeil DD, Dahn JR (2003) J Electrochem Soc 150:A21–A28
8. Yamaki J-i, Takatsuji H, Kawamura T, Egashira M (2002) Solid state ionics 148:241–245
9. Wang Q, Sun J, Yao X, Chen C (2006) J Electrochem Soc 153:A329–A333
10. Yang H, Bang H, Amine K, Prakash J (2005) J Electrochem Soc 152:A73–A79
11. Watanabe I, Yamaki J-i (2006) J Power Sources 153:402–404
12. Shu J, Shui M, Huang F, Xu D, Ren Y (2011) Ionics 17:183–188
13. Täubert C, Fleischhammer M, Wohlfahrt-Mehrens M, Wietelmann U, Buhrmester T (2010) J Electrochem Soc 157:A721–A728
14. Lu W, Belharouak I, Liu J, Amine K (2007) J Power Sources 174:673–677
15. Belharouak I, Sun Y-K, Lu W, Amine K (2007) J Electrochem Soc 154:A1083–A1087
16. Richard MN, Dahn JR (1999) J Electrochem Soc 146:2068–2077
17. Aurbach D (2000) J Power Sources 89:206–218
18. Andersson AM, Edstrom K (2001) J Electrochem Soc 148:A1100–A1109
19. Roth EP, Crafts CC, Doughty DH, James M advanced technology development program for Lithium-ion Batteries: thermal misuse performance of 18650 Li-ion cells; SAND2004–0584
20. Pasquier AD, Disma F, Bowmer T, Gozdz AS, Amatucci G, Tarascon JM (1998) J Electrochem Soc 145:472–477
21. MacNeil DD, Dahn JR (2001) J Electrochem Soc 148:A1205–A1210
22. Belharouak I, Vissers D, Amine K, Chemical E (2006) J Electrochem Soc 153:A2030–A2035
23. Bang HJ, Joachin H, Yang H, Amine K, Prakash J (2006) J Electrochem Soc 153:A731–A737
24. Ceder G (2010) Mater Res Bull 35:693–701
25. Belharouak I, Lu W, Liu J, Vissers D, Amine K (2007) J Power Sources 174:905–909
26. Delacourt C, Poizot P, Tarascon J-M, Masquelier C (2005) Nat Mater 4:254–260
27. MacNeil DD, Dahn JR (2001) J Electrochem Soc 148:A1211–A1215
28. Thackeray MM, Mansuetto MF, Bates JB (1997) J Power Sources 68:153–158
29. Hautier G, Jain A, Ong SP, Kang B, Moore C, Doe R, Ceder G (2011) Chem Mater 23:3495–3508
30. Theil S, Fleischhammer M, Axmann P, Wohlfahrt-Mehrens M (2012) J Power Sources
31. Kong W, Li H, Huang X, Chen L (2005) J Power Sources 142:285–291

Electrical safety

23

Heiko Sattler

Contents

23.1 Introduction

Due to continuous development in various areas of application for storage batteries, the demand for energy storage systems with high energy density is growing rapidly. The market requires energy supply systems that are increasingly based on renewable energies in order to make mobility less dependent on fossil fuels.

Lithium-ion batteries have been used in consumer electronics, communication, and IT technology for many years and are also being implemented in industrial applications with a growing frequency. In order to meet all requirements, energy storage systems must store large amounts of energy and need to have high voltages to be an economically viable solution.

The individual cells' voltage is severely limited by their basic chemical elements. For this reason, connecting individual cells in series to form batteries with high total voltages will be the leading market of the future (Fig. 23.1). This applies to both stationary batteries and batteries used in electric mobility applications. To

H. Sattler (✉)
VDE-Pruef- und Zertifizierungsinstitut, Merianstrasse 28, 63069 Offenbach, Germany
e-mail: heiko.sattler@vde.com

R. Korthauer (ed.), *Lithium-Ion Batteries: Basics and Applications*,
https://doi.org/10.1007/978-3-662-53071-9_23

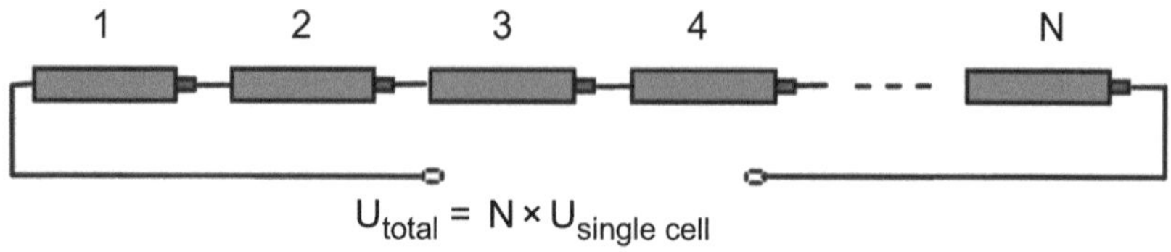

Fig. 23.1 Serial connection of individual cells in a battery with high total voltage

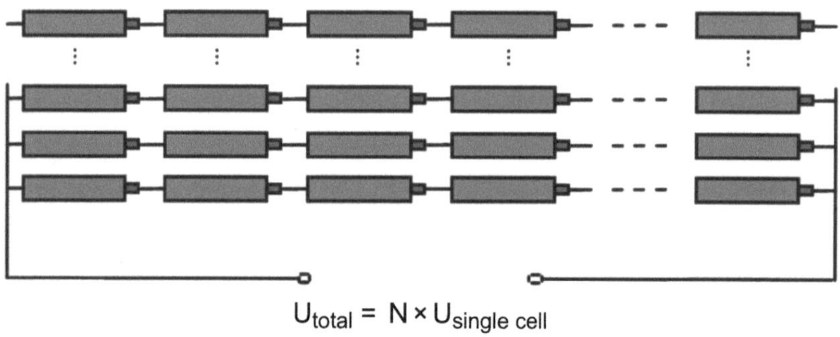

Fig. 23.2 Parallel connection of cell strings in a battery with high total voltage and high capacity and/or high energy content

obtain higher capacities and/or higher energy contents, such serial connections of cells (also known as cell strings or cell strands) are connected in parallel (Fig. 23.2).

The automotive industry uses hybrid batteries with voltages up to 180 V and energy contents of 0.6 to 2 kWh. Batteries for purely electric vehicles currently have voltages up to 600 V.

The energy content exceeds 15 kWh and is significantly limited by the battery weight. A plug-in hybrid battery ranges in between these energy classes with an energy content of around 5 to 15 kWh; its voltages are no higher than 400 V.

Stationary batteries are classified according to the energy they store. While stationary storage systems for non-commercial photovoltaic systems can store 2.5 to 10 kWh of energy, systems for commercial use store up to 1 MWh or more. Voltages can be as high as 500 V for small stationary home-use batteries and 1,500 V for commercial-use batteries.

23.2 Electrical safety of lithium-ion batteries

Electrical safety requirements for batteries used in electric vehicles depend on the given operating conditions and/or the situation in which safety is reviewed. The following operational conditions and situations must be considered for electric vehicles:

- normal operation (standby)
- faulty operation (considering a first fault)
- operation when driving
- operation when charging the battery while stationary
- safety while carrying out service work
- safety after an accident

Requirements vary greatly and partly depend on the systems connected to the electric vehicle. Additionally, different groups of people have to be protected. The general imperative is that personal injuries must be prevented as a matter of principle, and damage to objects and the environment must be avoided.

Fewer operational conditions need to be taken into account for stationary batteries that are used to immediately store energy, which has been generated by photovoltaic systems, for instance, because these batteries are not moved after installation. Therefore, only the following operating conditions must be considered:

- normal operation (standby)
- faulty operation (considering a first fault)
- operation when charging the battery from the photovoltaic system or power grid
- safety while servicing the system

In addition to the different operational conditions that need to be considered, the following hazards must be avoided for both applications:

- electric shocks
- overheating caused by damage to isolation (in contrast to hazards that can arise from overheating caused by chemical reactions)
- short circuits (between the battery's terminals)
- excessive currents when charging and discharging
- overcharging
- overdischarging
- unbalanced individual battery cells

23.2.1 Electric shock protection

These battery systems' voltages are considered "hazardous live"[1] and can cause electric shocks; therefore, all energized parts must be sufficiently isolated. There are several isolation methods.

[1]DIN VDE 0411 Part 1, EN 61010-1:2010: Hazardous live: capable of rendering an electric shock or electric burn

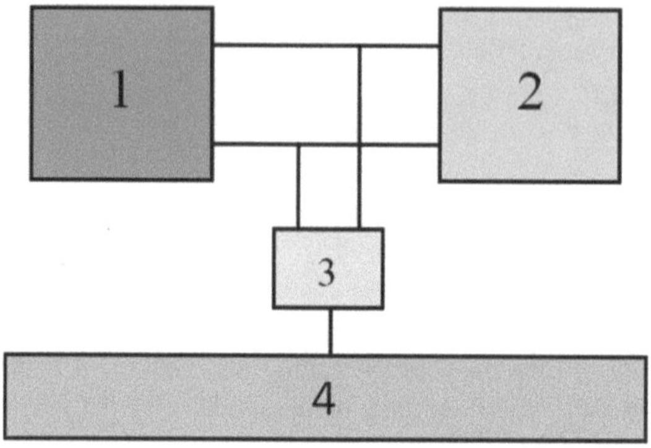

Fig. 23.3 Basic principle of a high-voltage on-board electrical system of an electric vehicle with an isolation monitor

The electric power circuit system (high-voltage network) of electric vehicles must be electrically isolated in accordance with ISO 6469-3[2]. The vehicles must have an isolation monitoring device (IMD) which permanently monitors the isolation from the chassis (Fig. 23.3).

Isolation is achieved through fixed isolation, on the one hand, and clearance and creepage distances[3] must also be observed, on the other hand. They prevent live electrical connections between the two battery terminals even if surfaces are contaminated. Such connections would create unwanted and, at worst, dangerous electrical current flow.

In electric vehicles, the chassis is isolated from the on-board electrical system. When wiring the entire vehicle, this should not be canceled out by inappropriate isolation or – with respect to battery voltage – insufficient clearance and creepage distances. The suitability of fixed isolation materials is determined by purely electrical characteristics such as electric strength and by mechanical characteristics such as resistance to high and low temperatures, resistance to fluids commonly used in vehicles such as oil and brake fluids, or to other fluids such as salt water.

It is also necessary to take into account the battery cell components inside the battery cells that could possibly escape from the cell in gaseous form and then condense within the housing. The most critical component is electrolyte, which nearly always consists of organic fluids. Since the battery is hermetically sealed to avoid external influences, electrolyte can accumulate inside the battery housing over the

[2]ISO 6469-3:2011 Electric road vehicles – Safety specifications – Part 3: Protection of persons against electric hazards

[3]Clearance and creepage distances should be measured in accordance with IEC 60664-1, which all standardization committees must observe.

entire service life of an electric vehicle's battery, which spans across more than ten years, and can thus compromise the fixed isolation within the battery.

If all electric shock safety measures are observed and intact, battery terminals can be safely touched. However, service personnel should exercise caution when handling tools to avoid inadvertently causing short circuits as they invalidate all isolation safety measures. For this reason, automotive manufacturers usually implement a special maintenance mode for battery systems that is designed to prevent high voltages at the terminals when performing service work. The mode reduces the risk of accidents caused by electric shock to a minimum.

In the event of an accident, however, this is not the case. Depending on the severity of an accident, conductive materials can enter the battery and invalidate all isolation safety measures. In this case, a potentially hazardous live voltage could occur in the entire high-voltage on-board electrical system. Should the safety shutdown of the battery fail in the event of an accident, the majority of automotive manufacturers provide so-called rescue data sheets (Fig. 23.4), which contain important vehicle construction information for emergency services. They help protect both rescue workers and passengers.

All rescue data sheets contain vital information about the layout of high-voltage wires in the on-board electrical system and describe where to cut open the car body in order to swiftly rescue trapped passengers.

Stationary applications are assumed to have battery voltages related to the protective earth potential (Fig. 23.5). In most cases, this is due to inverters connected to the battery that transform the energy stored in the battery into AC voltage, which is then either fed back into the power distribution network or made available for internal consumption.

While voltages related to the protective earth potential have one terminal connected to a potential that can safely be touched, the other terminal must be rendered safe through double-isolation[4] or reinforced isolation[5] in order to protect the operator and/or service personnel. For non-isolated parts, sufficient clearance and creepage distances must be observed for all surfaces touched by the operator. These distances must be in accordance with the maximum battery voltage.

Cables and other equipment connected to the battery voltage must be able to withstand all stresses, strains, and loads that can occur. In stationary batteries, these are purely electrical characteristics such as electric strength and current capability as well as resistance to temperatures that may occur during operation. In contrast to electric vehicles, there is usually no mechanical stress. Isolation material does not need to be resistant to chemicals because the battery is not hermetically sealed.

[4]Isolation consisting of basic and additional isolation

[5]Isolation of hazardous live parts that prevents electric shocks as effectively as double isolation

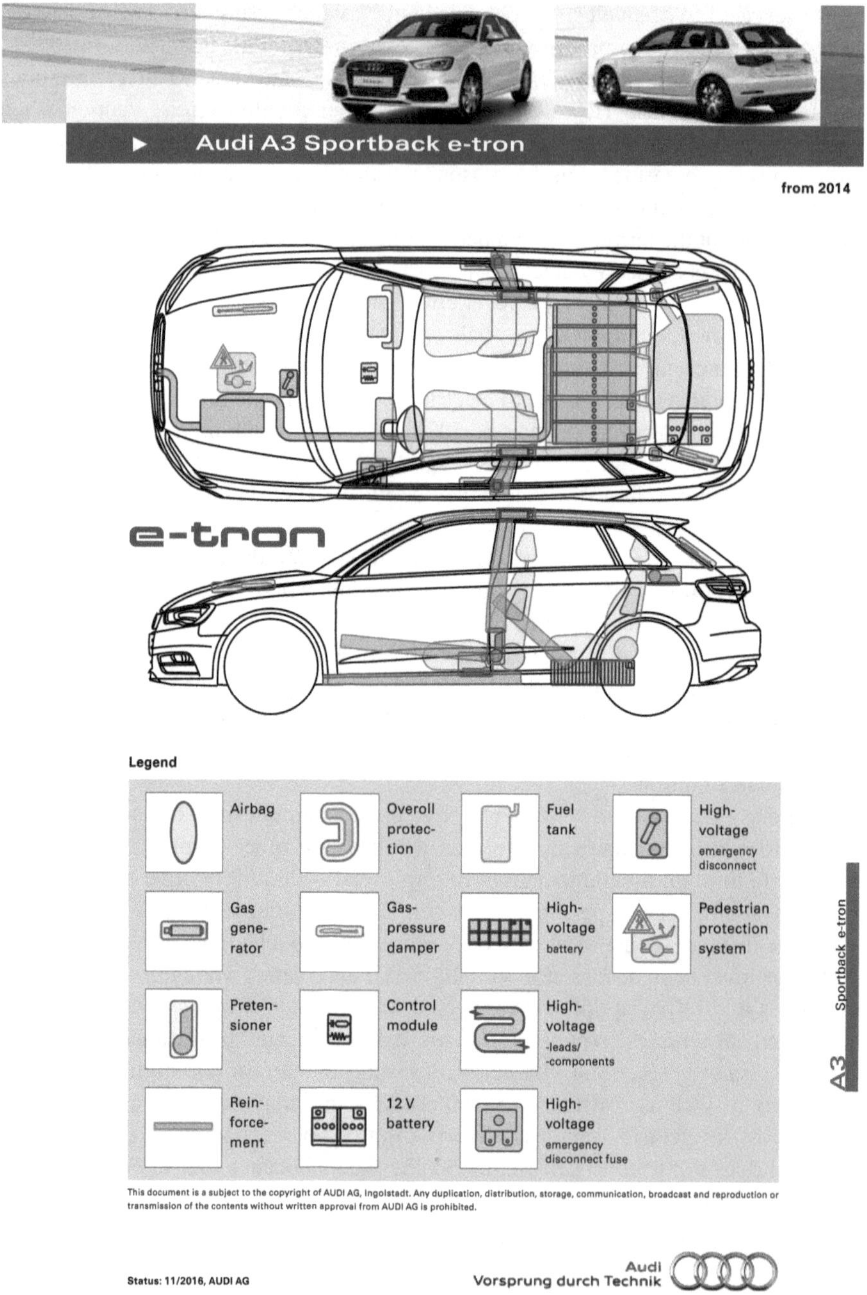

Fig. 23.4 Excerpt from Audi AG's rescue data sheets

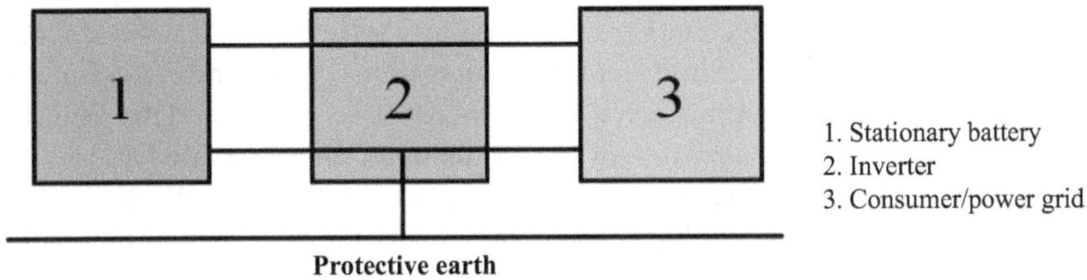

Fig. 23.5 Basic principle of stationary storage with battery voltage based on protective earth potential

23.2.2 Short circuits

In large batteries for industrial use, short circuits can cause high currents exceeding 1,000 A. These high currents, in turn, cause the temperature of live parts to rise substantially and can also increase the battery's temperature. This can result in the separator material within the cell failing, leading to a short circuit in the cell between the anode and cathode. Some of the materials currently used for lithium-ion cells are prone to thermal runaway. Therefore, batteries must include overcurrent protective devices, or the use of such components must be mandatory if batteries are to be integrated in a system. In any case, the battery terminals must be designed to prevent short circuits if a flat metal piece (e.g., a wrench) comes into contact with both terminals. Overcurrent protective devices must be suitable for the battery voltage and for use in direct voltage circuits.

23.2.3 Battery management systems

Battery management systems (BMS) prevent different states that are critical for lithium-ion batteries. States that have to be prevented include discharging and charging at high currents, overcharging, and overdischarging. These states are collectively called battery mishandling. In all of these cases, the battery can enter a critical state that could harm the environment and/or the operator. When certain material combinations are used, it is possible that the above-mentioned thermal runaway occurs. However, materials not prone to thermal runaway are also dangerous.

These batteries and their cells can also heat up when mishandled as described above, causing pressure to build up in the cells, which is usually blown off through safety valves. All materials leaking from the battery, especially organic material electrolytes, are usually toxic or flammable and potentially hazardous. For this reason, the functionality of the BMS preventing such situations is relevant to safety. Hardware protective functions, e.g., discrete or highly-integrated circuits, must be single-fault protected. Errors in hardware wiring should never endanger the battery. Due to increasingly complex algorithms, safety-relevant functions such as these are often transferred to software-based systems. This facilitates the parameterization of the different requirements for the available battery types.

On the other hand, software routines that guarantee safety must be programmed according to a methodology that enables such functions in order to ensure that the software can fulfill the respective tasks reliably. This is called functional safety. The basic standard for functional safety is IEC 61508[6]. It is implemented if no specific variant exists for a certain area of application that takes into account its specific conditions. For automotive engineering, ISO 26262[7] is applied. This relatively new standard adapts the basic requirements stipulated in IEC 61508 to the special conditions of vehicle technology. Consequently, the principles of ISO 26262 must be complied with in automotive engineering if safety-relevant software functions should or must be implemented.

23.3 Outlook

Current research projects are expected to result in more innovative energy storage systems with energy densities higher than those of lithium-ion batteries available today. Their main focus is on developing lithium-air batteries.

Other novel materials will possibly soon be available and could face new electrical and chemical safety challenges, necessitating ongoing adjustments of safety requirements.

[6] IEC 61508 Functional safety of electrical/electronic/programmable electronic safety-related systems

[7] ISO 26262 ("Road vehicles – Functional safety") consists of ten parts relating to safety-relevant electrical/electronic systems in vehicles. ISO 26262 defines a procedural model along with required activities and work products as well as methods to be used in development and production. The standard consists of the following parts:

1. Vocabulary
2. Management of functional safety
3. Concept phase
4. Product development at the system level
5. Product development at the hardware level
6. Product development at the software level
7. Production and operation
8. Supporting processes
9. Automotive Safety Integrity Level (ASIL)-oriented and safety-oriented analysis
10. Guideline (for information only)

Functional safety in vehicles

24

Michael Vogt

Contents

24.1 Introduction

ISO 26262 – Road vehicles – Functional safety In the past fifteen years, the number of electronic systems in vehicles has increased considerably. In the 1980s, most vehicles did not even have airbags, let alone electronic stability programs (ESP). Today, highly complex electronic safety systems have become increasingly widespread. Systems relevant to safety such as antilock bracking systems (ABS), ESP, or active steering have demonstrably reduced the number of road fatalities and

M. Vogt (✉)
SGS-TÜV GmbH, Hofmannstrasse 51,
81379 Munich, Germany
e-mail: michael.vogt@sgs.com

R. Korthauer (ed.), *Lithium-Ion Batteries: Basics and Applications*,
https://doi.org/10.1007/978-3-662-53071-9_24

injuries. However, although these systems are clearly beneficial, if they do not function according to the specifications, they are not only rendered ineffective, but can even cause dangerous driving situations. "Functional safety" is an instrument that identifies and prevents dangerous situations on the road.

Functional safety for automotive applications is regulated by the international standard ISO 26262, which came into effect in November 2011. The standard aims to reduce malfunctions of electrical and electronic systems to an acceptable minimum. It was developed by safety experts from eight countries. The project, under the mandate of the International Organization for Standardization (ISO), was initiated by the standardization committee for automotive technology of the German association of the automotive industry (VDA, Verband der Automobilindustrie).

ISO 26262 is derived from the basic functional safety standard IEC 61508 and describes requirements for the safe development of electrical and electronic systems (E/E systems).

According to the definition in Section 1 of ISO 26262, E/E systems consist of electrical and/or electronic elements (hardware/software), including programmable electronic elements. Consequently, ISO 26262 is the standard for safety-related development in the automotive industry.

24.2 Functional safety overview

Scope of ISO 26262 The ISO standard is currently limited to E/E systems that are installed in series-produced passenger cars with a maximum gross vehicle mass of 3,500 kg, with the exception of special-purpose vehicles such as vehicles designed for drivers with disabilities. The latest (first) edition does not apply to commercial vehicles and motorcycles. This area of application will be addressed in the second edition of the standard, which is scheduled for publication in 2018.

The ISO standard contains, among other things, requirements for functional safety management (FSM), set out in Part 2, and the corresponding supporting processes (Part 8) such as the development interface agreement (DIA), requirements management, change management, and configuration management. Furthermore, ISO 26262 addresses possible hazards caused by malfunctions of E/E safety-related systems. This is not related to the stability of such systems, but serves to analyze the hazards and assess the risks of malfunctions in certain driving situations.

For example, a vehicle is driving along a winding country road at moderate speed. One of the E/E systems that is relevant for driving dynamics malfunctions at a bend, causing the vehicle's driving dynamics to enter a critical state, and it swerves off the road. On the basis of this situation, the risk is assessed and the necessary safety objectives and requirements for the system are derived.

Requirements of ISO 26262 The requirements and methods set out in ISO 26262 correspond to the state-of-the-art in functional safety of road vehicles at the date of publication. Consequently, the standard serves to ensure that state-of-the-art E/E systems are implemented in the interest of product and manufacturer liability. It is

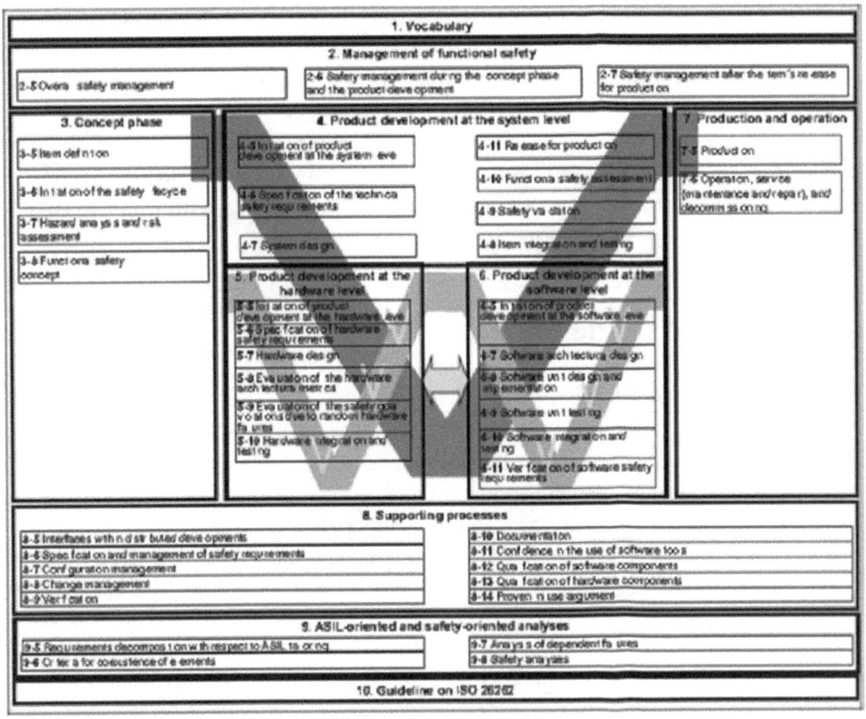

Fig. 24.1 Overview of ISO 26262

also possible that the burden of proof is reversed in a product liability case on the basis of the documentation obtained during development.

The structure of the standard and the V model [1] as the reference process model, which is widely used in the development of E/E systems in the automotive industry, is illustrated in Fig. 24.1. The V model facilitates a structured, iterative development process across all development stages.

ISO 26262 consists of ten parts. Parts 1 to 9 describe the standard; Part 10 is a guideline for information purposes only.

24.3 Functional safety management

The requirements for functional safety management described in Part 2 refer to a generic, comprehensive approach that can be tailored to each individual project. Safety-related development activities are defined at the start of each project by determining the developmental category, i.e., by distinguishing between either a new development or modification of an existing product by means of an impact analysis. All activities of the entire project are recorded in the safety plan, which is the central document for safety management. General functional safety management consists of specific requirements for all company organizations involved, processes for safety-related development, competence management of personnel, and the different roles and responsibilities during the entire safety life cycle.

The safety life cycle encompasses the development on system, hardware and software level as well as production, operation, service (maintenance and repair), and decommissioning. To integrate functional safety management into organizational processes, a quality management system (e.g., in accordance with ISO TS 16949) must be in place and functioning.

24.3.1 Safety life cycle

Parts 3 to 7 of the ISO standard describe the safety life cycle of safety-related development. Because the standard is intended for the automotive industry, the safety life cycle is based on the product development process of the automotive industry.

The concept phase (Part 3) serves as the foundation for all downstream stages of the safety life cycle. A hazard analysis and risk assessment are the core elements of this phase. They are used to evaluate malfunctions of E/E systems and then classify them according to their Automotive Safety Integrity Level (ASIL).

S0, E0, or C0 classifications result in assessment abortion

		CONTROLLABILITY C		
SEVERITY S	EXPOSURE E	C1	C2	C3
S1	E1	QM	QM	QM
	E2	QM	QM	QM
	E3	QM	QM	ASIL A
	E4	QM	ASIL A	ASIL B
S2	E1	QM	QM	QM
	E2	QM	QM	ASIL A
	E3	QM	ASIL A	ASIL B
	E4	ASIL A	ASIL B	ASIL C
S3	E1	QM	QM	ASIL A
	E2	QM	ASIL A	ASIL B
	E3	ASIL A	ASIL B	ASIL C
	E4	ASIL B	ASIL C	ASIL D

Fig. 24.2 ISO 26262 risk graph

ASIL measures the requirements to ensure that risks are adequately reduced. The individual parameters of ASIL are S (Severity), E (Exposure), and C (Controllability). The corresponding risk graph to determine the ASIL is depicted in Fig. 24.2. Quality management necessitates that development is carried out in compliance with aspects related to QM (quality management), followed by safety-related development in accordance with ASIL A to the highest class ASIL D.

In addition to hazard analysis and risk assessment and the document resulting from these analyses, which the standard refers to as the work product, there are several upstream and downstream activities involved until the concept phase is concluded. Section 24.4.2 describes the concept phase in detail with regard to electric mobility and especially battery systems and the related framework conditions. The concept phase is followed by product development at system level (Part 4), product development at hardware level (Part 5), product development at software level (Part 6), and production and operation (Part 7).

24.3.2 Acceptable risk

An acceptable risk describes the boundary between quality management and safety-related (ASIL) development.

The resulting risk acceptance threshold represents the socially acceptable risk and depends on the probability and extent of damage. This is illustrated in Fig. 24.3.

24.3.3 Legal background of ISO 26262

In principle, standards are neither laws nor regulations. However, they do have a presumption of conformity since their compliance ensures a sufficient degree of safety. ISO 26262 is based on the scientific fact that there is no absolute safety.

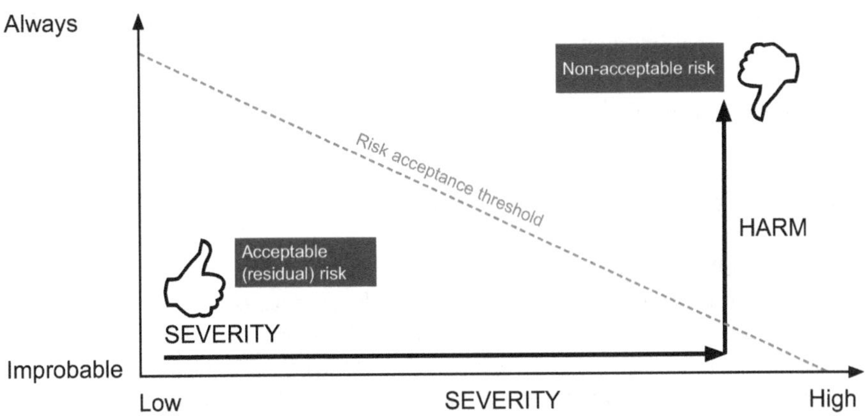

Fig. 24.3 Risk acceptance threshold

The standard's objective is to prove that safety-related E/E systems do not have any unacceptable risks.

According to European Regulation EC No. 661/2009 [2], the safety of vehicles must comply with the state-of-the-art in science and technology. Because the standard is a framework to obtain functional safety when using complex E/E systems in vehicles, it forms part of the current scientific and technological state-of-the-art. Functional safety is characteristic for such systems and can be achieved by using the methods described in ISO 26262.

Consequently, ISO 26262 is of legal relevance for the contractual and liability-related relationship of all contractual parties in the value chain, such as suppliers and manufacturers.

It stipulates, for instance, that a development interface agreement (DIA) needs to be concluded in order to contractually determine the responsibilities of vehicle manufacturers (OEM) and component suppliers (TIER). This is achieved by defining and documenting all safety activities in the concept phase, during development, and production. Compliance with the standard's requirements therefore also determines the manufacturers' responsibility for safety-related systems in civil and criminal law. This applies especially to vehicle manufacturers.

To clarify the legal relevance of ISO 26262, it is necessary to look at the European laws related to product liability and manufacturer liability. According to the German National and European Product Liability Acts, product liability constitutes the non-contractual, statutory liability without fault (tort product liability) for defective (sub-)products. This covers all cases related to the liability for defective products. Even products manufactured to specifications can be faulty. "The producer's liability obligation is excluded if the state of scientific and technical knowledge at the time when the producer put the product into circulation was not such as to enable the defect to be discovered." [3].

Manufacturer's liability under German law – especially as set out in § 823 of the German Civil Code (BGB) – is non-contractual, fault-based liability. It is applied in the following cases: breach of duty, infringement of legally protected interests, damage, and fault. No further information will be provided here on manufacturer liability. For legal interpretation, it is strongly recommended that legal experts for the kind of law concerned who are also familiar with country-specific laws and regulations are consulted.

24.4 Safety of electric mobility

ISO 26262 reflects the fact that, in recent years, the share of E/E systems in vehicles has increased continuously. Especially due to developments in electrification, this trend will continue because many components and systems can only be implemented by using E/E systems. The components and systems used in electric vehicles are either not present in combustion engines or are currently technologically implemented as mechanically operated systems. This applies in particular to the power train, brakes, steering, charging system and/or method (conductive vs. inductive), and the energy storage system. Additionally, systems for energy conversion, air conditioning, and heating of the vehicle's interior are required and necessitate new technical ways of implementation.

When using new and innovative technologies, one question inevitably arises: Which risks are involved and what needs to be done to ensure safety? In electric vehicles, hazards and risks arise in different technological areas. This applies particularly to energy storage systems used in electric vehicles. The hazards and risks that stem from battery systems are mainly found in the underlying electrochemical system. Consequently, these electrochemical systems must be protected at different levels. The battery system must be protected from external influences such as temperature increase and deformation. Therefore, modern battery systems have a housing. It serves to protect passengers from escaping gases and fluids in the event of a crash and to dissipate them in a controlled manner. The housing also shields the battery from external mechanical impact during an accident. Additionally, the vehicle's high-voltage technology must be designed in such a way that it protects persons. EMC influences caused by high-voltage technology can impact various E/E systems and must be reduced to a minimum.

This requires a methodological approach in which all aspects are considered in order to assess the hazards and risks. The hazards and risks arising from different technologies can only be minimized by using an interconnected safety concept. The methodological approach needs to enable the identification of all hazards and risks and the various technologies used must be able to mutually support each other. The safety of the entire vehicle depends on the safety measures identified in the different technological areas as well as on passive and active safety. The following section is devoted to describing how functional safety helps to achieve this objective.

24.4.1 Energy systems in electric vehicles

The integration of a battery system to provide electric drive energy is becoming increasingly established in the automotive industry. Typically, lithium-ion batteries are used in electric vehicles. This technology has been adopted successfully for many years in the consumer industry, for instance in cell phones, notebooks, and power tools. With regard to safety, the difference in electrical energy vehicles is the number of battery cells used. As a result, a larger amount of electrical energy and especially chemical energy content is installed in them. Another difference are the operating and environmental conditions, which fundamentally differ from those in the consumer market. Hazards can occur, for instance, due to external short circuits or overcharging the battery. Also, live and active components can be exposed and thus be touched in an accident. The DC voltages usually range between 400 and 800 V. This is another potential danger for vehicle occupants, rescue forces, and first responders.

Currently, major investments are being made in battery development to make batteries suitable and safer under automotive operating conditions. This goal can

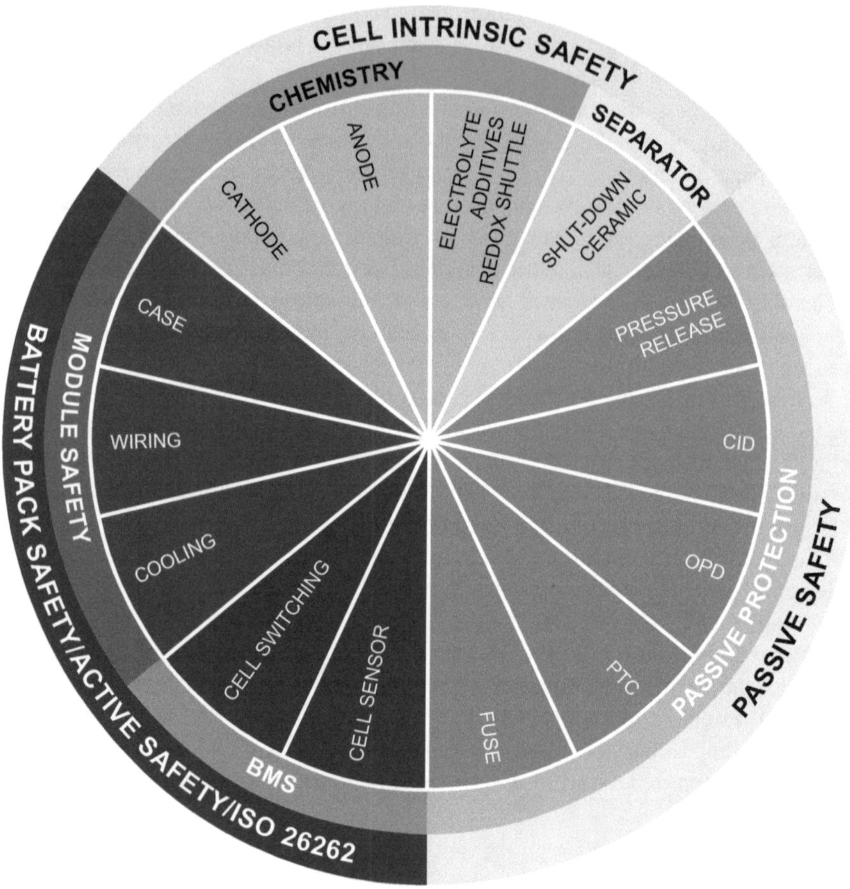

Fig. 24.4 Battery system safety

be achieved at several levels. The cell's safety is already determined when selecting its geometry, design (cell housing, passive cell protection measures), and chemistry (electrolyte, cathode material, anode). This is illustrated in Fig. 24.4.

Heeding this interrelation can significantly increase the safety of the technology used. Such measures are intrinsic and increase safety at cell level. Furthermore, the installation location and intelligent integration in the vehicle architecture can help protect the battery against external impact. In addition, hazards that cannot be minimized by intrinsic measures or installation location can be further reduced by intelligent E/E systems.

All this suggests that battery safety is influenced by different aspects. Hazards originating from battery systems can be divided into four classes (Fig. 24.5).

These are: hazards due to chemical characteristics (electrolyte spillage, outgassing, fire, explosion), electrical behavior (high voltage), mechanical stress, and

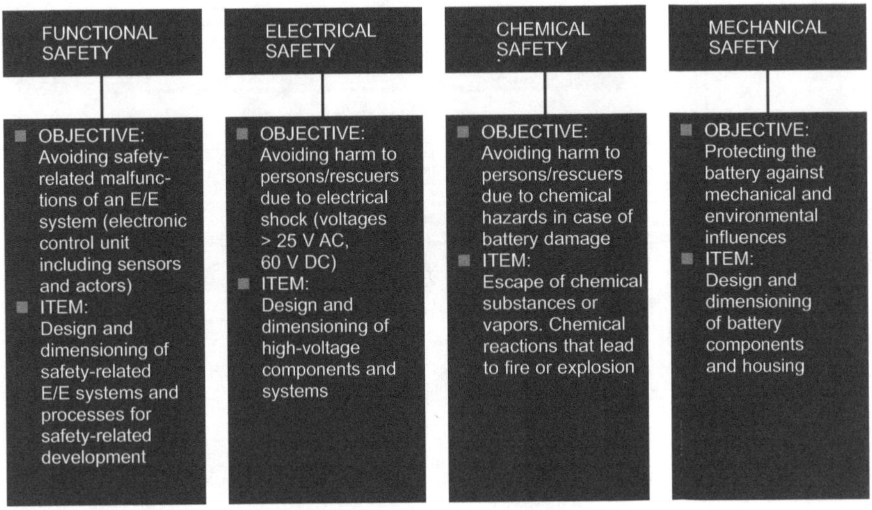

Fig. 24.5 Functional safety and other technologies

functional hazards due to malfunctioning E/E systems (in this case, the battery management system, BMS).

Battery systems not only contain electrochemical cells, but also sensors and actuators that are connected with the BMS, which is responsible for all functions that control and manage the battery system with regard to performance, service life, availability, and safety. It also serves as an interface to the entire vehicle. The battery system enables vehicle functions such as driving, energy recuperation, energy management, and power supply for electrical consumers. The BMS is therefore highly cross-linked with other vehicle systems.

The chemical hazards of battery cells and, consequently, battery safety is determined by the cell chemistry used (cathode, anode, electrolyte). Intrinsic measures inside the cell ensure the battery cells' safety. Such measures include ceramic separators, temperature-stable cathode materials, and electrolyte additives that raise the flash point. The battery system and individual cells are also monitored by the BMS with regard to temperature, pressure, voltage, power, SOC (state of charge), and other parameters. Critical conditions such as a temperature increase caused by external circumstances (such as overcharging), which in turn causes thermal runaway of individual cells, can be prevented through early detection and measures such as disconnecting the battery from the charging system or controlling the cooling power using the BMS.

Thermal runaway is caused by an exothermal reaction inside the cell, which can be triggered by external influences and/or internal short circuits. This could result in a cell fire or explosion. Potential influencing factors are illustrated in Fig. 24.6.

The heat released during exothermal reactions drastically speeds up the reaction, which in turn causes the temperature in the cell to rise even further. Once specific

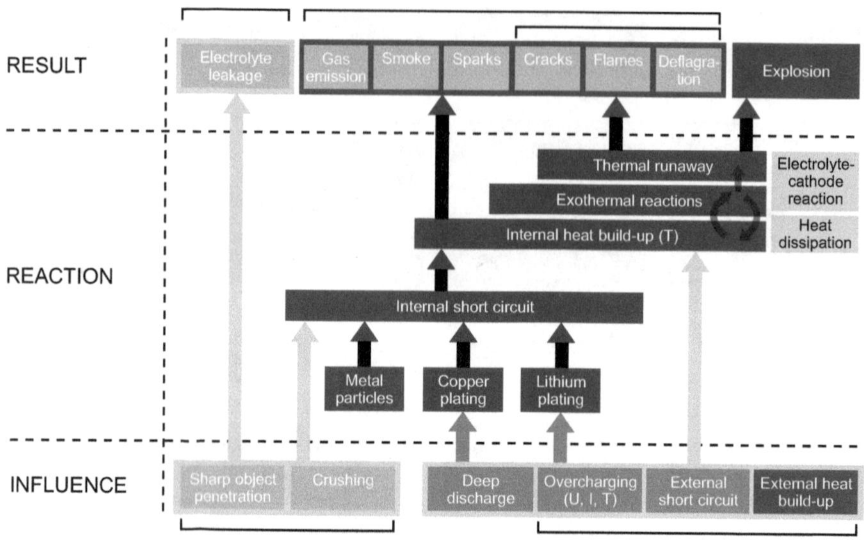

Fig. 24.6 Thermal runaway of lithium-ion cells – influencing factors

cell parameters (e.g., dT/dt) are exceeded, this process cannot be stopped. Moreover, neighboring cells in a cell system can be affected by a thermal runaway cell and cause it to propagate within the module.

Other cells in close proximity to the affected cell can be compromised and undergo the same chemical process.

The cell's and/or battery system's mechanical integrity is determined by its cell geometry, module and system design, and the battery system's integration into the vehicle system as a whole. Vibration, mechanical shock, or battery cell deformation constitute significant mechanical influences. Safety can be increased by carefully choosing the installation location, battery module, and battery housing design.

Electrical safety plays a crucial role for voltages exceeding 60 V DC or 25 V AC. Persons must be protected from active and live components. This can be ensured by implementing measures that prevent direct or indirect contact with these components. Direct contact can be prevented by housings, covers, and protected connectors.

Measures for preventing indirect contact are far more extensive. The aim is to prevent fault-related hazards such as live housing parts. To this end, the isolation between active components, systems, and housings must be designed, scaled, and tested in accordance with all relevant norms and standards.

This ensures adequate protection from isolation faults during the entire service life. Also, all metallic housings must have a low-impedance connection for other safety measures such as fuses to work properly. Ultimately, a comprehensive safety concept is needed that takes into account aspects such as the discharging duration of the intermediate circuit, isolation monitoring devices (IMD), or interlock loop configuration.

24.4.2 Integral safety examination

Functional, electrical, chemical, and mechanical safety can be examined in an integral way according to the measures described in ISO 26262. The safety life cycle is depicted in Fig. 24.7. It describes all necessary activities, requirements, and detection methods for all E/E system phases of life down to the hardware and software level. These phases of the safety life cycle can be roughly broken down into concept phase, product development, and start of production (SOP).

The concept phase is described in Part 3 of ISO 26262 in Sections 5 to 8. It essentially contains the following work products: item definition, hazard analysis and risk assessment, and functional safety concept (Fig. 24.8). The documents resulting from the "Initialization of the Safety Life Cycle" section will not be described in detail here. They are needed for safety-related development, but are not necessary for understanding the concept phase.

Item definitions, i.e., descriptions of the unit or system in question, take into account the system functionality, interfaces with other systems, operating and environmental conditions, legal requirements and framework conditions, known system hazards, etc. Such hazards can be identified by conducting technology-based risk analyses or can be based on other available data. Hazards determined and documented in this manner can be minimized by implementing different measures.

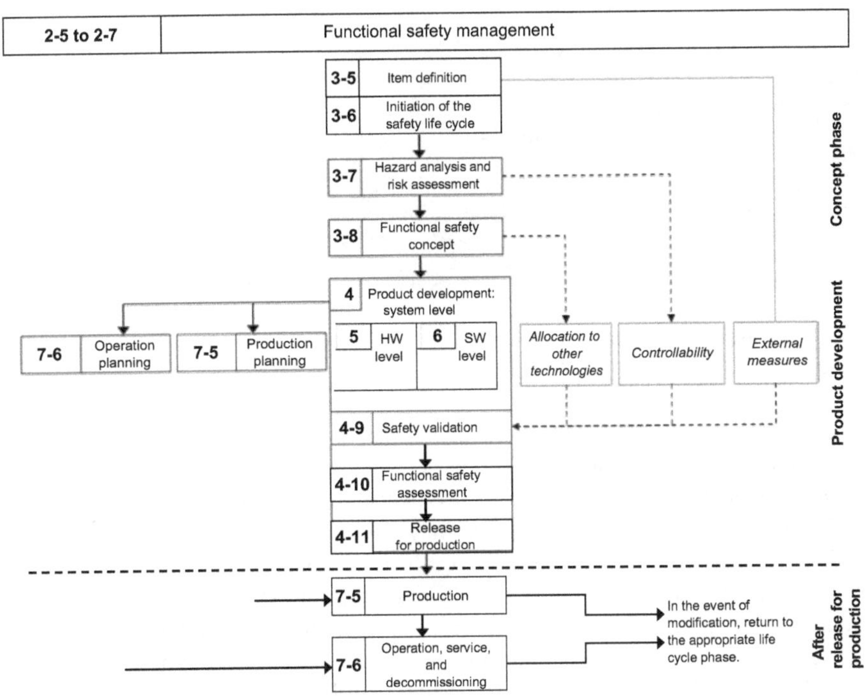

Fig. 24.7 Safety life cycle

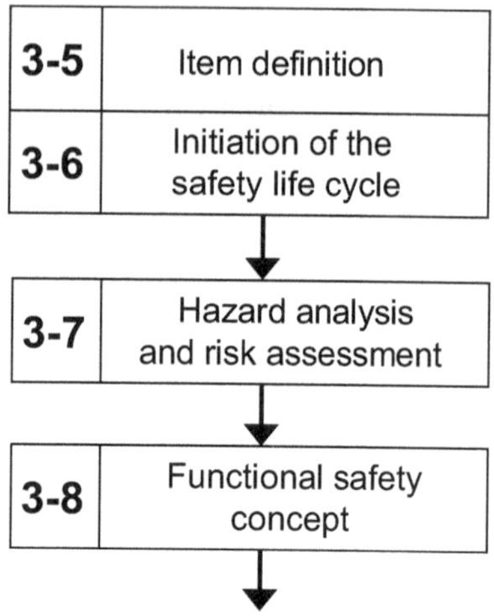

Fig. 24.8 Concept phase work products

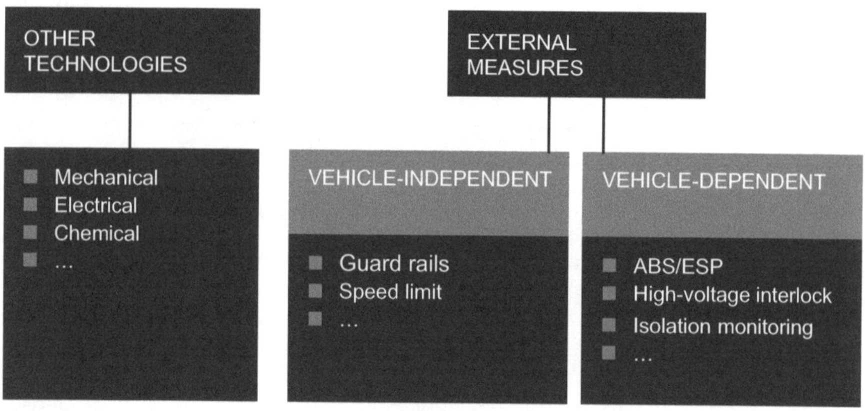

Fig. 24.9 Distinction of external measures

These include functional safety measures for E/E systems and other technologies as well as external measures. For battery systems, other technologies include electrical, chemical, and mechanical safety. External measures can be dependent on or independent of the vehicle (Fig. 24.9). They also increase the entire system safety.

Hazard analysis and risk assessment evaluate all hazards and risks related to E/E systems. Each danger caused by an E/E system malfunction is assigned a safety goal based on a hazard analysis and risk assessment that reduces the hazard and/or risk in accordance with ASIL. Functional safety is assessed according to Severity (S), Exposure (duration/frequency (E), and Controllability (C). To define the ASIL hazards and classify risks, the individual parameters (S, E, and C) are weighted.

Systems with a risk that is below the risk acceptance threshold are classified as QM. All safety-related components are classified from ASIL A to ASIL D.

The next stage involves describing the functional safety concept, which details the safety requirements obtained from the hazard analysis and risk assessment, the safety measures, and safety mechanisms to be developed. This includes handling systematic errors or system behavior when there is a hardware failure as well as handling the E/E system's transition to a safe state. All non-functional aspects are allocated to other technologies and examined. This is illustrated in Fig. 24.5.

The methodological approach of ISO 26262 can be applied to the development of innovative battery systems for use in road vehicles to ensure overall safety. Also, different safety-related measures can be compared and selected based on their effectiveness and costs.

Weighting of the E parameter (duration or frequency while driving) is adapted to combustion engines only because of the ISO 26262 publication date and therefore needs to be reassessed. This is due to the additional driving situations arising from electric mobility. Moreover, situations and potential dangers mutually interact. For example, when refueling a hybrid vehicle (= situation), an isolation fault (= malfunction) can occur in the high-voltage on-board electrical system. This is just one of the issues that are currently being intensively discussed. In this context, the potential hazards caused by the interaction between malfunction and situation must be taken into account. The ISO standard does not contain any information on this topic. However, it is urgently needed to ensure a more uniform development approach and to reduce competitive and cost pressure.

24.5 Practical application

Organizations interpret the norm differently and have varying development philosophies. Therefore, using ISO 26262 and implementing its requirements in practice have resulted in different applications. Prior to developing ISO 26262, the processes and practices in organizations were well established, which caused difficulties when implementing the new standards in the corporate culture. The concept phase is the foundation of safety-related development. All activities at this stage and all definitions resulting from this phase hugely impact downstream development phases and are thus crucial. For this reason, the most important topics of the concept phase will now be presented.

24.5.1 External measures vs. other technologies

External measures are measures that do not form part of an item, but reduce and/or prevent the potential risk caused by an item. For this reason, they are not taken into account in hazard analyses and risk assessments. These measures are only considered when drawing up the item definition and, later on, in the functional safety concept which is then created for use in the hazard analysis and risk assessment. If the external measure is an E/E system, it must be assigned an ASIL.

Fundamentally, measures from other technologies are only taken into account in the functional safety concept. They do not receive an ASIL classification because they are not an E/E system as defined by the standard, but rather measures related to electrical safety, design considerations, mechanical safety, etc.

However, the definition makes it possible to interpret other technologies as external measures. For this reason, it is difficult to make a clear distinction. For instance, depending on the item, the question is whether electrical safety is an external measure or another technology. If electrical safety is defined as an external measure as per the item definition, this can have adverse effects on subsequent projects. If external measures are forgone in subsequent projects, the safety concept may have to be revised, which would cause additional outlay because the safety implemented by external measures now needs to be realized by using other measures in the safety concept.

24.5.2 Hazard analysis and risk assessment

There are several challenges when conducting hazard analyses and risk assessments. A facilitator and an expert team are always needed. Ideally, the facilitator has some experience in hazard analysis and also has substantial know-how about the system (item) in question as well as the methodological approach. The facilitator can be an external person or a company employee, e.g., the functional safety manager. The expert team may only consist of employees from the organization and its affiliates. These are the only people that have the necessary professional and technical expertise to assess hazards and risks and interpret them. The facilitator guides the expert team during the hazard analysis and risk assessment and helps it to reach a consensus in the event of conflict. The outcome thus depends on the facilitator and the expert team. Hazard analyses and risk assessments also provide protection from product liability claims. Therefore, the expert knowledge should be very broad and supported by further investigations, if need be. This is absolutely essential for new and innovative technologies such as battery systems.

Experience and diligence are needed in order to prepare a hazard analysis and risk assessment. The operating modes, situations, and environmental conditions of the system to be assessed must be combined in a transparent manner. E/E system malfunctions are combined in this way and then subjected to a hazard analysis and risk assessment. With regard to battery management systems, this can involve monitoring the cell temperature, for instance. If a failure occurs during monitoring, different risks can arise depending on the current operating mode (driving, parking, charging), situation (country road, highway), and environmental conditions (high or low outside temperature).

Assessing the risk of hazards quite often is very complex. Frequently, there is no experience from previous development projects that can be used as a basis for assessment. It is crucial for the assessment to evaluate the system without safety measures. If such measures were considered in the analysis, it would be difficult to obtain a clear argument for safety-related development. This can be a decisive factor with regard to external measures and other technologies. Depending on the argument, it must be considered at different times in the safety life cycle. This brings us to the requirements management set out in the standard. Even if hazard analysis and risk assessment are carried out successfully, requirements management is still mandatory. It serves to ensure that, based on the hazard analysis and risk assessment, all requirements and resulting evidence (verification and validation) are carried out during the entire development phase.

24.6 Outlook

Currently, electric mobility and thus the development of battery systems are characterized by a strong momentum and ever-changing framework conditions. In addition to costs, range, and consequently customer acceptance, the success of battery systems hugely depends on their performance and safety during the entire service life. Moreover, new topics are emerging such as the "second use" of battery cells for stationary applications or smart battery cells. Safety is also a challenge in this area. The downstream use of smart battery cells in stationary applications also requires functional safety to be considered during development. However, the resulting requirements of the standard differ substantially. For this reason, a method is needed to adapt these requirements to enable the development of a smart cell for all areas of application.

Thoroughly assessing all risks related to the respective application is the first challenge that must be overcome. To this end, established processes must be refined and/or adapted while ensuring that they are still accepted.

This requires more intense collaboration between manufacturers (OEM) and system suppliers (TIER). Interfaces and functional safety management are already hot topics in development and will gain importance in the future.

Bibliography

1. http://www.v-modell-xt.de/
2. Regulation (EC) No 661/2009 of the European Parliament and of the Council of 13 July 2009 concerning type-approval requirements for the general safety of motor vehicles, their trailers and systems, components and separate technical units intended therefor
3. Act on Liability for Defective Products (German Product Liability Act, *ProdHaftG)*, Section 1, Paragraph 2, Item 5

Functional and safety tests for lithium-ion batteries

25

Frank Dallinger, Peter Schmid, and Ralf Bindel

Contents

25.1 Introduction

Functional and safety tests for lithium-ion batteries used in automotive and industrial applications are essential. Fig. 8.3 shows a diagram of such batteries. The battery consists of individual cells interconnected to form modules. Several modules are then also interconnected inside the battery itself. The battery also features a cooling circuit. A deaeration system evacuates gases during cell outgassing. The battery housing contains the battery management system (BMS) and the circuit interrupter.

During the battery's production cycle and service life, different components, assemblies, and the overall system are to be tested with varying objectives. The dimensions used for test structuring are shown in Fig. 25.1:

F. Dallinger (✉) · P. Schmid · R. Bindel
Robert Bosch GmbH, Wernerstrasse 51,
70469 Stuttgart, Germany
e-mail: Frank.Dallinger@de.bosch.com; PeterK.Schmid@de.bosch.com; ralf.bindel@zeiss.com

R. Korthauer (ed.), *Lithium-Ion Batteries: Basics and Applications*,
https://doi.org/10.1007/978-3-662-53071-9_25

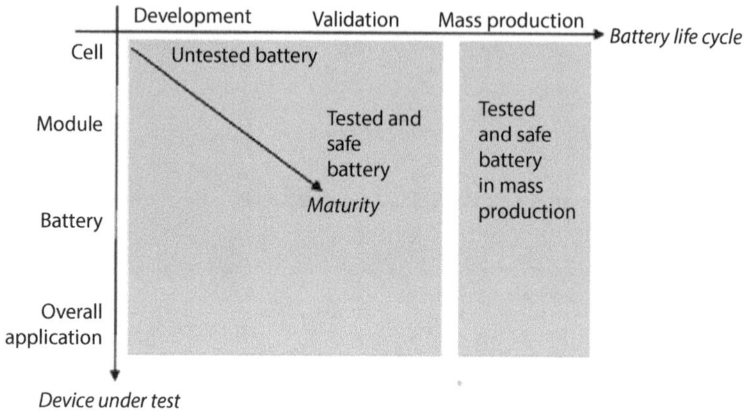

Fig. 25.1 Parameters determining scope of battery testing

- Battery life cycle: development – validation/endurance test – mass production
- Device under test (DUT): cell – module – battery – overall application
- Battery maturity: Battery maturity is to be taken into account during the tests in the development and validation phases due to the potential hazards posed by batteries. The maturity determines the required safety measures for the tests.

Table 25.1 shows an overview of the resulting tests. This overview is not complete due to the fact that the technology is still in its infancy.

As of March 2009, there are two specification definitions for lithium-ion battery reactions: the American FreedomCAR specification [1] and the European EUCAR specification [2]. As a general rule, EUCAR is also used in the US (Table 25.2).

25.2 Using EUCAR hazard levels for the test facility

The operator defines the hazard level based on the tests to be conducted and the hazard posed by the test specimens. The test system is then designed in accordance with the applicable safety class. Suitable safety technologies for the listed hazards and risks are installed and defined in a tool box. The tool box can then be used to create test systems for hazard levels 0 through 7 (Fig. 25.2) by choosing the appropriate safety technology modules (Table 25.3).

Tests with hazard levels 0 through 3 require alarm lamps, safety door locks, and safety interrupters. A safety control device already operates the safety signals and processes for these hazard levels. From hazard level 4 on, suitable safety elements need to be implemented to prevent blowing-off, fire or flames, bursting, and explosion.

The test system is amended accordingly for battery system tests when battery systems with hazard levels 4 through 7 are being tested.

Table 25.1 Overview of functional and safety tests for lithium-ion batteries

	Development	Validation, testing	Mass production
Cell	Impedance spectroscopy for cell chemistry optimization, testing electrical parameters	Time and cycle-based aging tests; testing electrical parameters; abuse tests; shake tests, if required	Forming; OCV, internal resistance measurement, testing tightness of cell
Module	Application tests with different test contents	Time and cycle-based aging tests; testing electrical parameters	Testing isolation, resistance; testing contacts
Battery	Application tests with different test contents	Time and cycle-based aging tests; testing electrical parameters; abuse tests; environmental tests	Testing electrical parameters; testing communication; testing tightness
Overall application	Application and development tests to verify improvement to application overall and interactions between individual components		

Table 25.2 Comparison test specifications EUCAR and FreedomCar

FreedomCAR Abuse Level	EUCAR Hazard Level	Hazard[a] of battery failure
Level 1	0	No hazard
	1	Reversible battery damage
Level 2	2	No hazard [b]
	3	Irreversible battery damage, leakage without hazard, as the case may be
Level 3	4	Hazard
	5	Irreversible battery damage
	6	
	7	

[a]Failure-induced hazards for people or things; use of personal protection equipment such as protective goggles and gloves is required
[b]Hazard is fully eliminated thanks to personal protection equipment (gloves, clothing, face protection [shield]).

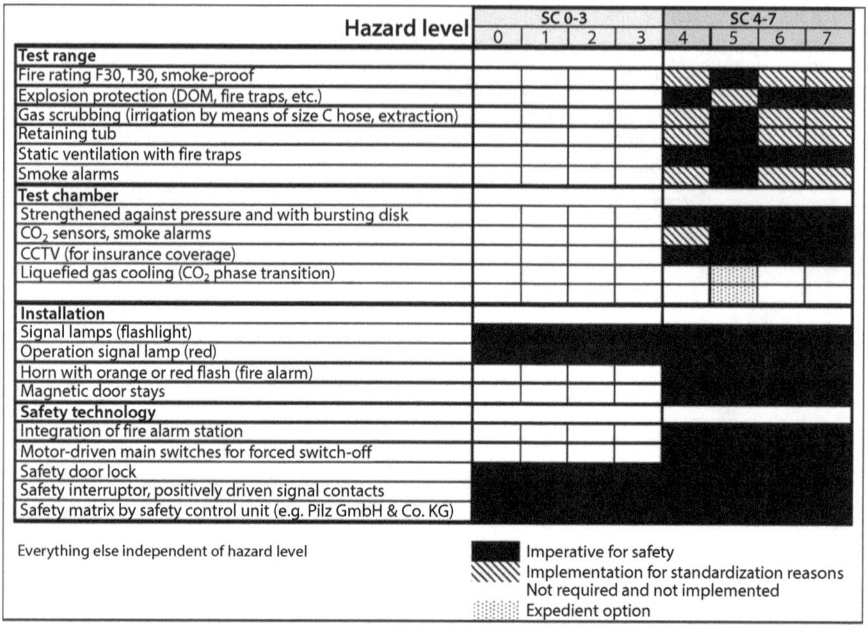

Fig. 25.2 Modular safety technology for battery test systems

Table 25.3 Hazard levels

Hazard level	Description	Classifying parameters and impact	Permitted hazard
0	No impact	No impact, no function impairment	
1	Passive safety equipment is tripped	No defect, no leakage, no blowing-off, no fire, no flames, no bursting, no explosion, no exothermal reactions, no thermal runaway. Cell still usable, safety devices and fuses must be repaired.	
2	Defect, damage	Similar to hazard level 1 but the cell is damaged irreversibly and must be replaced.	
3	Leakage, Mass loss < 50 %	No blowing-off, no fire, no flames, no bursting, no explosion. < 50 % electrolyte solution weight loss (solvent + conducting salt)	No hazardous or poisonous substances must be released into the work area during blowing-off.

Table 25.3 (Continued)

Hazard level	Description	Classifying parameters and impact	Permitted hazard
4	Blowing-off, mass loss > 50 %	No fire, no flames, no explosion. > 50 % electrolyte weight loss (solvent + conducting salt)	No hazardous or poisonous substances must be released into the work area during blowing-off.
5	Fire or flames	No bursting, no explosion (e.g., no debris ejected)	No hazardous or poisonous substances must be developed or released into the work area during blowing-off and burning.
6	Bursting	No explosion, but active electrode material ejected	No hazardous or poisonous substances must be developed or released into the work area during blowing-off, burning, and bursting.
7	Explosion	Explosion (e.g., cell destruction)	No hazardous or poisonous substances must be developed or released into the work area during blowing-off, burning, bursting, and exploding.

Personal protection equipment (PPE) for operating personnel is not sufficient for handling test specimens at higher hazard levels. Additional safety equipment is required in the test facility to protect against chemical hazards (considerable outgassing or outgassing of hazardous substances: level 4) and physical hazards (fire, bursting, explosion: levels 5, 6, and 7).

A testing room (safety box) constitutes the fire compartment according to DIN 4102, Part 2 (F30/T30), fire rating for Germany. Other European countries require up to 120 minutes; the US and Canada up to 240 minutes, depending on the local authority's requirements.

The safety box contains the following: washing system with a fire water supply, retention basin and drainage, door lock, smoke alarm, fire traps. A temperature chamber is installed in the testing room. It has the standard features required for battery testing: pressure resistance higher than the minimum triggering level for an overpressure drain. This is achieved by means of pressure relief equipment with an exhaust pipe (stack), gas and/or temperature sensors to detect fires (preferably CO sensors due to their cross-sensitivity toward hydrocarbons and thus toward opened cells), inertization, CO_2 cooling, ducts, and connection adapters for electrical and fluid media. A safety control device monitors and controls the entire system.

It also uses safety sensors to monitor the equipment status and feedback from test chambers and testers. It activates the relevant safety systems in the event of a hazard. It also prevents operating conditions hazardous to personnel from arising during regular, failure-free test system operation. The safety control also connects to a higher-level detection system.

25.3 Functions and modules for battery testing

Battery testing systems consist of many modules (Fig. 25.3) that cover different battery testing functions to ensure safe, repeatable tests, assist the operator in creating test procedures, and manage specifications and results.

Battery charging and discharging Batteries are submitted to repeated stress by current sink (discharging) and current source (charging) during standard operation. In test facilities, this stress is simulated with power supplies, what are referred to as battery testers. They function as a source during charging and as a sink during discharging. Energy recovery during discharging is highly important in this respect. It can be fed back into the intermediate DC circuit, the testing circuit, or into the public grid. It is worth investing in power supplies with a good energy efficiency rating since energy costs are lower and no large cooling equipment needs to be installed. The battery tester voltage and current range need to be suitable for the tested battery. Table 25.4 shows typical performance data.

The instrumentation installed in the equipment is not only used to control current and voltage; it also handles the evaluation and output of measured values. It must therefore be sufficiently accurate for the planned task.

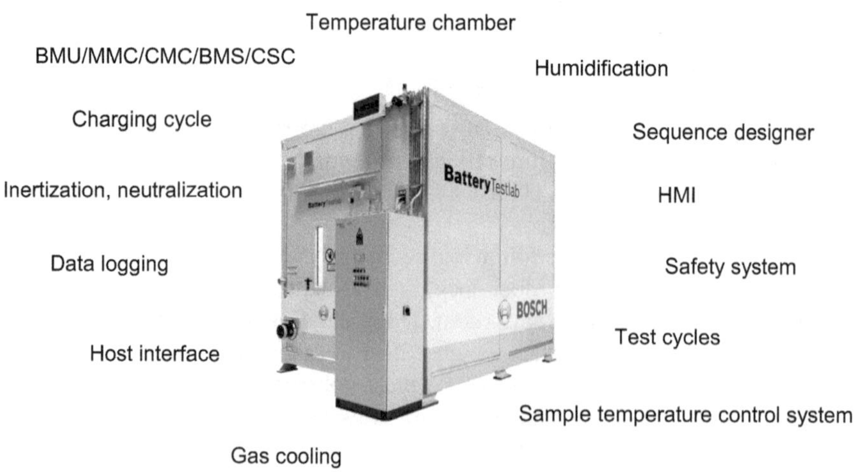

Fig. 25.3 Functions and modules of battery testing systems *(Source* Bosch)

Table 25.4 Typical battery tester performance data

	Cells	Modules	Packs
Output voltage	0 V to 6 V*	0 V to 60 V	50 V to 600 V and 900 V, respectively
Output current (max.)	± 400 A	± 300 A to 600 A	± 600 A to 900 A
Power	± 2.4 kW	± 18 kW	± 120 kW to 350 kW

*Also some negative voltages for diffusion testing (e.g., forming)

Table 25.5 Typical battery tester accuracy requirements

	Cells	Modules	Packs
Slew rate	300 A/ms	600 A/ms	1,200 A/ms
Current ripple factor	0.02 %	0.2 %	0.2 %
Measurement accuracy Current	± 80 mA to ± 400 mA	± 80 mA to ± 0.4 A	± 120 mA to ± 0.6 A
Measurement accuracy Voltage	± 2 mV to ± 10 mV	± 20 mV to ± 0.1 V	± 200 mV to ± 1 V

Development and validation projects require high accuracy and dynamics. Qualities similar to those of the charging systems built into the vehicles suffice for end-of-line tests (Table 25.5).

Ambient temperature management In test systems, the temperature around the car batteries or in the application concerned is simulated by temperature-controlled test chambers in which the tests are conducted. The temperature range is – 40 °C to + 85 °C. Since lithium-ion battery technology is still in its infancy, there is still a need for suitable standards. This is why standards from electronics testing are currently being used for temperature gradient testing, e.g., ISO 12405-1, IEC 60068-2-38, and ISO16750-4. Demands for steeper gradients up to 4 K/min need to be discussed for lithium-ion batteries. Heat transfer from the test chamber air into the individual battery cells via the battery housing and the space underneath is too slow for these batteries.

Battery temperature control The temperature in individual battery cells is safety-relevant and considerably influences the batteries' maximum power output and aging behavior (Fig. 25.4).

Only a fraction of the nominal output is available at low battery temperatures. Power drain also influences a battery's service life. This is why batteries have a temperature control system. Powerful liquid-based systems are state of the art in this respect, as are direct temperature control systems in which the cooling plate installed in the battery (DUT) functions as an evaporator for a suitable compression-type

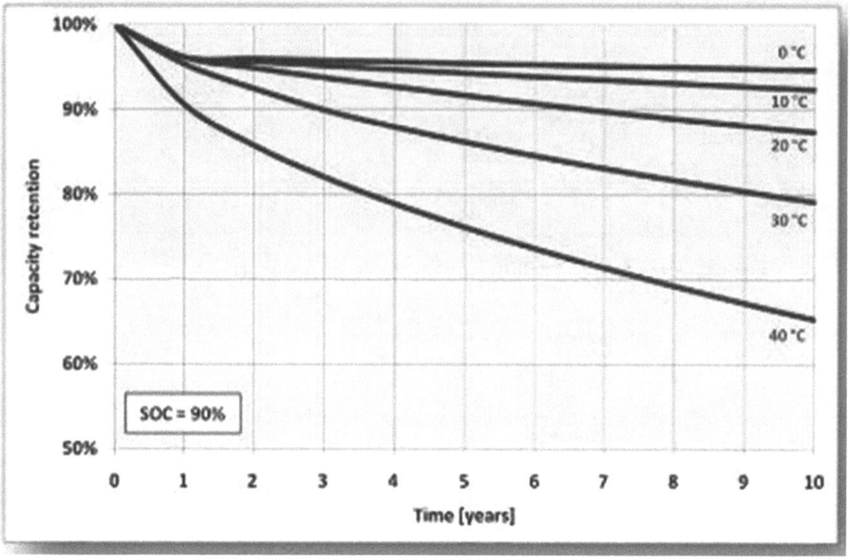

Fig. 25.4 Aging behavior as a function of temperature (*Source* Pankiewitz, AABC 2012, Orlando)

cooling system, and air-based temperature control systems. End-use scenarios are simulated in battery testing systems using temperature control systems. Control of these systems is either static (test program specifications) or dynamic (battery management system communication with the test facility control).

Specifications are derived from the temperature control medium's temperature or the flow rate or both. High-performance battery testing systems also control power input and output in relation to the battery temperature.

Data management Data management performs different tasks in a battery testing system. Test procedures need to be recorded as a prerequisite for repeatable tests with comparable results. This can be done directly in the battery testing system control. However, this means that test procedures are not automatically available to other battery testing systems. This is why individual battery testing systems in a test facility are ideally linked via a higher-level manufacturing execution system (MES). Recording test procedures in the MES enables valid and validated test procedures to be correlated unambiguously with the corresponding test specimen. The test specimen type is typically identified by scanning when it is introduced into the battery testing system. The current version of the appropriate test procedure is then loaded automatically from the MES. This procedure eliminates operator errors. Recording results in a higher-level MES also brings advantages. It enables statistics from the collected data to be produced that could point toward systematic alterations in the battery production process chain. It also enables the operators to easily compare the production plants within a manufacturing network.

Safety control The safety control has an important task within the battery testing system. Lithium-ion batteries store high-density energy. Hazards are thus not only the high voltages and currents but also a sudden and unscheduled release of such energy. Possible causes include manufacturing errors, incorrect module and pack assembly, battery management system faults, or incorrect handling.

The safety control uses suitable sensor technology to monitor limit values for temperature, voltage, and current. It provides safety device queries and protection against contact with live parts through an interlock. In the event of a malfunction, it also interrupts test procedures, triggers protection measures, and alerts help, thus providing protection for the operator.

Automation system The automation system provides communication between the individual components in the test system (sub-systems) and with the test specimen (Fig. 25.5). The communication channels are bidirectional and target values can be sent and measurement readings received. Safe, convenient test system operation requires the sub-systems to be placed under a higher-level control. Operating individual sub-systems with an individual control would be too complex and error-prone. The automation system also is a prerequisite for implementing automatic and dynamic test procedures. It dynamically amends the test procedures in accordance with the values received from the test specimen and the sub-systems.

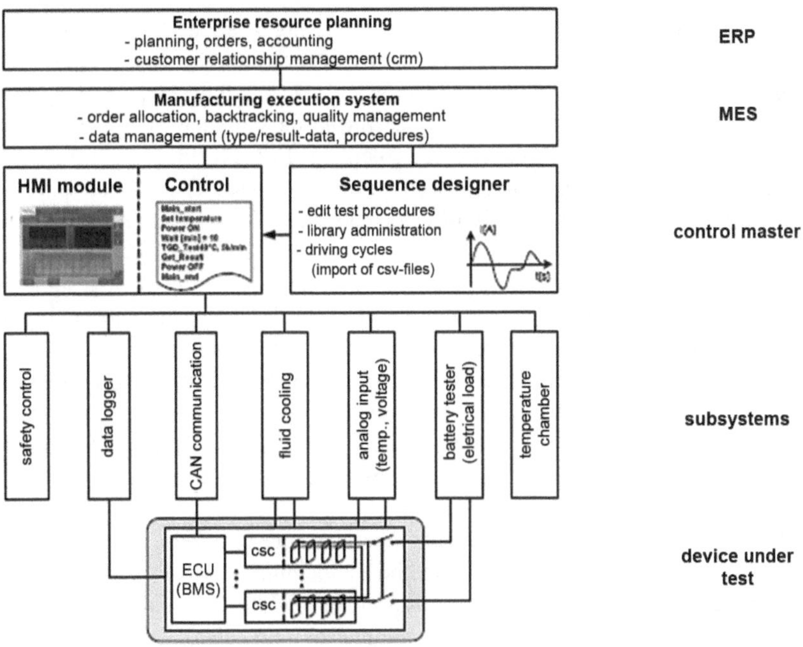

Fig. 25.5 Control diagram of the test facility (*Source* Bosch)

It also provides tools for the operator with which complex test procedures can systematically be developed and tested for plausibility.

The communication interfaces in the automation system depend on the test assignment concerned. Cell tests require bidirectional interfaces with the test system's sub-systems. Module tests also require communication with the CSC (cell supervisory circuit). As a general rule, CAN bus systems are used for this purpose. Battery packs also require communication with the BMS via different bus systems (e.g., CAN, LIN, OBD2).

25.4 Battery testing system examples

Testing facilities are used for many applications: from individual cells and cell components in research and development to module test systems to overall systems for cars, commercial vehicles, and stationary plants. Research still concentrates on chemical materials, which are actually only used in small quantities. In regard to battery packs, however, the focus is increasingly set on electronics and temperature management. The cycler that charges and discharges the cells is very important in terms of cells, but less important for modules, and least important for battery packs. Signal quality, signal rise speed, and current noise on cell level are still decisive parameters. However, in the battery pack, they are merely the trigger for necessary cell heating during thermal dimensioning. As the test object size increases, the influence of temperature control devices on larger test specimens decreases. This is due to the surface of a three-dimensional object increasing less than the volume of the same object, hence cooling impact also decreases. Nevertheless, these two test facility components still have an impact on volume and costs. Safety systems for systems engineering feature an appropriate design. During development, early prototypes with their associated hazard potential are tolerated. In mass production, however, the material is technically mature and well known from a safety perspective.

Fig. 25.6 shows a battery test facility for batteries with a volume up to 25 m^3. Typical load is around 2 t.

As a general rule, the test examines service life, temperature-control design, and data exchange with the battery control unit (BCU) or the BMS, the latter being the main control component for each pack unit. The testing current reaches 900 A. The multi-chamber design helps to meet the strict requirements for the North-American market and obtain approval to test low maturity prototypes. If fire protection requirements are lower, a one-chamber system is used. Its outer wall features high-quality thermal isolation and functions as a fire protection-certified temperature-control chamber.

Suitable devices are used to load the chambers. As a general rule, a forklift cannot be used because of potential damage to the temperature-control chamber's inner walls. Specially modified lifting apparatuses are used instead. Usually, test systems are not operated as individual systems. A number of units, up to 50, are operated in a factory building. This enables higher numbers of tests to be conducted and reduces infrastructure costs for transformers, lead frames, coolants, and other material.

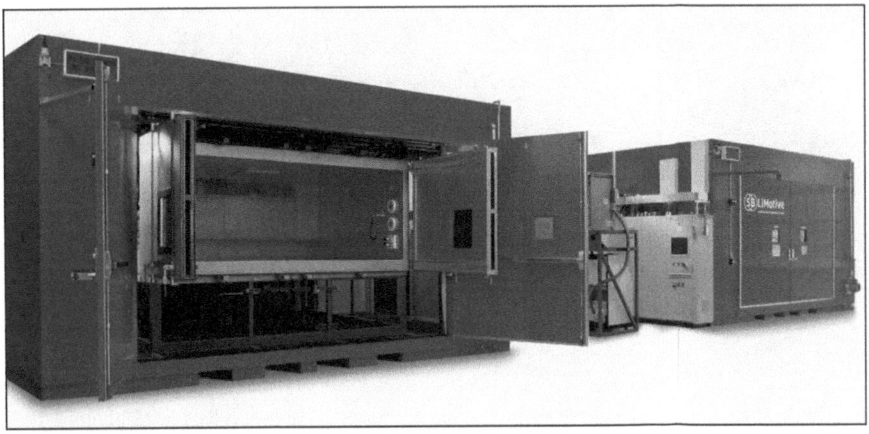

Fig. 25.6 Pack test system for commercial vehicles, stationary plants, and batch tests (Cobasys Research Laboratory, Michigan, US, *Source* BOSCH)

Fig. 25.7 Research laboratory test facilities for cell tests (*Source* Bosch)

Fig. 25.7 shows a research test facility. Smoke exhaust pipes to be used in the event of an emergency can clearly be discerned on the roof, as can the red fire bulkheads near the floor. Floor pipes are used for "shower" cleaning and subsequent draining of the cleaning agent.

Individual cell tests are conducted for research as a general rule. Compact single-chamber systems are used for this purpose. Impedance measurement and half-cell measurement devices are already installed in such systems. Fig. 25.8 (left) shows a highly compact module.

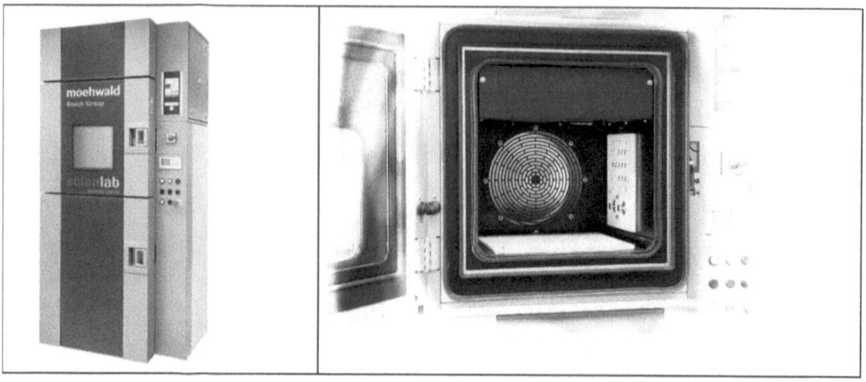

Fig. 25.8 *Left* Lithium-ion battery single test facility for up to 100 Ah cells; *right* Lithium-ion battery cell production test facility (*Source* Bosch)

Fig. 25.9 Cell test facility with a disposal storage unit (moehwald/scienlab)

What are known as forming testers are used for mass cell testing during production (Fig. 25.8 right). These testers do not come with individual test chambers. They are part of the manufacturing plant and are already installed in a safety area.

Outdoor facilities (Fig. 25.9) with double-chamber systems are generally used for critical cell material testing in research and for abuse tests. One-chamber systems have also become available recently. Permission requirements stipulate the installation of gas scrubbing or contaminant separation devices, because malfunction actually forms part of the test.

25.5 Outlook

Lithium-ion battery technology for automobile and industrial applications is a technology that is still in its infancy. Many parameters and interdependencies in batteries are still not fully understood at present. It can be assumed that the progressive development of battery technology and the associated production technology will have a positive influence on the evolution of testing strategies and test beds. Multi-purpose battery test systems are used at present, but more specific test contents and increasing lot sizes will probably require improved, more cost-effective testers in the future. At this stage, researchers are already looking into decreasing test duration because of the extensive time-based and cycle-based aging tests. Stress test systems will probably be developed which will produce safe batteries in considerably less time. The BMS (battery management system) is also being increasingly tested in battery pack test systems. The battery testing system control thus needs to be connected to simulation systems and record result values with a high frequency.

Current long-term service life tests show that charging and discharging rates have a considerable influence on capacity loss and, consequently, battery service life. This parameter needs to be taken into account when stress tests for realistic service life forecasts are developed. It could also increase validation test duration and therefore the expense for validation under life-like conditions.

Bibliography

1. FreedomCAR: Electrical energy storage system abuse test manual for electric and hybrid vehicle applications; SAND 2005–3123
2. Josefowitz W et al (2005) Assessment and testing of advanced energy storage systems for propulsion, European Testing Report. In: Proceedings of the 21 worldwide battery, hybrid and fuel cell electric vehicle symposium & exhibition, Monaco, 2–6 Apr 2005

Transportation of lithium batteries and lithium-ion batteries

26

Ehsan Rahimzei

Contents

26.1 Introduction

Lithium batteries were introduced on the market in the seventies and were produced for special applications in relatively small lot sizes. Today, billions of rechargeable lithium-based batteries are transported every year.

Lithium batteries, lithium cells, and devices equipped with them are considered to be dangerous goods based on their basic properties. Repeated incidents regularly

Author's note: At the time of printing, the 2015 or 2016 systems of rules and standards including some extensive amendments already applied with regard to lithium batteries as discussed below.

E. Rahimzei (✉)
Battery Production, VDMA Verband Deutscher Maschinen- und Anlagebau e.V.,
Lyoner Str. 18, 60528 Frankfurt am Main, Germany
e-mail: ehsan.rahimzei@vdma.org

R. Korthauer (ed.), *Lithium-Ion Batteries: Basics and Applications*,
https://doi.org/10.1007/978-3-662-53071-9_26

give rise to discussions on tightening their transportation conditions. Air traffic is particularly critical about the commercial and private transportation and application of lithium batteries because these incidents lead to very much uncertainty.

Today's regulations were created first and foremost for batteries used in devices. Large-scale batteries such as those in partially or fully electric vehicles or large electrical storage systems were not sufficiently taken into account at the time. Internationally coordinated efforts are currently under way to ameliorate this situation. The commercial transportation of lithium batteries and cells is subject to the regulations for the transportation of dangerous goods. National and international laws and regulations must be applied and heeded for all carrier types. Air traffic has additional regulations that govern and restrict individual use and private transportation of lithium batteries and lithium battery-powered devices. The requirements differ for on-board and registered baggage.

The system of rules and standards and the resulting legislation are subject to continuous amendments. This is why the transportation and requirement details in this Chapter concerning lithium batteries and lithium cells only constitute a snapshot at the time of writing. As a rule, they are amended annually or biannually. This Chapter focuses on the transportation of lithium batteries as dangerous goods. The information provided here cannot be used to replace state-required training or official examinations.

26.1.1 Fundamentals of regulations for the transportation of dangerous goods

Many everyday goods are subject to the regulations for the transportation of dangerous goods. These regulations describe the prerequisites for the safe transportation of goods that in principle can present a danger. Such goods are substances and objects with properties that present a danger to important public property or to the health of humans, animals, and the environment, in particular when they are being transported.

Transportation of dangerous goods must take into account all necessary steps from planning to implementation. To this end, the goods that are to be transported need to be identified and classified; suitable packaging material needs to be chosen, and the goods need to be appropriately packaged. The packages need to be marked and identified and the dangerous goods documents have to be drawn up. Loading, transportation, receipt, and unloading are also all regulated by the dangerous goods transportation legislation. The term "dangerous goods" should not be confused with the term "dangerous substance". The system of rules and standards concerning dangerous goods exclusively governs the goods that need to be transported and the dangers resulting therefrom.

According to the framework governing dangerous goods transportation, responsibility is shared between the contractors or owners of companies involved in packing, loading, shipping, unloading, receiving, or unpacking dangerous goods and the manufacturers of packaging, containers, or vehicles intended for these purposes.

26.1.2 The origin of the UN system of rules and standards

The principles of the dangerous goods regulation are developed and determined by an international commission, the United Nations Economic Commission for Europe (UNECE) in the so-called UN Model Regulations. Based on these principles, regulations for specific carrier requirements are developed and implemented in national law.

The expert committee for the transportation of dangerous goods affiliated with the UN has its registered headquarter in Geneva, Switzerland. It is divided into the sub-committee for the transportation of dangerous goods, on the one hand, which develops the UN Model Regulations, the so-called Orange Book, and the UN Manual of Tests and Criteria. On the other hand, there is the sub-committee concerned with the Globally Harmonized System of Classification and Labeling of Chemicals, GHS.

The ICAO, the International Civil Aviation Organization, is directly affiliated with the UN, as is the IMO, the International Maritime Organization.

These organizations have affiliated work groups; similarly to the UNECE work groups, they elaborate carrier-specific systems of rules and standards (Table 26.1).

26.1.3 General classification

Correct identification and classification are important steps for handling dangerous goods. The specific subsequent steps are derived from them. Dangerous goods are grouped depending on their properties and divided into nine classes (Table 26.2).

Lithium batteries, lithium cells, and devices equipped with them are listed in Class 9. In addition to general regulations, the listing under the respective UN number

Table 26.1 Work groups for carrier-specific systems of rules and standards for transportation

Carrier	Organization/agreement	System of rules and standards
Air traffic	International Civil Aviation Organization (ICAO)	ICAO Technical Instructions (TI)
	International Air Transport Organization (IATA)	IATA Dangerous Good Regulations (DGR) (IATA equivalent to ICAO TI)
Maritime traffic	International Maritime Organization (IMO)	International Maritime Dangerous Goods (IMDG)
Road traffic	(UN Economic Commission for Europe [UNECE])	Accord européen relatif au transport international des marchandises Dangereuses par Route (ADR)
Rail traffic	Intergovernmental Organization for International Carriage by Rail (OTIF)	Regulations concerning the International Carriage of Dangerous Goods by Rail (RID)
Inland navigation	(UNECE)	Accord européen relatif au transport international des marchandises Dangereuses par voies de Navigation interieures (ADN)

Table 26.2 Dangerous goods classification

Class	Substance class
1	Explosives
2	Gases
3	Flammable liquids
4	Flammable solids; substances liable to spontaneous combustion; substances which, on contact with water, emit flammable gases
5	Oxidizing substances and organic peroxides
6	Toxic and infectious substances
7	Radioactive material
8	Corrosive substances
9	Miscellaneous dangerous substances and articles, including environmentally hazardous substances

contains concrete information: name and description, class, allied perils, packaging group, hazard label/identification, special provisions, limited and exempted quantities, packing instructions, and several details concerning transportation.

26.1.4 Carrier-specific and national implementation

The UN Model Regulations constitute the foundation for concrete carrier-specific provisions.

The dangers resulting from transportation of substances and objects are weighted differently depending on the carrier. Accordingly, the provisions for different carriers vary in terms of content and structure.

In addition, the manner in which they are implemented nationally can vary. For example, the content and structure of the ADR are very similar to the UN Model Regulations, while many details of the American Code of Federal Regulation, Title 49, Transportation (49 CFR), are different in terms of structure and content. For specific issues, multilateral agreements exist in addition to country-specific provisions. Also, authorities might have their own interpretations. The UN Manual of Tests and Criteria describes in detail many tests for classifying substances and objects as well as the respective criteria to be used and tests for transportation equipment.

26.2 Transportation of lithium batteries and lithium cells

Commercial lithium battery transportation is generally subject to the regulations for the transportation of dangerous goods; there are some exemptions. Preparing transport and transportation must be performed exclusively by appropriately trained personnel and/or the process must be supervised by experts or qualified companies.

Apart from a few exemptions, transportation requires that the batteries or cells have been successfully tested in compliance with sub-section 38.3 of the UN Manual of Tests and Criteria. This should be taken into consideration as early as possible during cell and battery design as well as for the devices powered by the batteries, ensuring that the design complies with the test requirements from the very beginning. Cell and battery capacity limits should be taken into account already in the planning stage to enable exemptions.

The cells' or batteries' weight can decisively influence the transportation possibilities and package sizes, especially for air traffic. It also influences test parameters.

Unfortunately, many parameters vary greatly as they are often changed for revised tests or carrier-specific quantity thresholds and limits. Nevertheless, creative definitions are possible. A difference of a few grams only can determine whether a certain carrier is suitable for a certain transportation situation or not. Relatively small system alterations require the repetition of all tests. The upcoming version of the UN Manual of Tests and Criteria will apply to changes to both hardware and software, e.g., to battery monitoring and management systems.

Thus, dangerous goods regulations influence the design and production process a long time before the batteries reach the logistics chain. Knowledge about the transportation of dangerous goods and the required prerequisites is a must for every product designer and planner. The decision as to whether a product can ever be transported as wished is already made at the design and planning stage. Moreover, decisions made at this stage extensively influence the considerable testing costs. Fig. 26.1 shows the example of a decision tree for lithium-ion battery and

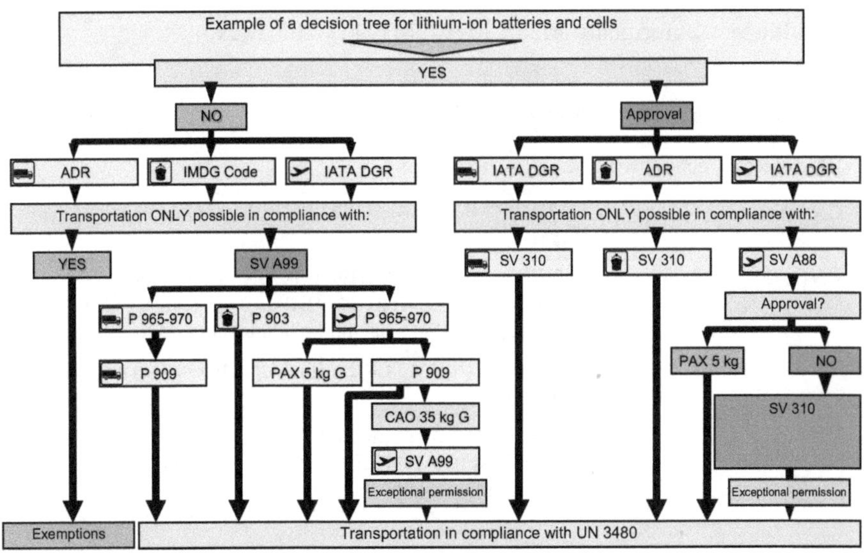

Fig. 26.1 Example of a decision tree for the transportation of lithium-ion batteries or cells (classified in accordance with carriers)

cell transportation with different carriers and different prerequisites. The necessary requirements are discussed in abbreviated form in the following chapters.

26.2.1 Classification and transportation provisions

This section will briefly present the relevant entries in the different provisions. The lithium battery cells of Class 9 and the devices powered by them are subject to the valid versions of the dangerous goods regulations and their exemptions (Table 26.3).

Although not explicitly specified, the term "lithium metal batteries" refers to non-rechargeable systems and the term "lithium-ion batteries" refers to rechargeable systems.

Road traffic and rail traffic The following is a list of the relevant special and packaging provisions as well as peculiarities regarding road and rail traffic. The full text can be found in the respective system of rules and standards:

- ADR, RID
  Special provisions: 188, 230, 310, 348, 360, 376, 377, 636
  Packing instructions: P903, P908, P909, LP903, LP904
- Exemptions in accordance with ADR sub-sections 1.1.3.6 and 1.1.3.7:
  Mitigations for transporting limited amounts per transport unit can be granted in compliance with general exemptions according to ADR sub-section 1.1.3.6. ADR sub-section 1.1.3.7 states that lithium batteries are exempted if they are installed in and power a vehicle or equipment that is used during transportation.
- Multilateral agreements: M285, M292, M294, M295, M296

Table 26.3 UN dangerous goods list (excerpt for lithium batteries)

UN dangerous goods list	Name and description
UN 3090	LITHIUM METAL BATTERIES (including lithium alloy batteries)
UN 3091	LITHIUM METAL BATTERIES CONTAINED IN EQUIPMENT or LITHIUM METAL BATTERIES PACKED WITH EQUIPMENT (including lithium alloy batteries)
UN 3480	LITHIUM ION BATTERIES (including lithium ion polymer batteries)
UN 3481	LITHIUM ION BATTERIES CONTAINED IN EQUIPMENT or LITHIUM ION BATTERIES PACKED WITH EQUIPMENT (including lithium alloy batteries)

Air traffic The following is a list of the relevant special and packaging provisions as well as peculiarities regarding air traffic:

- IATA DGR, ICAO T.I.
  Special provisions: A48, A88, A99, A154, A164, A181, A182, A183, A185, A201
  Packing instructions: P965, P966, P967, P968, P969, P970
  Peculiarities: State Variation US 2 & US 3

Maritime traffic The following special and packaging provisions apply to maritime traffic:

- IMDG Code:
  Special provisions: 188, 230, 310, 348, 360, 376, 377, 957
  Packing instructions: P903, P908, P909, LP903, LP904

26.2.2 UN Manual of Tests and Criteria

Chapter 25 already explained the relevance of tests for the work of product designers and planners. Apart from a few exceptions and in compliance with the dangerous goods regulations for lithium batteries, all new cells or battery types must pass all tests stipulated in the UN Manual of Tests and Criteria before they can be transported. This applies when several cells or batteries are connected to form new batteries or battery packs, in particular. Design measures and safety mechanisms are also prescribed.

In essence, today's tests comprise the following: simulation of elevation, changes in temperature, vibration, shock, external short circuits, impact, overcharging, and deep discharge. A large number of cells and batteries are tested with different states of charge; rechargeable batteries are tested after multiple charging and discharging cycles. Each battery component must be subjected to the tests. First of all, the cells must pass. Then, if applicable, sub-assemblies such as modules and finally the battery itself must be tested. An exception is made for batteries consisting of very large modules. Additional criteria must be met to qualify for exemption. Much stricter provisions for new qualifications must be fulfilled, and a quality management system with suitable quality assurance measures must be in place.

26.2.3 Transportation of untested lithium batteries and cells

Transportation of batteries and cells not tested in compliance with the UN Manual of Tests and Criteria is possible. This exemption applies when batteries are transported to be tested externally. For this reason, it only applies to small series of up to 100 cells and to pre-production prototypes. Depending on the carrier, additional measures are required to ensure adequate transportation safety.

Such measures can include packaging with higher safety standards. Special permits or even exceptional permissions from authorities are required in some cases. For the UN Model Regulations and the ADR, special provision 310 is important. Peculiarities concerning suitable protection and increased packaging requirements must be taken into account.

In aeronautics, special provision A88 is the equivalent of provision 310. It prescribes special permits from the authorities. The Federal Office of Civil Aeronautics (Luftfahrtbundesamt, LBA) is the responsible authority in Germany. Before such a permit is granted, the German Federal Institute for Materials Research and Testing (Bundesanstalt fuer Materialforschung und -pruefung, BAM) may have to carry out an expertise on the measures required to achieve comparable safety standards. If a special permit is not granted, transportation in compliance with ICAO TI Section 1.1.2 is possible for a good cause. Exceptional permission from all countries involved are required in this situation. Transportation to or within the USA in compliance with 49 CFR almost always requires a special permit issued by the relevant authority for all carriers.

26.2.4 Transportation of used lithium batteries and cells

Used batteries are also subject to the dangerous goods regulations. The provisions for new batteries can generally be applied to functioning and undamaged used batteries. Defective or damaged batteries are subject to stricter provisions, in some cases even to a total transportation ban.

Since January 2015, new provisions have applied to defective or damaged batteries, irrespective of the carrier. In compliance with ADR 2015, damaged batteries can be transported in compliance with SV 376 and batteries for recycling and disposal can be transported in compliance with SV 377. Total transportation bans apply to air traffic (ICAO TI IATA DGR: special provision A154).

In addition, the relevant special provision 636 and packaging provision 909 of the ADR apply to transportation of undamaged used batteries. Waste batteries and batteries for recycling and disposal are prohibited in air traffic in compliance with IATA DGR special provision A183. Exemptions must be granted by the relevant national authority of the dispatching country and the carrier's country.

Currently, an international work group is striving to develop the basic requirements for the transportation of defective batteries on UNECE level. There are also efforts toward drawing up UN provisions for the transportation of used lithium batteries from devices for recycling and disposal purposes based on the ADR. However, it will take several years for an applicable procedure to be developed.

26.2.5 UN entries relating to lithium batteries and cells

The following UN entries can be of importance especially for road traffic and, to a certain extent, for maritime traffic:

- UN 3166 VEHICLE, FLAMMABLE GAS POWERED or VEHICLE, FLAMMABLE LIQUID POWERED or VEHICLE, FUEL CELL, FLAMMABLE GAS POWERED or VEHICLE, FUEL CELL, FLAMMABLE LIQUID POWERED
- UN 3171 BATTERY-POWERED VEHICLE or BATTERY-POWERED EQUIPMENT

Here, road traffic is completely exempt from the ADR provisions.

Bibliography

1. Recommendations on the transport of dangerous goods Vol. I/II, 2011, Rev. 17, ISBN 78-92-1-13914-1
2. ADR 2011, 2010, ISBN 978-3-609-69338-5
3. Technical instructions for the safe transport of dangerous goods by air, 2011–2012 edition, ISBN 978-92-9231-605-1
4. ADR: http://www.bmvi.de/SharedDocs/DE/Artikel/G/Gefahrgut/gefahrgut-recht-vorschriften-strasse.html
5. IATA Guidance Document concerning lithium batteries: http://www.iata.org/publications/Pages/lithium-battery-guidelines.aspx
6. UN Manual of Tests and Criteria: http://www.unece.org/fileadmin/DAM/trans/danger/publi/manual/Rev.6/1520832_E_ST_SG_AC.10_11_Rev6_WEB_-With_corrections_from_Corr.1.pdf German translation provided by BAM: UN Handbuch Prüfungen und Kriterien www.bam.de/de/service/publikationen/publikationen_medien/handbuch_befoerderung_gefaehrlicher_gueter.pdf
7. Multilateral agreements: www.unece.org/trans/danger/multi/multi.html
8. Compendium Li-Ionen-Batterien: https://shop.vde.com/de/kompendium-li-ionen-batterien

Lithium-ion battery recycling 27

Frank Treffer

Contents

27.1 Introduction and overview

After a long hiatus, research on electric vehicles with battery energy storage systems was taken up again in the early 1980s. Developing low-cost batteries with a high power and energy density was and still is an important focus of such research. Lithium-ion batteries, which currently dominate in many consumer electronics applications such as laptops, represent a battery type that is able to achieve an acceptable range of up to 250 km in electric vehicles with a "still acceptable" battery weight of around 300 kg. For several years now, intensive research to further improve this technology has been promoted in both the private and the public sector.

This has thus led to greater awareness of electric mobility over recent years. Both large car manufacturers and energy suppliers have been working to develop transport concepts based on electrical and hybrid vehicles. An important prerequisite for success in this respect is to ensure medium-term and long-term availability of the special metals used in such batteries. Environmentally sound, low-cost battery recycling is the right approach to provide mass production in the automotive industry.

F. Treffer (✉)
Umicore AG & Co. KG, Rodenbacher Chaussee 4,
63457 Hanau-Wolfgang, Germany
e-mail: frank.treffer@eu.umicore.com

R. Korthauer (ed.), *Lithium-Ion Batteries: Basics and Applications*,
https://doi.org/10.1007/978-3-662-53071-9_27

A series of industrial processes already exist to recover special and precious metals from scrapped cars. Recycling precious metals from catalytic converters is a good example of such a process.

As early as 2003, Umicore established a battery recycling process that has now achieved recognition worldwide. This process primarily recovers cobalt, nickel, and copper from used lithium-ion, lithium-polymer, and nickel metal hydride batteries using an environmentally friendly approach. The lithium and aluminum contained in batteries enter the process slag and are forwarded to the industry together with other slag minerals. Recovery of Li is meanwhie available and part of the existing recycling scheme and these solutions have become part of the existing recycling processes.

One of the first commercial pilot battery recycling plants was built in Sweden in 2004. Its nominal throughput was 2,000 t of batteries per year. Today, there are industrial plants with an overall capacity of 7,000 t. Even large battery systems from hybrid and electric vehicles are recycled in this plant, where they are mechanically pretreated (disassembled) in special plants.

Recycling concepts have the potential to generate considerable profit: 10 million electric vehicle batteries weighing between 10 and 100 kg alone correspond to a future capacity of several hundred thousand tons per year, and that is just Europe. If recycling requirements from other applications are also taken into account, there is a need to start building up the capacities required in 20 to 30 years' time now.

Most high-performance batteries for hybrid and electric vehicles are currently still under development. They feature manifold designs and represent a network of components such as electronic systems, housings, and cooling devices. Due to this highly heterogeneous assembly, only suitable integral recycling concepts and modern technologies are currently able to extract recyclable materials, achieve effective process stability, and operate at a low cost.

Large quantities of these strategic metals are required for use as energy storage systems in hybrid and electric vehicles on a massive scale. This, in turn, creates a need for industrial scale recycling of cobalt, nickel, copper, lithium, and manganese ("economies of scale"). Secondary recycling of these materials must be further developed into an overall process chain to recover them as they are of great strategic value for Europe. This recycling process chain is crucial for introducing electric mobility onto the mass market and ensuring a medium-term and long-term supply of the aforementioned strategic metals.

27.2 Lithium-ion battery recycling

27.2.1 International state of the art

Recycling batteries from small devices has been possible for years. Pyrometallurgical and hydrometallurgical processes have become established as the main methods used. In this respect, it is important to distinguish between refining and actual recycling. Not all recycling companies perform the final recovery steps (refining) that ultimately recover valuable materials to the level of purity required for reuse.

Some recycling companies prefer a combination of pyrometallurgy and hydrometallurgy and use a special process control to achieve a high recovery rate for nickel, cobalt, and copper. This method can handle a wide variety of source materials; it can process all standard rechargeable NiMh and lithium-ion batteries, for example. This flexibility or unimportance of battery type is not without difficulties, but is essential for effective, economical battery recycling. The battery recycling process is also geared toward the high volumes expected in the future. Some companies have many years of experience in this respect and are in a position to increase capacity within a short period of time. Should the market situation make recovering lithium profitable, the necessary process concepts are already in place.

Availability and demand dictate which metal recycling services will be developed and offered. Such decisions thus depend on market value and, ultimately, on the cost and effort required for recovery. Most companies carefully monitor availability and demand of raw materials. This has ultimately promoted process development and manufacturing of noble earth concentrates from used NiMH batteries.

In contrast to portable application batteries, the much larger batteries in electric vehicles (EV) or hybrid vehicles ([P]HEV) require special preparation for reasons related to processing. Mechanical treatment is primarily required due to the batteries' dimensions and weight (50 to 450 kg). Each battery's particular mechanic and electrochemical end-of-life state requires special attention to eliminate any storage, transport, and handling hazards.

27.2.2 Lithium-ion battery recycling technologies

One of the world's most modern recycling plants in the Belgian city of Hoboken, Antwerp, (Fig. 27.1) handles a wide variety of secondary materials. Every year, more than 350,000 t of source material (such as catalytic converters, printed circuit boards, cell phones, intermediate industrial products and residues, slags, and flue

Fig. 27.1 Umicore in Hoboken, Belgium. The largest recycling plant of its kind worldwide, it handles more than 350,000 t per year of source material [1]

dust) are handled by a complex, yet efficient combination of different processes to recover valuable metals effectively.

Process control has been optimized to treat precious and special metals, providing short cycle times and high (precious) metal yields. Throughput is further increased by combined processing of a great variety of complex materials containing noble metals. Flexibility and sensitivity to impurities are also improving.

The copper contained in the slag binds the precious metals before being granulated immediately after tapping and then transferred to the subsequent electrolysis, where the precious metals are separated from the copper. In contrast, the primary slag is subjected to another blast furnace process, during which lead and other non-ferrous metals are separated and remaining noble metals recovered. The resulting side-stream material is also introduced into the process cycle and subsequently further processed. Aside from precious metals, many kinds of special metals can also be efficiently recycled to a high standard.

Emission limits are reliably met with the help of a highly efficient waste gas treatment system that completes the entire process. The enriched concentrates are then transferred to specific, hydrometallurgical processes to achieve high-purity elements. Years of experience show that combining pretreatment of end-of-life components, identifying and locating raw material components, the sampling process, pyrometallurgy and hydrometallurgy, and their integral optimization are important prerequisites for success in modern recycling processes.

As a result of improvements to recycling and refining processes in recent years, much expertise has been gained in logistics, metal management (trading of valuable metals), analytics, and sampling (sample recovery and analysis). This is now a significant basis for all kinds of future recycling of post-consumer uses. It has also ultimately produced the present-day battery recycling process, which is described below.

As part of a government-funded joint project on lithium-ion battery recycling (LiBRi [2]), the first plant for the pretreatment of industrial battery systems was developed, installed, and commissioned between 2009 and 2011 (Fig. 27.2), and was constructed in Hanau (Germany). This pretreatment process step perfectly complements battery recycling. It enables safe handling even of large battery systems (industrial batteries), especially those used in the automotive industry and electric mobility applications. It ensures that the services offered in industrial-size battery recycling plants are complete. This plant complies with all technical and legal requirements with regards to the European battery directive, not only for small device batteries, but also for larger batteries.

The EU directive requires a recycling quota of more than 50 %. The processes described above in combination with the pyrometallurgical and hydrometallurgical recycling processes discussed below consistently fulfill these requirements for lithium-ion and nickel metal hydride battery systems of different types and capacities.

During the pretreatment stage, the battery systems are first disassembled to cell or module level (Fig. 27.3) with the cells remaining unopened during this process.

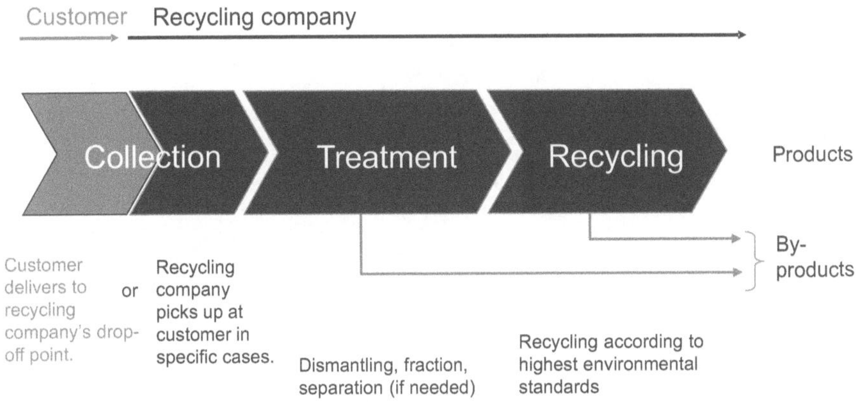

Fig. 27.2 Lithium-ion battery recycling process model with an integrated pretreatment process

Cells and modules represent the key material fractions that are then fed into the actual recycling plant. All other materials are separated and forwarded to enterprises in the metal and metallurgical industry in their purest possible state.

Small device batteries, especially lithium-ion batteries from cell phones, laptops, MP3 players, power tools, and e-bikes, are transferred directly to the recycling process step, i.e., without pretreatment (Fig. 27.4). This approach enables the operator to handle an extensive variety of battery types in a single process. This is why the process is highly efficient, especially when it nearly reaches its maximum workload (currently 7,000 tons per year).

Battery materials are collected into melting batches based on element analyses for the pyrometallurgical process steps. Pretreating feedstock in this way helps to

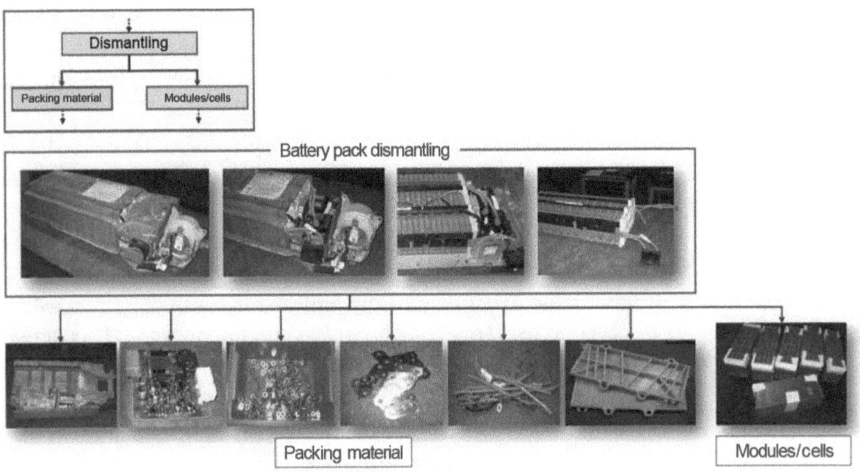

Fig. 27.3 Pretreatment/disassembly of battery systems into individual material fractions, such as metals, plastics, electronic components, composites, and battery cells or modules

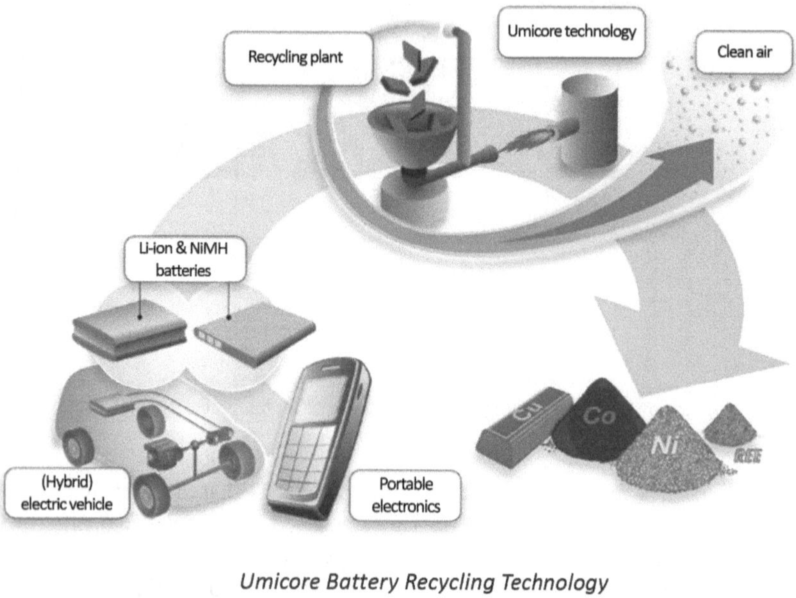

Fig. 27.4 Recycling process diagram for rechargeable end-of-life batteries

manage the process and increase the recycling quota of preferred elements. The pyrometallurgical process results in three initial material flows: alloys, slag, and flue dust.

The feedstock's metallurgic characteristics are used to concentrate the valuable materials in the slag. The fusible metal is perfectly conditioned for hydrometallurgical preprocessing using chemical separating processes (solvent extraction). This is where the individual elements, especially Co and Ni, are recovered in their purest form.

The easily oxidizable components are extracted from the slag. The slag itself is currently used as a raw material (mineral aggregate) in the ready-mix concrete industry. It is possible to recover oxidic lithium contained in the slag, as demonstrated by the work at Clausthal University of Technology as part of the LiBRi project [2].

The third material flow is the exhaust gas flow with the flue ash. The flue ash is transported in a closed system to a special exhaust gas treatment plant. This active treatment plant completely destroys all VOCs (volatile organic compounds) and captures other hazardous substances, which account for only 2 to 3 % of the raw material. It is also possible to treat flue ash using hydrometallurgical processes (LiBRi project [2]). However, this is not cost-effective due to the high expense.

The metals recovered through battery recycling, e.g., cobalt and nickel, can be converted into (pre-)products for battery materials, what are known as precursors.

Fig. 27.5 Closed-loop approach [2]

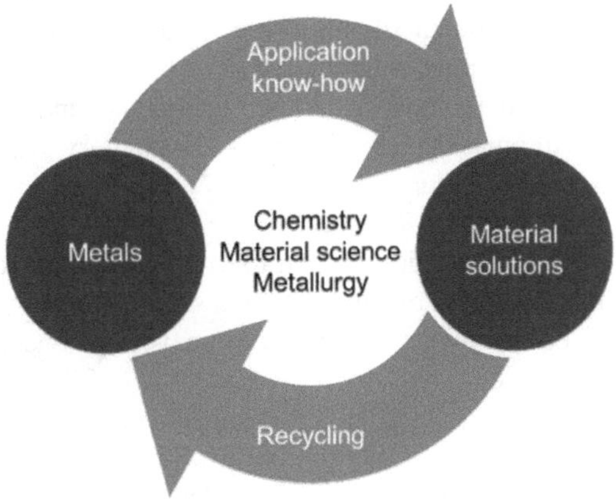

The precursors, in turn, are used to manufacture items such as new cathode materials for batteries. This closed-loop approach is deeply rooted in most companies' corporate philosophy (Fig. 27.5).

Besides this process chain, efforts must be made to incorporate recycling into early stages of product development in the future. This is why the sectors involved must work together more effectively.

27.3 Outlook

The respective recycling technology and the objectives of the individual companies strongly determine further development in battery recycling processes. As a general rule, current recycling technologies are well-established, validated, and successfully under control on a technical level. The processes differ with respect to pretreatment (concentration), exhaustive material preprocessing, focus on materials, environmental impact, and energy demand.

Other efforts, especially regarding battery recycling, are being made to resolve still unanswered logistic and safety-relevant problems, particularly handling the expected high material throughput. This includes solutions for efficient collection of used lithium-ion batteries.

Thus, the next steps in improving battery recycling are [2]:

- Studies on safe used-battery transportation, especially damaged systems and large quantities
- Development and implementation of a logistics chain for used batteries in large quantities, instead of cost-intensive individual solutions
- Extended business models which take into account recycling and potential reuse

- Research on used lithium-ion batteries (e.g., recording of life cycles and aging characteristics) to improve batteries and battery systems
- Development of removable cell and module connections to implement repair and disassembly-friendly design
- Rapid diagnostics for used batteries (standard)
- Automated pretreatment of xEV battery systems and preparing plants and processes for large quantities

Questions concerning centralized or decentralized solutions still remain unanswered, not only in the battery recycling sector. Because of currently low volumes, it can be stated that a centralized battery recycling concept has advantages over a decentralized system.

Centralized battery recycling concepts mean focusing metallurgic processing in one location or one plant worldwide. The goal would be to improve capacity utilization and increase efficiency. This is in contrast to a decentralized infrastructure, where several plants are built in strategic locations worldwide.

The following lists only the most important factors:

Advantages of a centralized battery recycling concept:

- Reduction in the overall energy demand, because 100 % capacity utilization cannot be achieved with current volumes → Excess capacities
- Overall reduction of process-related CO_2 emissions

Advantages of a decentralized battery recycling concept:

- Unaffected by export restrictions
- Less transportation and therefore less associated CO_2 emissions

At this stage, it is possible to provide customers with the best recycling service (on-site service) using an upstream network of rechargeable battery drop-off points.

Bibliography

1. Hagelücken C, Treffer F (2011) Beitrag des Recyclings zur Versorgungssicherheit – Technische Möglichkeiten, Herausforderungen und Grenzen, EuroForum-Konferenz in Stuttgart, Mai
2. Projektabschlussbericht Verbundprojekt "Entwicklung eines realisierbaren Recyclingkonzeptes für die Hochleistungsbatterien zukünftiger Elektrofahrzeuge" – Lithium-Ionen Batterierecycling Initiative – LiBRi

Vocational education and training of skilled personnel for battery system manufacturing

28

Karlheinz Mueller

Contents

28.1 Introduction

Modern qualification concepts enable battery manufacturers to react dynamically to technical challenges and to familiarize new staff with the latest company processes early on. Also, they can qualify their skilled personnel to work with new technologies and changed processes and tasks by implementing company-specific vocational training and continuing education, which is integrated into the company's processes and adapted to the personnel's individual talents and interests.

K. Mueller (✉)
Berufsbildungsausschuss, ZVEI – Zentralverband Elektrotechnik- und Elektronikindustrie e. V., EABB Consulting, Merckstrasse 7, 64342 Seeheim-Jugenheim, Germany
e-mail: mueller.zwingenberg@t-online.de

R. Korthauer (ed.), *Lithium-Ion Batteries: Basics and Applications*,
https://doi.org/10.1007/978-3-662-53071-9_28

Both vocational education and training largely involve providing staff with a qualification that is rigorously geared to the value added chain and its processes and integration. As a result, the staff share a comprehensive understanding of the company's processes, thus promoting communication and cooperation in intelligent production by all parties involved. Vocational education already lays the foundation for continuing education, enabling organizations to develop the competences of their skilled personnel together with product and process innovations [6].

The following chapter focuses on Germany's vocational training and continuing education system.

28.2 Qualified staff – versatile production

Electric mobility is Germany's opportunity and challenge to further enhance the country's top position as an industrial, economic, scientific, and technological location.

The German government and the economy have a common goal: "By the year 2020, Germany will be the lead supplier and key market for electric mobility" – one of the most ambitious technological transformation processes of the upcoming decades. This development will give rise to intersectoral cooperation, new value added chains, and revised business processes and operational procedures. It is only possible to master these tasks if all sectors have the personnel required to support and design the change [7].

Therefore, all companies must look ahead and verify whether they will have the necessary personnel with the know-how at their disposal at the right time. And this is exactly the problem; on the one hand, the sectors require more skilled personnel and demand better qualifications. On the other hand, the workforce is aging and new personnel is becoming scarce. Skills shortage and the continuous demographic decline in manpower can quickly pose an existential challenge for companies.

28.3 Innovative recruitment of new employees and skilled workers in the metal-working and electrical industry

When circumstances change, companies need new human resources development strategies for professional vocational education and training to ensure future competitiveness. In recent years, the metal-working and electrical industry has developed and implemented modern, forward-looking job descriptions that fulfill the quality requirements for this dynamic industrial sector. These new occupations have broad qualification profiles based on an integral professional understanding adapted to business processes and customer relations. The corresponding process-oriented job descriptions have advantages in areas where dynamic change, manifold innovations, or complex problems make regular work days challenging, e.g., in the electric mobility sector's fields of activity.

The vocational education regulations are open to interpretation; the training organizations can now recruit new employees very flexibly. Vocational education takes place on the job, with hands-on tasks. In contrast to the past, the examination requires apprentices to execute a company project or a complex work task. This proves that examinees are capable of making professional decisions in an individual operational context [4].

A process-oriented continuing education system has been set up, based on the respective professional training, to perfectly qualify skilled workers: Young people completing a vocational education in the electrical engineering sector have several possibilities. They can continue training to become Systems, Production, Assembly, or Service engineers and then acquire the vocational training diploma of the German Chamber of Industry and Commerce (Industrie- und Handelskammer, IHK) as a certified Process manager – Electrical engineering (Operative professional) (Fig. 28.1) [8].

This continuing education gives rise to skilled workers who are capable of mastering the technological and organizational challenges of innovative and dynamic technology sectors such as electric mobility. Certificates of this kind open doors to company departments that have always been restricted to technicians and engineers. The production technology sector offers comparable vocational education and training and continuing education, which is described in detail below [10].

28.3.1 Job descriptions in electric mobility

The qualification requirements for the electric mobility field of activity "Automotive engineering (eCar)" including battery systems are covered by the following vocational education:

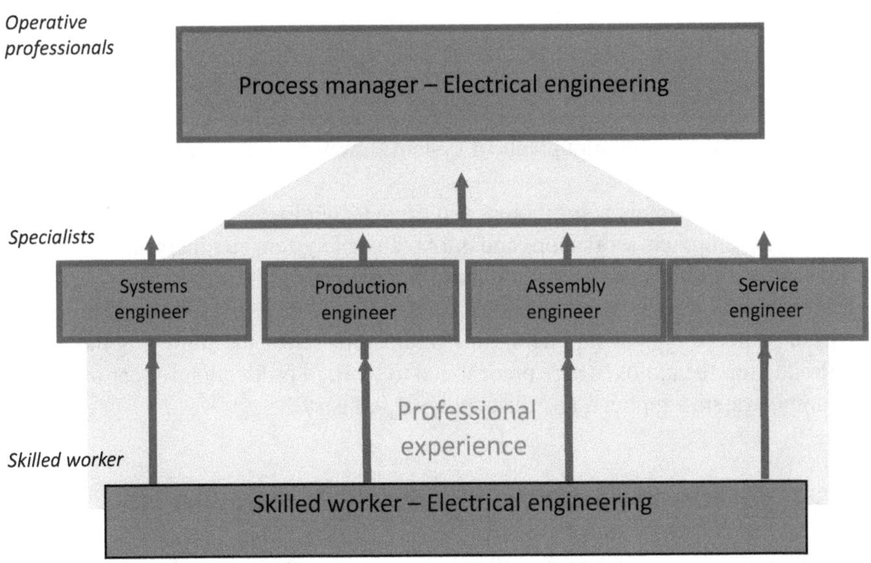

Fig. 28.1 Vocational training and continuing education in the electrical engineering sector

- Electronics technicians for devices and systems have comprehensive systems competency in electronics. They have detailed knowledge about automotive hardware and software components, both with regard to their technical design and systemic functionalities and including the related sensor technology and actuators.
- Electronics technicians for information and systems technology have comprehensive know-how of electrical engineering and software. They focus on linking hardware and integrated software components, so-called embedded systems.
  This includes developing hardware-oriented software, programming interfaces, and integrating them into the respective vehicle systems.
- Electronics technicians for motors and drive technology (industry and trade) are specialists in electric motors and their control and regulation systems. They know the different motor types, their coil parameters, and their operating behavior.
- Mechatronics fitters have systems competency in linking different mechanical, electrical, and electronic system components, and they understand their individual functions and how they interact in the overall system.
- Production technologists have comprehensive process skills. They are responsible for workflow stability, the quality of products, and flexible and efficient production processes.
- Electronics technicians for automation technology set up and operate automated manufacturing plants.

The duration of vocational education for all professions is 3 or 3½ years and the training is held in companies and vocational schools [2]. Their qualification focus predestines these professions for the following training organizations and fields of application:

- Electronics technicians for devices and systems as well as electronics technicians for information and systems technology are predestined to work in the development workshops and pilot plants of system suppliers, car manufacturers, and the supplying industry.
- Electronics technicians for motors and drive technology are predestined to work in the development workshops and pilot plants of system suppliers and car manufacturers and for mass-producing manufacturing plants.
- Mechatronics fitters are predestined to work in the development workshops and pilot plants of system suppliers, car manufacturers, and the supplying industry.
- Production technologists are predestined to work in production for car manufacturers, system suppliers, and the supplying industry.

28.3.2 Vocational education in battery system production

The following describes the areas of operation and the relevant professional qualifications of vocational education that are connected to battery production.

Production technologist Area of operation: Production technologists prepare production orders, create product samples and prototypes, test production facilities, and operate test facilities.

They commission and set up machinery and use software to simulate, control, and monitor processes. They operate, monitor, optimize, and ensure smooth production. Product technologists are in demand in "Automotive engineering (eCar)" to work on production lines for new cars or car parts, e.g., electric motors and batteries.

Their relevant professional qualifications are:

- setting up production facilities, starting new processes
- organizing logistic processes
- operating, optimizing, and monitoring production facilities
- securing quality standards and process flows

Electronics technician for devices and systems Area of operation: Electronics technicians for devices and systems manufacture components and build devices and systems. They commission systems and devices and maintain them. They build, for example, battery systems, electric motors, traction control systems, and inverters in the "Automotive engineering (eCar)" sector.

Their relevant professional qualifications are:

- designing circuitry and building prototypes
- integrating electronic modules and components
- installing and configuring software
- analyzing and testing technical functions
- testing and repairing systems and devices

Systems information technology specialist Area of operation: Systems information technology specialists develop, implement, and maintain industrial IT systems. For example, they implement software components, configure assemblies, and program embedded traction control systems and systems for battery management, safety, diagnostics, and driver aid systems in the "System services" area.

Their relevant professional qualifications are:

- implementing and testing IT components
- installing and configuring operating systems and networks
- creating software components, integrating interfaces
- integrating and testing system components
- supporting troubleshooting

28.3.3 High-tech qualifications for battery system production

Quality requirements with regard to electrode production, cell assembly, forming, and battery assembly are mainly geared toward production technology.

In contrast, the quality requirements with regard to battery system integration, prototype construction, and testing are related to electrical engineering, electronics, and IT. In addition, forming and battery assembly require electrical engineering know-how in the areas of inspection and testing as well as in handling high-voltage components (Fig. 28.2).

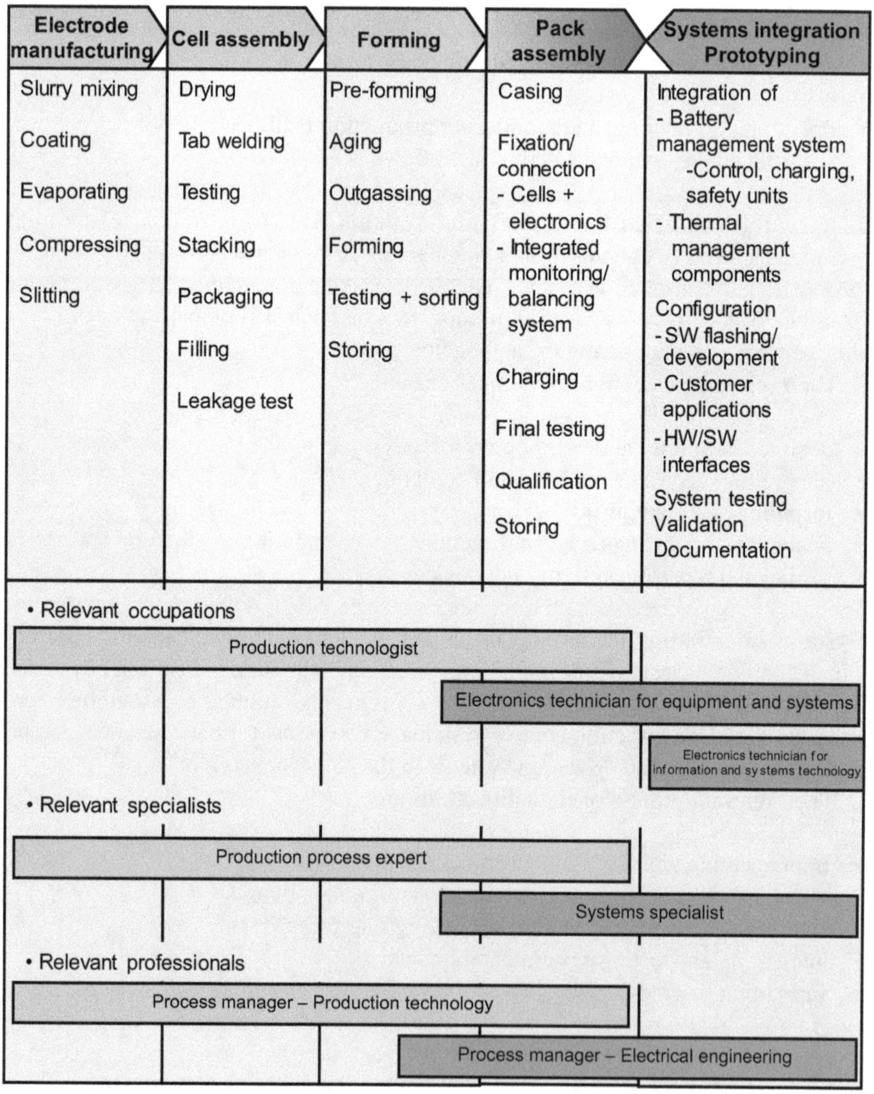

Fig. 28.2 Skilled personnel for battery system production

28.4 Integrated production technology qualification concept

The most important sector of the German economy is the production sector. More than two-thirds of the country's economic performance is connected directly or indirectly with the production sector. The keywords flexibility, adaptability, modularity, and system integration may not be new. However, in the field of electric mobility, they have a new meaning and much greater importance. Production technology focuses on production facilities that can flexibly implement different technologies and produce varying lot sizes in order to economically manufacture important components and systems. Organizing production involves managing changes in value added chains, business processes, and operational procedures through cooperation between the car manufacturers and their suppliers. Both issues are especially relevant for battery production [11].

Production technology is a good example of how pressure to innovate can emerge in an economic sector that has been very successful to date. The existing professions no longer meet the needs of the sector; therefore, solutions that systematically link the contents of vocational and continuing education are required. To achieve this, an integrated approach has given rise to new vocational training regulations for the profession "Production technologist" as well as new and custom-fit training examination regulations for the profession "Process manager – Production technology" [3].

28.4.1 Production technologist

Production technologists are a new breed of skilled workers with extensive decision-making capabilities with regard to standard production processes and innovative production technologies. Their potential is based on special professional characteristics and operational socialization that stems directly from the working processes and the related tasks [1].

Field of work and fields of activity Production technologists are employed in the workflow between development and production, in pilot and zero series, and in production start-up. They are competent networkers and cooperate with product and process developers, suppliers, and manufacturers. In battery production, this applies especially to the production process and production facility development.

When humans and process technology interact successfully, stable, flexible production systems can be achieved which are less susceptible to failure. Skilled personnel must have a high degree of expertise, as they have to communicate directly with the machinery during the production process to ensure high mass-production standards.

The start-up phase is of immense importance in this respect: It is the first time that newly developed or modified products and production technologies in a specific production system are combined with internal and external logistics processes as well as with IT tools, and everything is expected to operate flawlessly. Combining and optimizing components is the specialty of production technologists. For them, the start-up phase is concluded when the production process is "stable", i.e., when the intended lot size of the product can be produced with the planned cycle time, quality, and without exceeding the budgeted costs. The duration of the start-up phase is key in determining whether the product comes onto the market on time and whether it will be an economic success.

Qualification and task This profession requires three years of vocational education comprising content that is process-oriented and geared toward production technology and IT.

Production technologists set up production machines, test product samples and prototypes, determine process parameters, and program control software. They use up-to-date material and novel production processes. They team up with foremen, technicians, and engineers to analyze and simulate processes and to try out and optimize products and processes; they cooperate with suppliers and customers. They prepare production and assembly, set up the flow of materials, simulate production start-up, regulate and control processes, and monitor product quality (Fig. 28.3).

While the qualification of mechatronics fitters focuses on systems competency, the qualification of production technologists is characterized by process competency, integrating production technology, production organization, and IT skills. They do not have a special "role" but work as an "equal among equals", contributing their process competency, a key component for intelligent production, to the skilled worker teams.

This considerably increases all teams' competencies, making them more effective. Vocational education for production technologists is an attractive opportunity; companies offering such positions can win over motivated and well-suited youths for their professional training programs.

28.4.2 Continuing education in production technology

Process-integrated recruitment of skilled workers is supported by a two-tiered qualification concept for vocational training and continuing education following vocational education.

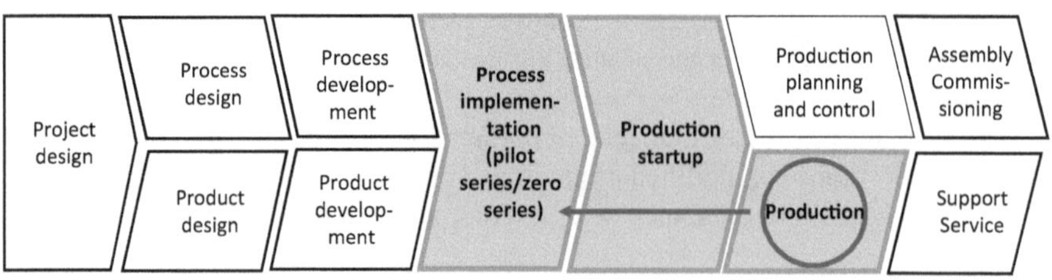

Fig. 28.3 Fields of activity – Production technologist

The first step of professional continuing education and job specialization consists of the "Process specialist" and "Application specialist" profiles. The second step of the qualification concept regulates the "Process manager – Production technology" certificate, a high-quality certificate as "Operative professional" which is a qualification that is equivalent to a bachelor's degree.

The qualification opportunities and measures can be geared toward improving innovation capabilities and fostering the learning of each individual on the job. The key concepts of this new and integrated qualification concept are therefore to establish qualification structures that are tailored to the work processes and can be flexibly implemented, and to ensure permeability of professional development [9].

Continuing education for process specialists and application specialists These specialized continuing education profiles connect professional vocational education with professional vocational training of the operative professionals. To qualify for this specialist profession, individuals must qualify in the work processes described in the regulations, which are considered the benchmark and determine the qualification standards.

Process specialists analyze process requirements, find technical solutions, calculate costs and lot sizes, estimate process times, create process descriptions, help design production plants, and plan and control production, find solutions for equipment availability improvements, and optimize processes (Fig. 28.4).

Application specialists handle customer requests, clarify technical requirements, costs, and deadlines, help develop solutions for customers, translate customer orders into design, production, or delivery orders, monitor goods and services and deadlines, create agreed-upon reference products, handle complaints, change requests, and warranty claims, and support customers (Fig. 28.5).

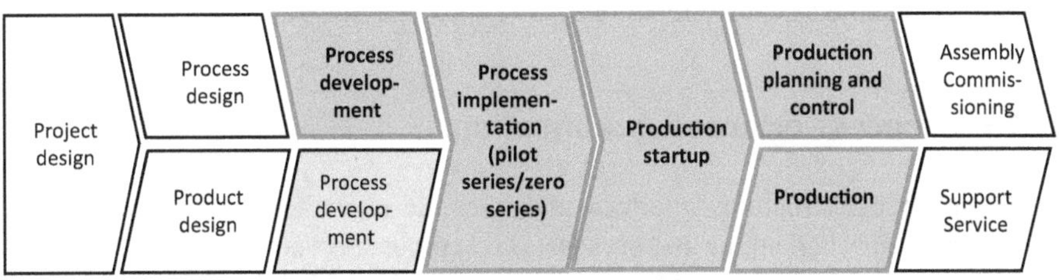

Fig. 28.4 Fields of activity – Process specialist

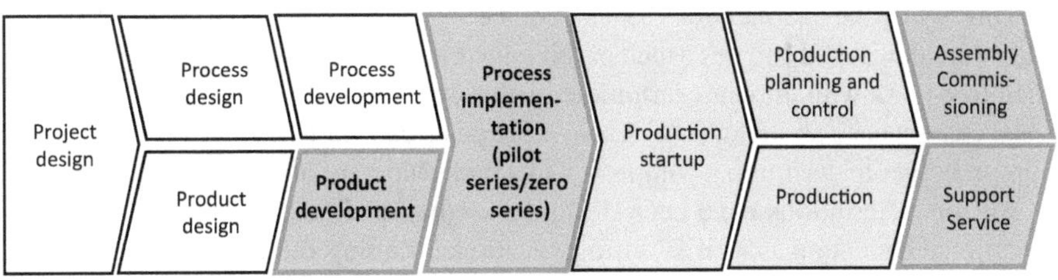

Fig. 28.5 Fields of activity – Application specialist

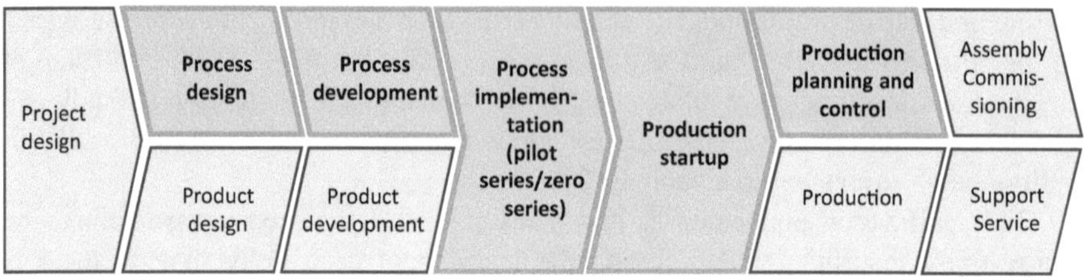

Fig. 28.6 Fields of activity – Process manager, Production technology

Process manager – Production technology (Operative professional) Operative professionals head organizational units and carry out technical engineering tasks. They are qualified to solve complex problems in specialized work areas, have decision-making responsibility, and promote their teams' professional development.

Process managers (production technology) manage production processes and implement the related innovation and improvement schemes, and they manage complex production projects. In this respect, they work in product and process design, process development and implementation, production start-up, or production planning and control (Fig. 28.6).

They require specific qualifications to take the exam, which can be acquired in the production technologist vocational education program and the subsequent qualifications to specialize as a process specialist or application specialist or they are considered to be acquired based on their pertinent professional experience. They are examined by an examination committee of the German Chamber of Industry and Commerce. The examination's objective is to prove that the examinee is able to plan, design, optimize, implement, and take on responsibility for production processes and handle executive functions.

28.5 Process-oriented qualification

Modern process structures interlock and integrate many activities. The tasks are becoming more challenging and broader; single procedures are combined and cross-linked, and decision-making responsibilities are delegated to the skilled employee level. This requires competent employees on all hierarchical levels who have a comprehensive and shared understanding of the processes.

This is why the qualifications for each of the vocational education and training profiles were developed and described based on work and business processes. The teams therefore work with the same terminology, descriptions, definitions, diagrams, drawings, etc. On each skilled worker level, individuals can learn to design the respective sub-processes and acquire a comprehensive understanding of the overall process.

All job descriptions have been designed so that qualification is possible during vocational education as well as during vocational training and continuing education in real-life work processes and projects. Their content is adapted to the specific company requirements and application areas. Qualification therefore is highly transferable and promotes extensive professional decision-making capabilities.

This on-the-job training does not happen without effort; it requires reflection on the work to draw the correct conclusions. By developing an awareness of what they have learned, employees move to a higher level of abstraction and transfer newly acquired knowledge to other, new situations. This practical on-the-job learning is more effective than learning in a seminar. It is based on the skilled personnel's experience; it answers questions; motivation and transferability come almost automatically. Even older employees learn more effectively in this way and can also contribute their experience [5].

28.6 On-the-job learning

Electric mobility has very dynamic transformation processes. It is those processes, which especially require future-oriented skilled personnel qualifications.

To this end, the qualification processes must be defined and interlinked with the new work processes to the greatest extent possible to ensure vocational training and continuing education efficiency and sustainability. The workforce must be able to handle the technology and the modified processes. Actually this goes without saying, but is often not connected.

Company management and executives must face up to the issue of employee qualification and create the necessary framework, especially in the context of technological and organizational transformation. Designing a framework that promotes learning does not require new "responsibilities". The executives are merely doing their job, in their original field of responsibility, and are responsible for promoting qualification. This is of paramount importance for the employees' decision-making capabilities and, ultimately, for the performance of the executives' area of responsibility. In this way, they can directly promote the employees' professional development and, at the same time, meet current requirements. The executives can also ensure that learning becomes an integral part of the work processes and that the learners contribute to their qualification themselves.

This new approach to qualification enables companies to find an efficient type of employee qualification and is an outstanding personnel development tool in the context of demographic change. Career changers and skilled personnel have the opportunity to further their competencies on the job and therefore have manifold employment opportunities in the innovative fields of electric mobility [12].

Bibliography

1. BiBB (ed.) (2008) Aus- und Fortbildung in der Produktionstechnologie. Bonn: BiBB– ARGE PT
2. BMBF (ed.) (2011) Ausbildungsberufe für die Elektromobilität. Broschüre/CD-ROM. Bundesministerium für Bildung und Forschung, Bonn

3. Borch H, Zinke G (2008) Aus- und Fortbildung aus einem Guss. Berufsbildung in der Produktionstechnologie. In: BiBB – BWP 4/2008, pp. 43–47
4. Diegner B (2010) Industrielle Elektroberufe: Einsatzgebiete für die Nachwuchssicherung und Fachkräfteentwicklung im Bereich der Elektromobilität. In: Korthauer (ed.) Handbuch Elektromobilität. EW Medien und Kongresse GmbH, Frankfurt/M
5. FhG – ISST (ed.) (2007) Weiterbildung mit System. Lernen im Prozess der Arbeit (APO IT). W. Bertelsmann Verlag, Bielefeld
6. GGEMO (ed.) (2010) Zwischenbericht der Nationalen Plattform Elektromobilität – AG 6 Ausbildung und Qualifizierung. BMWi, Berlin
7. GGEMO (ed.) (2011) Zweiter Bericht der Nationalen Plattform Elektromobilität. BMWi, Berlin
8. DIHK (ed.) (2011) Rahmenplan – Geprüfte Prozessmanager Elektrotechnik. DIHK Verlag, Meckenheim.
9. Müller K (2009) Enge Verzahnung von Aus- und Weiterbildung in der Produktionstechnologie. In: Loebe/Severing (ed.) Zukunftssicher durch flexible Ausbildungszeiten? Neue Metall- und Elektroberufe in der Diskussion. W. Bertelsmann Verlag, Bielefeld
10. Müller K, Schenk H (2011) Berufliche Karrieren mit System – Fort- und Weiterbildung in der Elektrotechnik. In: BIBB-BWP 1/2011, pp. 36–40
11. VDMA (ed.) (2007) Aus- und Weiterbildung: Produktionstechnologe/in. Fachkräfte für die intelligente Produktion. VDMA, Frankfurt am Main
12. ZVEI (ed.) (2009) Neue Weiterbildungsmöglichkeiten für die Elektroindustrie. In: ZVEI Mitteilungen 17/2009, pp. 14–15

Standards for the safety and performance of lithium-ion batteries

29

Hermann von Schoenau and Kerstin Sann-Ferro

Contents

29.1 Introduction

Today, standards are the foundation of almost all technical developments. They are the basis for companies that want to penetrate national and international markets and they also provide legal certainty. In our society, standards ensure trust and safety, provide a kind of legal certainty and bundle developers' resources in order to create safe and manageable products in a goal-oriented manner.

H. von Schoenau (✉)
Schoenau-Consulting, Hauptstrasse 1 a (Schlosshof),
79739 Schwoerstadt, Germany
e-mail: hermann.schoenau@t-online.de

K. Sann-Ferro (Dr.)
DKE, Stresemannallee 15, 60596 Frankfurt,
Germany
e-mail: kerstin.sann@vde.com

R. Korthauer (ed.), *Lithium-Ion Batteries: Basics and Applications*,
https://doi.org/10.1007/978-3-662-53071-9_29

As a result of the policy of moving away from fossil and nuclear fuels and the fast-paced technical development of lithium-ion batteries, the standards governing their use in different areas of application are mandatory in order to inspire confidence among consumers and provide legal certainty for manufacturers.

The rapid development of lithium technology – from cell phone batteries to large-scale energy storage systems – has led to standardization efforts worldwide. The intensive use of these batteries has made the development of safety standards inevitable. In comparison to previously widely-used lead-based systems, the energy density of lithium batteries is much higher; batteries are therefore smaller and can be implemented in completely novel applications. This is due to the lithium battery electrolyte's unique kinetic stability and the usage of intercalation materials.

Batteries are among the oldest products in electrical engineering and have thus been subject to a lengthy standardization process. Most likely, lithium-ion batteries will go through a standardization process similar to that of lead-acid batteries. Many standards are available for different applications. The most important standards are: dimensional and material standards, performance standards, and safety standards.

29.2 Standards organizations

The standardization committees of DIN and DKE (Deutsche Kommission für Elektrotechnik Elektronik Informationstechnik, German Commission for Electrical, Electronic & Information Technologies) cooperate on three levels: national level (DIN Standards, DIN SPEC and VDE AR), international level (resulting in international ISO and IEC standards and specifications), and European level (resulting in European CEN, CENELEC (CLC), and ETSI standards and specifications) (Fig. 29.1). It is in the interest of an industrial nation such as Germany to contribute to standardization efforts in order to facilitate export and import, remove trade restrictions, and ensure the international division of labor. The increasing economic integration of neighboring countries and country groups renders consistent and interoperable standards with a greater depth of standardization more and more important.

The members of ISO and IEC (international level) and CEN and CLC (European level) are the national committees, which are mostly the national standardization organizations. They appoint experts and delegates from industry, government bodies, associations, and academia to participate in the technical work.

ETSI, however, is a membership organization, which means that any company or organization with an interest in the creation of telecommunications and related standards can become a member of ETSI.

The DIN is the German member organization in ISO and CEN, whereas the DKE is the German member organization within the electrotechnical standardization organizations IEC and CENELEC.

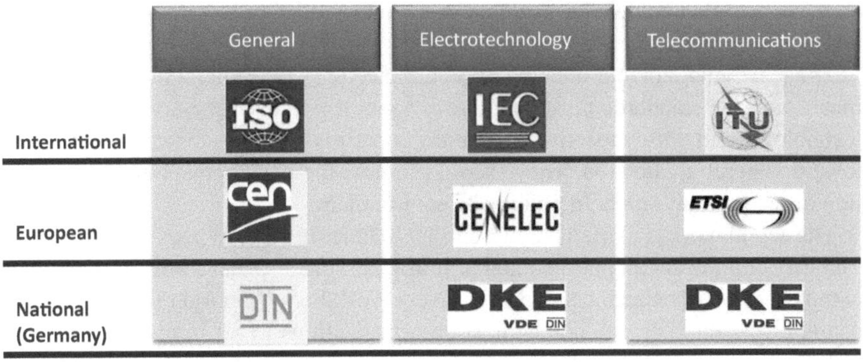

Fig. 29.1 Standardization organizations

The council board is the highest policy-level body. In the next hierarchical layer, the steering committees administrate the organization and manage all activities. The central secretariat organizes all general committees.

The task of developing the international and European standards is performed in a decentralized manner by technical committees (TC). The scopes of these committees are examined and authorized by the respective coordinating body.

Sub-areas of operation are distributed to subcommittees (SC), especially in ISO and IEC. All members are welcome to participate in the TCs and SCs by their experts. Official member delegations are sent to the meetings of the technical committees and subcommittees.

The national organizations work in so-called "mirror committees" in order to harmonize their efforts on a national level. Representatives are appointed in the national bodies to pursue national interests at international and European levels. The basic principle underlying participation in the national bodies of DIN or DKE is to involve all organizations with a vested interest in developing standards such as companies, commerce, trade, testing institutes, and authorities. They can authorize technical experts to cooperate in national bodies. The expert groups must be appropriately represented in the standardization bodies. This is especially important for developing safety standards. Furthermore, the results must be presented to the public. The public's response is part of standardization and must be given appropriate consideration.

29.3 Standardization process

The history of the development of technical standardization is characterized by different phases that have had varying importance over time. These are plant standardization, national standardization, and supranational (international and regional) standardization. International standardization has become more important with increasing global trade. For a global market to work, trade restrictions must be removed or prevented.

The standards' content must be in accordance with the general public's requirements, and the standards must take into account the state-of-the-art in science and technology and the respective economic conditions. They contain rules geared toward general application. They should promote the development and humanization of technology, and help protect the environment.

The actual work is carried out in a similar manner in all organizations (Fig. 29.2). The first step toward an international or European standard that a member, technical committee, or organization submits is a New Work Item Proposal (NWIP). Anyone can apply to the national standardization organizations to set in motion standardization efforts. If a proposal is accepted, the next step involves extending the work program of an existing technical committee or founding a new technical committee. Often, work groups work on the first drafts, resulting in a committee draft (CD) on a higher committee level. The CDV is then distributed by the central secretariat to the members as an international or European standard draft to put to the vote. If a sufficient number of members accept the draft, the final draft (FDIS) is finalized and passed in order to be publicized as international or European standard. The intended standard version must be presented as a draft to the public on national level.

The detailed standard text is usually finalized by work groups (WG) before being passed on to TC or SC level to receive an official opinion. The member countries appoint experts to the work groups. The standardization organizations ensure that the standards are the result of a voluntary consensus based on public participation

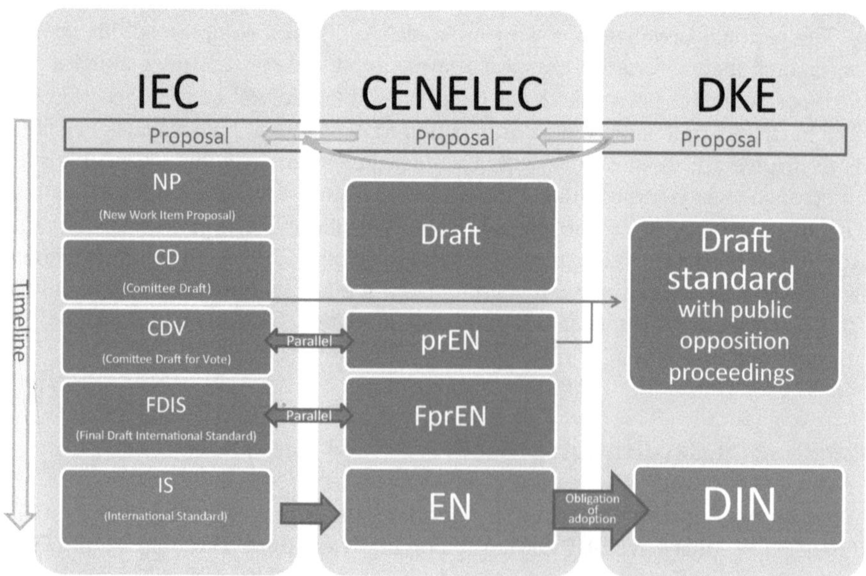

Fig. 29.2 Standard development steps on different levels in electrotechnical standardization

and involving entities with a vested interest such as companies, commerce, universities, consumers, trade, testing institutes, and authorities.

All standards are subject to a so-called maintenance cycle. They are regularly checked for updates that may be required or reasonable. Furthermore, standards may be revised if necessitated by technical innovation.

29.4 Battery standards application

Standards for batteries essentially distinguish between primary and secondary batteries. The information provided here is limited to secondary batteries because recharging is a key characteristic for novel areas of application.

Furthermore, battery standards depend on the application. These are mainly portable, stationary, and industrial applications as well as road vehicle and traction applications.

The next step is classification according to the battery's chemistry. There are batteries with acidic electrolyte versus non-acidic electrolyte and battery systems based on lead-acid versus nickel-cadmium and lithium-ion. Lead-acid and nickel-cadmium are usually considered to be a group in terms of safety requirements and kept separate from lithium batteries, which are characterized by different safety aspects.

There are regulatory requirements for batteries which are more important than standards; compliance with standards is voluntary. The regulatory requirements are:

All lithium-ion batteries are subject to the UN Recommendations on the Transport of Dangerous Goods. All tests described there in chapter 38.3 are mandatory from cell level upwards, with only a very few exceptions. They ensure that batteries are safely transported. The German battery law translates EU Directive 2006/66/EC into national law and also applies.

29.4.1 Standards for electric mobility

The German Federal Government intends to have one million electric vehicles on the roads by 2020. In recent years, this has spurred the development of lithium-based drive batteries. However, the necessary standards have not been developed as quickly, and many of the existing standards have only recently taken into account current developments. This situation is even more complex due to the fact that there are different concepts for electric mobility's infrastructure (Fig. 29.3).

Drive batteries have one major advantage: Automotive manufacturers – who are naturally avidly interested in selling safe products – already meet large areas of the safety requirements with the aforementioned industry standards. Standardization is very difficult due to the technical complexity, especially when an entire battery system is considered, including its thermal and electronic management systems. Table 29.1 shows several SAE standards, some of them quite recent.

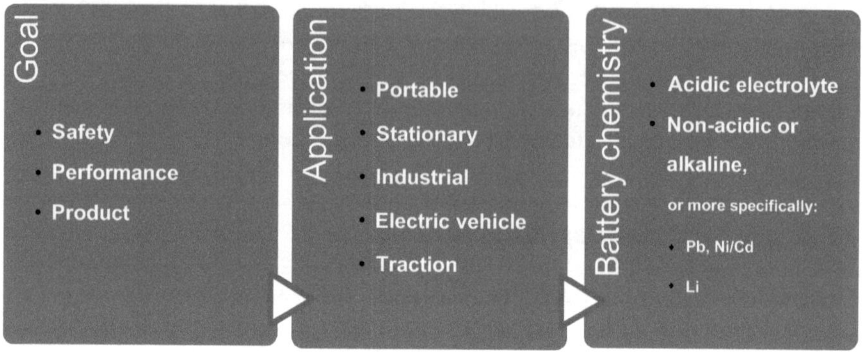

Fig. 29.3 Battery standards classification

Table 29.1 Battery standards

SAE Automotive Battery Standards	
SAE J240	Life Test for Automotive Storage Batteries
SAE J537	Storage Batteries
SAE J551	Performance levels and methods of measurement of electromagnetic radiation from vehicles and devices (30 to 1000 MHz)
SAE J1127	Battery Cable
SAE J1455	Recommended Environmental Practice for Heavy-Duty Trucks
SAE J1718	Measurement of Hydrogen Gas Emission From Battery-Powered Passenger Cars and Light Trucks During Battery Charging
SAE J1742	Connections for High Voltage On-Board Road Vehicle Electrical Wiring Harnesses-Test Methods and General Performance Requirements
SAE J1766	Recommended Practice for Electric and Hybrid Electric Vehicle Battery Systems Crash Integrity Testing (Work in progress)
SAE J1772	SAE Electric Vehicle Conductive Charge Coupler
SAE J1773	SAE Electric Vehicle Inductively Coupled Charging
SAE J1797	Recommended Practice for Packaging of Electric Vehicle Battery Modules
SAE J1798	Recommended Practice for Performance Rating of Electric Vehicle Battery Modules
SAE J1811	Power Cable Terminals
SAE J1939	The SAE specification for the CAN Bus
SAE J2185	Life Test for Heavy-Duty Storage Batteries
SAE J2288	Life Cycle Testing of Electric Vehicle Battery Modules
SAE J2289	Electric Drive Battery Pack System Functional Guidelines

Table 29.1 (Continued)

SAE Automotive Battery Standards	
SAE J2293	Energy Transfer System for Electric Vehicles
SAE J2344	Guidelines for Electric Vehicle Safety
SAE J2380	Vibration Testing of Electric Vehicle Batteries
SAE J2464	Electric Vehicle Battery Abuse Testing
SAE J2602	The SAE specification for the LIN Bus

29.4.2 Standards for stationary energy storage systems

Lithium-ion batteries have become increasingly important for stationary systems. This applies especially to stationary home storage systems installed in combination with photovoltaic systems, but also to large-scale systems providing intermediate storage. Operational safety is especially critical for domestic application and must be standardized and monitored.

The challenge of standardization in this case is to define basic requirements that ensure that batteries cannot harm the user even if mishandled or defective. The VDE has drawn up application rules to check functional safety requirements and foreseeable mishandling, which include drop tests, thermal mishandling, and overcharging.

29.4.3 Standards for other applications

As mentioned above, lithium-ion batteries are used in a great many areas such as portable applications (e.g. power tools), industrial applications as well as in railway and aviation applications.

Lately, efforts have been undertaken to close these gaps so that documents are available for most areas of application.

29.5 Current standardization projects and proposals for lithium-ion batteries

Currently, standards are being developed related to the following issues:

- Safety standards for stationary applications (home storage systems)
  - in the future: large-scale smart grid systems (MWh range)
- Safety standards for traction applications
- Labeling standard for recycling
- Series of standards for flow-battery systems
- Safety standards for high-temperature batteries

and many more.

29.6 Standards list

Table 29.2 contains standardization projects for lithium-ion batteries and projects currently under development.

29.7 Outlook

Standardization work is an important component of successful development in the battery industry. Reliable international standards are absolutely essential in order to make novel applications safer and controllable and to enable them to prevail in the market.

Table 29.2 Standardization projects for lithium-ion batteries

Standard	Title (English)	Title (German)
Stationary storage systems		
DIN EN 61427-1 Ed. 1	Secondary cells and batteries for renewable energy storage – General requirements and methods of test – Part 1: Photovoltaic off-grid application	
DIN EN 61427-2 Ed. 1	Secondary cells and batteries for renewable energy storage – General requirements and methods of test – Part 2: On-grid applications	
E DIN EN 62485-5 Ed. 1	Safety requirements for secondary batteries and battery installations – Part 5: Lithium-ion batteries for stationary applications	
E DIN EN 62932-1 Ed. 1	Flow battery systems for stationary applications – Part 1: General Aspects, Terminology, and Definitions	
E DIN EN 62932-2-1 Ed. 1	Flow battery systems for stationary applications – Part 2-1: Performance general requirements & methods of test	
E DIN EN 62932-2-2 Ed. 1	Flow battery systems for stationary applications – Part 2-2: Safety requirements	
VDE-AR-E 2510-2		Stationäre elektrische Energiespeichersysteme vorgesehen zum Anschluss an das Niederspannungsnetz
E VDE-AR-E 2510-50		Stationäre Energiespeichersysteme mit Lithium-Batterien – Sicherheitsanforderungen

Table 29.2 (Continued)

Standard	Title (English)	Title (German)
Road vehicles/LEV		
DIN EN 62660-1 Ed. 1	Secondary lithium-ion cells for the propulsion of electric road vehicles – Part 1: Performance testing	
DIN EN 62660-2 Ed. 1	Secondary lithium-ion cells for the propulsion of electric road vehicles – Part 2: Reliability and abuse testing	
DIN EN 62660-3	Secondary lithium-ion cells for the propulsion of electrical road vehicles – Part 3: Safety requirements for cells and modules	
ISO 12405-1	Electrically propelled road vehicles – Test specification for lithium-ion traction battery packs and systems – Part 1: High-power applications	
ISO 12405-2	Electrically propelled road vehicles – Test specification for lithium-ion traction battery packs and systems – Part 2: High-energy applications	
ISO 12405-3	Electrically propelled road vehicles – Test specification for lithium-ion traction battery packs and systems – Part 3: Safety performance requirements	
ISO/IEC PAS 16898	Electrically propelled road vehicles – Dimensions and designation of secondary lithium-ion cells	
ISO 6469-1	Electrically propelled road vehicles – Safety specifications – Part 1: On-board rechargeable energy storage system (RESS)	
ISO 6469-2	Electrically propelled road vehicles – Safety specifications – Part 2: Vehicle operational safety means and protection against failures	
ISO 6469-3	Electrically propelled road vehicles – Safety specifications – Part 3: Protection of persons against electric shock	
ISO 6469-4	Electrically propelled road vehicles – Safety specifications – Part 4: Post crash electrical safety	
E DIN EN 50604-1	Secondary lithium batteries for LEV (Light Electric Vehicle) applications – Part 1: General safety requirements and test methods	

Table 29.2 (Continued)

Standard	Title (English)	Title (German)
Portable applications		
DIN EN 61960 Ed. 2	Secondary cells and batteries containing alkaline or other non-acid electrolytes – Secondary lithium cells and batteries for portable applications	
DIN EN 61960-3 Ed. 1	Secondary cells and batteries containing alkaline or other non-acid electrolytes – Secondary lithium cells and batteries for portable applications – Part 3: Prismatic and cylindrical lithium secondary cells and batteries	
DIN EN 61960-4 Ed. 1	Secondary cells and batteries containing alkaline or other non-acid electrolytes – Secondary lithium cells and batteries for portable applications – Part 4: Coin type (button) lithium secondary cells and batteries	
DIN EN 62133-2 Ed. 1	Secondary cells and batteries containing alkaline or other non-acid electrolytes – Part 2: Safety requirements for portable sealed secondary cells, and for batteries made from them, for use in portable applications	
Industrial applications		
E DIN EN 62619 Ed. 1	Secondary cells and batteries containing alkaline or other non-acid electrolytes – Safety requirements for secondary lithium cells and batteries for use in industrial applications	
DIN EN 62620 Ed. 1	Secondary cells and batteries containing alkaline or other non-acid electrolytes – Secondary lithium cells and batteries for use in industrial applications	
Traction applications		
E DIN EN 62485-6 Ed. 1	Safety requirements for secondary batteries and battery installations – Part 6: Lithium-ion batteries for traction applications	
Transport		
IEC 62281 Ed. 2	Safety of primary and secondary lithium cells and batteries during transport	

Apart from current topics such as electric mobility and stationary applications, other areas of application will emerge in the future, such as shipping and aerospace. All areas of expertise require experts who are willing to work on standards without remuneration in order to obtain and maintain a complete catalog of standards.

Frequently, the world of standards lags behind technical developments. Private enterprises should provide more opportunities for experts to participate in standardization bodies as this is an opportunity to set standards. Courts often consult standards when dealing with legal disputes and settling claims because they reflect the technical state-of-the art, providing there is an economic interest to comply with the standards.

Fields of application for lithium-ion batteries

30

Klaus Brandt

Contents

Due to ever-growing energy requirements, the world's population is dependent on fossil energy sources, which will result in a shortage of these resources and possible climate changes. It is widely recognized that energy production increasingly needs to be covered by renewable energy sources. This trend has been fueled by the rapid economic growth of the so-called emerging countries and the decision by some industrial nations to phase out nuclear energy production. The increased use of renewable energies, such as solar and wind power, has ultimately resulted in the necessity to store electric power temporarily in order to bridge the gap between the time of producing energy and consuming it. Batteries can provide part of the required capacity, ranging from battery units for homes with several kWh of storage capacity through to large batteries with capacities in the MWh range suitable as grid storage systems. Other systems such as pump storage systems or compressed-air storage systems already exist, but can only be expanded to a limited extent because they require a specific geographical environment. In addition, other electrochemical

K. Brandt (✉)
Lithium Battery Consultant, Taunusstrasse 43,
65183 Wiesbaden, Germany
e-mail: akkubrandt@aol.com

R. Korthauer (ed.), *Lithium-Ion Batteries: Basics and Applications*,
https://doi.org/10.1007/978-3-662-53071-9_30

storage systems are currently being applied or developed in the stationary battery system segment such as high-temperature batteries and redox-flow systems.

A significant share of the planet's oil reserves are currently used for mobility. Electrically powered vehicles with battery-based energy storage systems (BEVs = battery-electric vehicles) have existed for niche applications for more than 100 years. Their use is essentially limited because the amount of energy they can carry per kg storage weight for conventional batteries is more than two orders of magnitude below that of the chemical energy sources gasoline and diesel (lead-acid battery around 40 Wh/kg, gasoline around 10,000 Wh/kg). The power and range of electric vehicles are thus severely restricted.

Battery systems with high energy densities and specific energies are required to advance electric mobility. Among the current technologies, lithium-battery systems are the technology that meets these requirements, even if they are still far from achieving the energy density of fossil fuels. Hybrid vehicles (HEV) offer an interim solution, as their drive power is provided either by the combustion engine or by the electric motor or by both simultaneously. These vehicles save energy partly by recuperating energy via the electric motor, which serves as a generator. Fuel cells with hydrogen are also being developed as an energy source to power vehicles. However, because abrupt load changes and peak loads result in low energy efficiency, they are combined with batteries and also constitute a hybrid solution.

Stationary energy storage systems (ESS) and all types of electrically powered vehicles (xEV) are in all probability the main future lithium-battery system applications. Nevertheless, there are other applications, e.g., in the industrial sector, where it could be beneficial to harness the technology in order to recover energy or replace fossil fuels. Port operations are a good example of such an application. Operations that involve raising and lowering heavy loads, e.g., with mobile container cranes, can benefit from energy recovery. In addition, electric or hybrid drives could be implemented in harbor tugs to make them more environmentally friendly. The same applies to airport operations. Lithium-battery systems will also capture a share of the conventional market for stationary lead-acid batteries, which are mainly implemented in emergency power supply systems.

30.1 Stationary applications

The areas of application can be divided up according to three criteria. First of all, a distinction is made between batteries that are part of the electricity grid and standalone solutions. The latter are self-sufficient energy production systems which provide a reliable energy supply in regions that have no or insufficient connection to the grid, for example to telecommunication installations located in remote areas. The two other criteria are the length of time during which the stored energy has to be delivered (discharge time required) and the size of the storage unit (Table 30.1).

The growing share of electricity generated from renewable energy sources is leading to increasing fluctuations in supply and, in turn, to instabilities in the grid. ESS based on lithium-ion batteries are capable of providing electric power at short

Table 30.1 Classification of ESS applications according to discharge time required and size

	Local storage systems Battery size ≤ 100 kWh	Central storage systems Batteries ≥ 1 MWh
Short discharge time required (less than 1 hour)	Uninterrupted power supply (UPS) Compensation of load peaks	Grid stabilization
Long discharge time required (1 hour to several days)	Photovoltaic systems on buildings Telecommunications Arbitrage in the industrial sector	Central photovoltaic systems Wind parks Arbitrage in the grid

notice and can therefore be deployed to stabilize the grid [1]. There are a growing number of projects that generate power ranging from several MW to over 10 MW in the USA, Asia, and Europe [2, 3]. When installed locally, they mainly protect critical systems such as data storage systems and telecommunication installations from brief interruptions in the mains power supply through an uninterrupted power supply (UPS). Longer disruptions are largely bridged by diesel generators and, in some cases, already by fuel cells. For these applications, lithium-ion batteries face strong competition from cheaper lead-acid batteries (Table 30.2).

Photovoltaic systems are implemented both locally and centrally to produce power. Whereas energy production peaks around midday, consumption is typically at its highest around late afternoon. Batteries can be used to bridge the roughly four-hour gap. If installed in an individual building, the power supply of the building can be self-sufficient providing there is enough sunshine and storage capacity. There are a growing number of systems on the market that provide homes with power by combining self-generated solar energy, battery storage, and connection to the grid. Systems of this kind are subsidized by the state in countries such as Germany [4, 5]. In addition, electric vehicles can be integrated into ESS. Depending on the situation, the vehicle's traction battery can be charged with solar power, or the traction battery can be implemented as an additional storage medium to supply the household with power. It is also possible to quickly charge a traction battery with the energy stored in the stationary battery [6]. This concept is termed "vehicle-to-home" (V2H).

Table 30.2 Competing electrochemical storage systems for ESS applications

	Local storage systems Battery size ≤ 100 kWh	Central storage systems Batteries ≥ 1 MWh
Short discharge time required (less than 1 hour)	Lead-acid battery Lithium-ion battery	Lithium-ion battery
Long discharge time required (1 hour to several days)	Lithium-ion battery Fuel cell Lead-acid battery	Lithium-ion battery High-temperature batteries Redox-flow systems

A wind park's capability to produce a regular, constant supply of energy strongly depends on its geographical situation and the time of year. Parks with capacities exceeding 100 GW require high storage capacities, largely to compensate for fluctuations in production over longer periods. Lithium batteries can only be a part of the storage solution for large-scale parks.

Arbitrage involves storing power from the mains grid when energy is being produced in abundance and is cheap and then releasing it back into the grid when demand is high and energy is therefore expensive. An example of this is the storage of off-peak energy produced at night. For this purpose, batteries can be installed locally and connected up to the grid in larger units.

A prerequisite for using a local storage system is the smart grid [7, 8], which enables the operator to control the storing and releasing of energy back into the grid and to offset it from energy costs. In the smart grid, the capacity of batteries that are primarily connected to the grid for other purposes can be used for arbitrage. This can also include the batteries of electric vehicles that are connected to a charging station. This concept is known as "vehicle-to-grid" (V2G) [9]. Industrial systems often have short-term peak loads, which have to be factored in when designing the grid connection. If battery storage units are able to bridge these peaks in demand, a less powerful connection suffices, thus reducing costs [10]. The advantage for the grid operator is that expanding the grid can possibly be foregone in spite of rising demand.

30.2 Technical requirements for stationary systems

The technical requirements for ESS batteries largely depend on the specific application and, above all, on the amount of energy that has to be stored and the discharging duration (Table 30.2). Safety is the number one priority, in particular when these batteries are located inside buildings. The service life for ESS installations is 20 years. Depending on their application, this can mean very high cycle numbers, for example a total of around 8,000 cycles for PV systems, with one cycle each per day. While, for most applications, energy density does not play an important role, power density is of importance for applications with a short discharging duration.

The history of large battery systems being implemented to support the electricity grid goes back relatively far, starting with lead-acid batteries and high-temperature batteries [2]. Lithium-battery systems have several characteristics that distinguish them from other storage systems. They have a very high storage efficiency; i.e., a high percentage of the stored energy – over 95 % depending on the charging and discharging rates – is released again. Depending on the chemistry of the electrodes and the cell design, they are capable of storing energy over long periods of time without high levels of power loss (low level of self-discharging). On the downside, they are more expensive than lead-acid batteries.

The costs of large-scale lithium-ion storage units will decisively influence their market share in the ESS segment. Researchers are therefore currently trying to determine whether the batteries of electric vehicles (EV) are not suitable for a second life as stationary storage units after they have reached the end of their useful service life in vehicles. As lithium-ion batteries age, their capacity decreases continuously. Batteries in EV applications are designed so that they still have 80 % of their original capacity at the end of their service life, which could be sufficient for them to be used as ESS batteries. However, the usability of used batteries is not only restricted by their lower capacity but also by the fact that their high-current capability (power) decreases [12].

Lithium batteries, like other batteries, use one and the same active material for chemically storing energy and for converting energy between electric power and the chemical storage unit.

Energy conversion and storage are separated in other storage systems, for example in redox-flow systems [13, 14] and fuel cells. When the storage capacity of redox systems is increased, a higher storage capacity is needed for the reactances but a larger energy conversion unit, the more expensive component, is not required. It can therefore be assumed that these systems will have a cost advantage, in particular when large amounts of energy have to be stored over long periods of time. Sodium-sulfur is a high-temperature system [15], which is deployed in Japan with systems up to 200 MWh [16] and is also being tested by American energy supply companies.

From an economic point of view, electrochemical storage systems such as lithium-ion batteries are not suitable for storing large quantities of power – such as the capacity needed for offsetting seasonal fluctuations of power generated by wind turbines or photovoltaic systems – for long periods. For these applications, power-to-gas or power-to-liquid fuel concepts are more suitable. They involve transforming electric power into gaseous or liquid fuels such as hydrogen or methanol [17].

30.3 Automotive applications

There are two motivations behind the application of batteries in automobiles. On the one hand, the aim is to boost the energy efficiency of vehicles with conventional combustion engines and, on the other hand, to use renewable energies indirectly through battery storage to transport persons and goods.

Essentially, the energy efficiency of a conventionally powered vehicle can be increased by three measures: by reducing the vehicle's weight, recuperating break energy, and downsizing the combustion engine. Reducing the weight of a conventional vehicle by replacing the lead-acid battery with a lithium battery only decreases the weight by around 10 kg [18]. Recuperating break energy with a generator (mostly identical to the starter motor or the traction motor) and storing it in the vehicle's battery makes it possible to cut fuel consumption. In a micro-hybrid vehicle, this recuperated energy is used to switch off the vehicle's motor when it is at a standstill (e.g., at traffic lights) and to subsequently start it again [19]. In hybrid vehicles with a dual power train – with both a combustion engine and an electric motor – the electric power is also used to support the combustion motor with acceleration. The combustion engine can therefore be downsized (Table 30.3).

Table 30.3 Characterization of different xEV (cars)

	xEV	Battery size	Power/ energy ratio (P/E)	Energy saving	Primary energy source
Start-stop		≤ 1 kWh	10	Idling	Gasoline/diesel
Micro-hybrid	HEV	≤ 1 kWh	30	Recuperation	Gasoline/diesel
Hybrid	HEV	1 to 2 kWh	20	Recuperation/ downsizing	Gasoline/diesel
Plug-in hybrid	PHEV	5 to 15 kWh	5 to 15	Recuperation/ downsizing/ power grid	Gasoline/ power grid
Full EV	BEV	20 to 80 kWh	2 to 3	Recuperation/ power grid	Power grid
EV with range extender		20 to 40 kWh	2 to 3	Recuperation/ power grid	Power grid/ gasoline/diesel

Fully electric driving is possible with increasing battery size. A battery energy content of 1 kWh corresponds to a 5 to 10 km electric driving radius for a car. Plug-in hybrids typically have a battery with 10 kWh of usable energy, which can also be charged from the mains grid, enabling them to cover short distances of 50 km or more without the support of a combustion engine. Pure electric vehicles need batteries with a storage capacity of 20 kWh to 100 kWh to cover everyday distances. Higher energy contents and, as a result, distances of over 300 to 500 km are not possible with the specific energies attainable today because the weight of the battery cannot be raised indefinitely.

The battery charging concepts for electric vehicles play a key role in making them feasible for everyday use. A socket in the garage at home can generally provide a maximum of 3 kW of power. The battery of an electric vehicle is charged over night. To raise the acceptance of electric mobility, a public charging infrastructure is required with charging stations that provide significantly higher power outputs. In several countries and regions, an infrastructure of this kind is currently being developed. Charging quickly (< 1 hour, > 20 kW charging capacity) is technically demanding and needs additional infrastructure.

Concepts that involve exchanging the discharged battery for a charged battery at battery-changing stations require considerable investment in infrastructure [20] and have not yet been implemented today. A solution to the limited range of vehicles is the range extender: a small generator powered by a combustion motor that recharges the traction battery when needed.

The higher the share of energy needed for driving that comes from the socket, the greater the possibility of indirectly using renewable energies provided by the power grid for automotive applications. Nevertheless, longer distances also require larger and more expensive batteries. The costs per storable energy unit of batteries for xEV are still higher than those for batteries used in portable applications. Larger production units and optimization across the entire value chain are, however, leading to steadily declining costs [21, 22].

Another group of vehicles that lends itself to electric drive is commercial vehicles, in particular buses [23]. Implementing HEV concepts, for example in city buses and refuse collection trucks, can save much more energy than in cars because they start and stop more frequently. Driving fixed routes is an advantage for BEVs because the vehicles' limited range can be dealt with by installing charging stations or battery-changing stations. The batteries for this application are much larger, up to 100 kWh for BEVs. In China, there are already many vehicles of this kind on the roads [25], and they are currently being tested in many cities in other countries. The principal aim of electrifying transport is often to improve the air quality in cities.

30.4 Technical requirements for automotive applications

The safety requirements are higher for automotive applications than for stationary applications because of additional provisions in the event of vehicle batteries being mechanically damaged in accidents.

Based on the ratio between the amount of power required and the stored energy (Table 30.3), batteries can be split into two types: high-power batteries ($P/E \geq 10$) or high-energy batteries ($P/E < 10$). Different types of cell chemistry as well as cell and battery designs are required for these two types.

The most important characteristic of the BEV is its range and, therefore, the battery's specific energy and energy density. This requirement is instrumental in shaping the choice of cell chemistry in lithium-ion batteries, which are the sole battery type used in BEVs. The motivation to match the ranges covered by a full tank of diesel or gasoline is the main driver behind the development of new systems such as lithium-sulfur or lithium-air, which could eventually replace today's prevailing lithium-ion battery technology [24].

Historically, one of the milestones on the path toward the lithium-ion battery was the development of the rechargeable lithium-metal battery [25, 26]. The appeal of this system is that the metal anode offers a high specific capacity in comparison to the carbon anode. However, this technology, which combines this metal anode with liquid organic electrolytes, has not proved successful. A variant of the technology with a solid polymer electrolyte is implemented in a fleet of electric vehicles in Paris [27]. Other concepts integrating solid electrolytes are currently being developed.

The costs of the battery and fossil fuels together with the application will determine how economic the BEV solution is. Today, most BEV applications are not a cost-effective solution for the consumer unless they are subsidized. The costs of the battery systems therefore have to be reduced considerably to facilitate broad market penetration. The purchase of xEVs is currently being subsidized in many countries in order to provide this technology with the necessary impetus [28].

The first HEV applications have been launched with NiMH batteries, which have a sufficient P/E ratio [20]. For PHEV applications, however, the electric range is of importance so lithium-ion batteries are deployed in them. Today, lead-acid batteries are still implemented for start-stop operations and in HEVs. Although they are significantly more cost-effective, they only recuperate a limited amount of energy. The ever-growing pressure on the automotive industry to cut CO_2 emissions of vehicles will probably result in lithium-ion batteries becoming sufficiently attractive for these applications in the medium term due to their lower weight and additional energy-saving potential.

As reflected by discussions about the advantages and disadvantages of different vehicle concepts, it is clear that there are two main obstacles slowing down the widespread electrification of mobility: the relatively low energy density compared to fossil fuels and the high costs in comparison to combustion engines. Electric mobility will become more attractive and its use more widespread with the continuous improvement of the energy density of lithium-ion batteries and with cost reductions brought about by economies of scale and the development of rapid charging networks.

30.5 Further applications

ESS and xEV are indisputably the most frequently discussed applications.

Any other applications of electrochemical storage systems generally arise, firstly, when there is a possibility of significantly recuperating energy that has been already used and, secondly, when hybrid vehicles can be deployed to cut fuel consumption, and, thirdly, when the higher output of lithium-ion batteries compared to lead-acid batteries makes it possible to implement electric traction in a field where it was not previously possible.

An example of the first application is the raising and lowering of heavy loads, for example by port cranes, providing they are not stationary, in which case feeding the energy back into the grid is generally the more straightforward solution [29]. This also applies to any mobile applications that involve frequent acceleration and breaking, for example transporting containers in ports [30]. Another possible application of this is in the rail sector, in particular when there are very short distances between stops. In Great Britain, a battery-powered local train is already in operation [31]. If, however, an electric system is already in use, feeding break energy back into the grid is probably preferable.

There are multiple applications for diesel-electric drives, e.g., locomotives, cranes, vessels. The diesel engine has to be designed so that it can accommodate peak loads. If this load can be buffered by a battery, the motor can be downsized and, as a result, a more consistent load and fuel savings are possible [32].

Electric boat engines with lead-acid batteries are already being deployed on a small scale in some inland waters to reduce noise and pollution of these waters. Today, ferries are powered by hybrid engines and lithium-ion batteries in order to avoid emissions near the coast [33]. While in port, vessels can draw their power supply from batteries, further reducing the environmental impact [34]. The higher capacity of lithium-ion batteries will open up other applications, for example harbor tugs or towing vehicles at airports [35].

Bibliography

1. Kamath H (2011) Integrating batteries with the grid. In: 28th International battery seminar & exhibit, Fort Lauderdale, 14–17 March 2011
2. Doughty DH, Butler PC, Akhil AA, Clark NH, Boyes JD (2010) Batteries for large scale stationary energy storage. Electrochem Soc Interface 19(3):49–53
3. AES Energy Storage: Very Soon, Utility-Scale Storage Will Always Be On The Resource Menu, Forbes, 12 November 2015, http://www.forbes.com/sites/peterdetwiler/2015/11/12/aes-energy-storage-very-soon-utility-scale-storage-will-always-be-on-the-resource-menu/#3fba0eee41b9, accessed on: 3 June 2016
4. VDI Nachrichten (2012) Lithium-Ionen-Speicher sollen Stromversorgung im Niederspannungssektor stabilisieren. http://www.vdi-nachrichten.com/artikel/Lithium-Ionen-Speicher-sollen-Stromversorgung-im-Niederspannungssektor-stabilisieren/56916/2, accessed on: 2 June 2016
5. Kyocera news releases: Kyocera to start exclusive sales in Japan of new residential-use energy management system combining solar power with Li-ion battery storage unit. http://global.kyocera.com/news/2012/0102_qpaq.html, accessed on: 2 June 2016
6. Denso Corporation news release: Denso develops vehicle-to-home power supply system for electric vehicles. http://www.globaldenso.com/en/newsreleases/120724-01.html. accessed on: 2 June 2016
7. Harris C, Meyers JP (2010) Working smarter, not harder: an introduction to the "Smart Grid". Electrochem Soc Interface 19(3):45–48
8. Electric Power Research Institute (ed.) EPRI smart grid demonstration update, April 2012. http://smartgrid.epri.com/doc/EPRI_Advisory_Update_April_2012_Issue.pdf, accessed on: 2 June 2016
9. Kempton W, Marra F, Anderson PB, Garcia-Valle R (2012) Business models and control and management architecture for EV electrical grid integration; prior publication, Chapter 4.
10. Reducing Energy Costs with Peak Shaving in Industrial Environments, Schneider Electric, http://www.schneider-electric.com.hk/documents/energy-efficiency-forum/Reducing-Energy-Costs-with-Peak-Shaving.pdf, accessed on: 3 June 2016
11. Norman S (2012) Demand for large scale batteries and alternatives. AABC Europe, Mainz, 18–22 June
12. C.K. Narula, R. Martinez, O. Onar, M.R. Starke, G. Andrews, Economic analysis of deploying used batteries in power systems, Oak Ridge Natl. Lab. (2011). Report ORNL/TM-2011/151.
13. Nguyen T, Savinell RF (2010) Flow batteries. Electrochem Soc Interface 19(3):49–53
14. Gibbard HF (2011) Redox flow batteries for energy storage. In: 28th International battery seminar & exhibit, Fort Lauderdale, 14–17 March 2011
15. Sudworth J, Tilley AR (1985) The sodium sulfur battery, Springer
16. NAS Energy Battery Storage Systems, NGK Insulators, Ltd, October 2013, http://www.eei.org/about/meetings/meeting_documents/abe.pdf, accessed on: 3 June 2016
17. Electricity Storage in the German Energy Transition, Agora Energiewende, December 2014, https://www.agora-energiewende.de/fileadmin/Projekte/2013/speicher-in-der-energiewende/Agora_Speicherstudie_EN_web.pdf, accessed on: 3 June 2016
18. Eger U (2011) Dual-battery system with lithium battery for the 12-V powernet of a vehicle. AABC Europe, Mainz, 6–10 June
19. Kessen J (2012) Lithium-ion advances in micro-hybrid applications. AABC Europe, Mainz, 18–22 June
20. How electric cars swap batteries. MIT Technology Review. http://www.technologyreview.com/demo/425889/how-electric-cars-swap-batteries/. accessed on: 2 June 2016
21. Pillot C (2012) Battery and material market outlook. AABC Europe, Mainz, 18–22 June
22. Nelson PA, Gallagher KG, Bloom I, Dees DW (2011) Modeling the performance and cost of lithium-ion batteries for electric-drive vehicles. Report ANL-11/32, Argonne National Laboratory, Sept 2011. http://www.ipd.anl.gov/anlpubs/2011/10/71302.pdf, accessed on: 2 June 2016

23. Sauer DW (2012) Full electric buses for public transport – markets and technology options for energy supply by lithium-ion batteries. AABC Europe, Mainz, 18–22 June
24. Thielmann A, Isenmann R, Wietschel M (2010) Technologie-Roadmap Lithium-Ionen-Batterien 2030. Fraunhofer-Institut für System- und Innovationsforschung ISI, Karlsruhe. http://www.forum-elektromobilitaet.de/assets/mime/c6ef10b72e9f2b1588821ed9baae7ba0/Lib_Road[1].pdf, accessed on: 2 June 2016
25. Brandt K (1986) A 65 Ah rechargeable lithium molybdenum disulfide battery. J Power Sources 18:117–125
26. Brandt K (1994) Historical development of secondary lithium batteries. Solid State Ionics 69:173–183
27. Marginedes D, Planchais E (2011) Lithium metal polymer: performance and design for Paris EV project Autolib. AABC Europe, Mainz, 6–10 June
28. Helveston JP, Liu Y, McDonnell E, Fuchs E, Klampfl E, Michalek JJ, Will subsidies drive electric vehicle adoption? Transportation Research Part A 73 (2015) 96–112
29. Green Batteries for RTG Cranes; Green Port 15 January 2016, http://www.greenport.com/news101/energy-and-technology/green-batteries-for-rtg-cranes, accessed on: 3 June 2016
30. Hamburger Hafen: Batteriebetriebene Schwerlastfahrzeuge fahren Container, Windkraft-Journal, 25 April 2014, http://www.windkraft-journal.de/2014/04/25/hamburg-hafen-batteriebetriebene-schwerlastfahrzeuge-fahren-container/51620, accessed on: 3 June 2016
31. Low carbon battery-powered train carries first passengers, The Guardian, 13 January 2015, http://www.theguardian.com/environment/2015/jan/13/low-carbon-battery-powered-train-carries-first-passengers, accessed on: 3 June 2016
32. Emissionsfrei im Hafen; Schiff & Hafen. http://www.schiffundhafen.de/news/schiffbau/single-view/view/emissionsfrei-im-hafen.html. accessed on: 2 June 2016
33. Erste Hybrid-Fähre verbindet Deutschland und Dänemark, Wirtschaftswoche, 25 May 2016, http://www.wiwo.de/technologie/green/tech/keine-abgase-in-kuestennaehe-erste-hybrid-faehre-verbindet-deutschland-und-daenemark/13644656.html, accessed on: 3 June 2016
34. Driving change; Airport World. http://www.airport-world.com/home/item/1460-driving-change. accessed on: 2 June 2016
35. Bernardus Hybrid Tug, Netherlands, Ship Technology.com http://www.ship-technology.com/projects/bernardus-hybrid-tug/, accessed on: 3 June 2016

Part V

Battery Applications – Sectors and Requirements

Requirements for batteries used in electric mobility applications

31

Peter Lamp

Contents

31.1 Introduction

Future mobility needs novel concepts that strike a balance between individual mobility needs, sustainable use of resources, and protection of the environment (Fig. 31.1). Also, owing to climate change and the limitation of fossil fuels, more efforts are required to decrease CO_2 emissions. In this regard, the entire automotive industry has already achieved considerable success by optimizing engine technology and introducing automatic engine start/stop systems and brake energy recuperation. These efforts in conventional drive technology are ongoing.

However, new drive technologies are required to enable the long-term transition from fossil fuels to fuels based on renewable energies. From today's perspective,

P. Lamp (✉)
BMW AG, 80788 Munich, Germany
e-mail: peter.lamp@bmw.de

R. Korthauer (ed.), *Lithium-Ion Batteries: Basics and Applications*,
https://doi.org/10.1007/978-3-662-53071-9_31

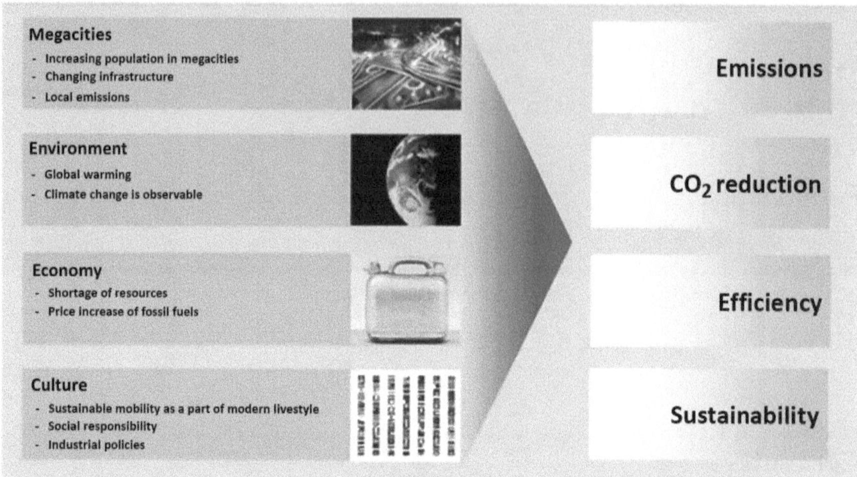

Fig. 31.1 New framework conditions require new mobility concepts

power train electrification in hybrid vehicles, plug-in hybrid vehicles, and electric vehicles is the technically practical means of achieving this goal, which is approved by society, politics, and industry.

Notwithstanding the degree of electrification, electrical energy storage systems – i.e., batteries – play a crucial role in this technological change. Storage and power density essentially define the power train and/or vehicle characteristics and thus also determine the CO_2 savings potential and, to a considerable extent, customer acceptance.

Introducing electric vehicles has often been unsuccessful in the past because of the battery technologies available at the time. Especially lithium-ion technology has been further developed in recent years, resulting in a different perspective and intensified development of electric vehicles. The requirements for batteries used in vehicles are high, and modern battery technologies still do not fully meet them. This Chapter discusses the status quo and necessary future developments.

31.2 Requirements for vehicle and drive concepts

It is important to distinguish between the different basic vehicle concepts for electric vehicles and the use of batteries in them. These concepts are essentially: hybrid, plug-in hybrid, and electric vehicles. However, they all have special forms that vary depending on the manufacturer and/or vehicle model (even from the same manufacturer).

The basic concepts will be briefly described below. However, this Chapter does not provide a universal definition but helps distinguish between the concepts mentioned here.

Micro-hybrid: Engine start/stop systems and brake energy recuperation by using starter generators, which are modified if necessary, in combination with a battery or an additional battery. No drive support by an electric motor, no electric driving.

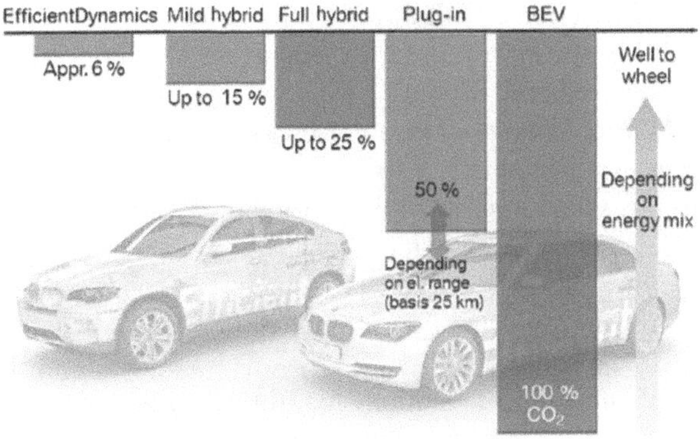

Fig. 31.2 CO_2 savings as a function of the electrification concept and/or the degree of electrification

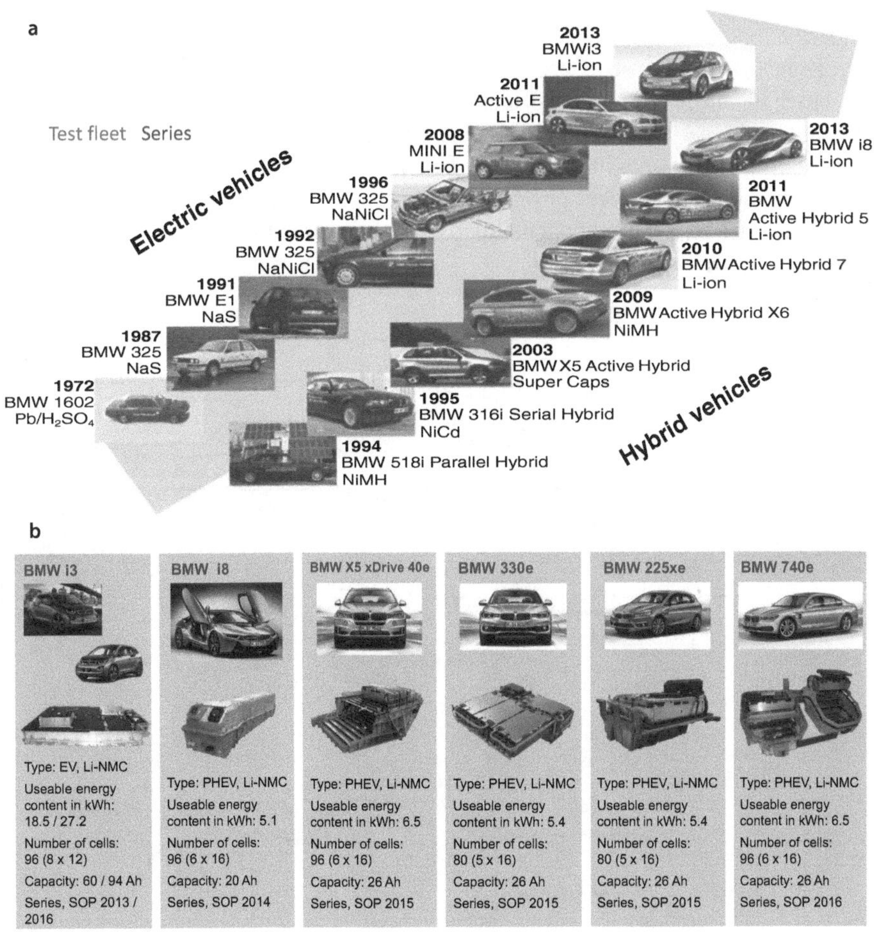

Fig. 31.3a+b Development history of hybrid and electric vehicles (*Source* BMW AG)

Mild-hybrid: All functionality of the micro-hybrid concept plus an electric motor integrated into the power train in order to support the combustion engine in certain driving situations. No electric driving.

Full-hybrid: All functionality of the mild-hybrid as well as an electric power train for purely electric driving (usually 1 to 3 km).

Plug-in hybrid: All functionality of the full-hybrid as well as battery charging via both a vehicle generator and an external charger, e.g., socket or charging station. More electric driving due to larger battery (usually up to approx. 100 km).

Electric vehicle: Purely electric power train. 100 % electric driving. Vehicle range depends on battery capacity.

There are different technical solutions for all these technologies that will not be set out in detail here. The amount of CO_2 savings essentially depends on the degree of electrification, i.e., electric driving. This is illustrated in Fig. 31.2. German manufacturers already offer mild-hybrid and full-hybrid vehicles as part of their standard portfolios. At the end of 2013, plug-in hybrid and electric vehicles entered the market.

BMW has been developing the aforementioned vehicles based on decades of experience with electrical power trains (Fig. 31.3). This experience is necessary in order to select concepts and technologies for every application and to design requirements that result in the ideal outcome in terms of costs and technology and that deliver the best-possible customer benefits.

Developing an electrical power train is always a complex, multistage process. Specifications for sub-components such as batteries are only determined right at the end of this process (Fig. 31.4).

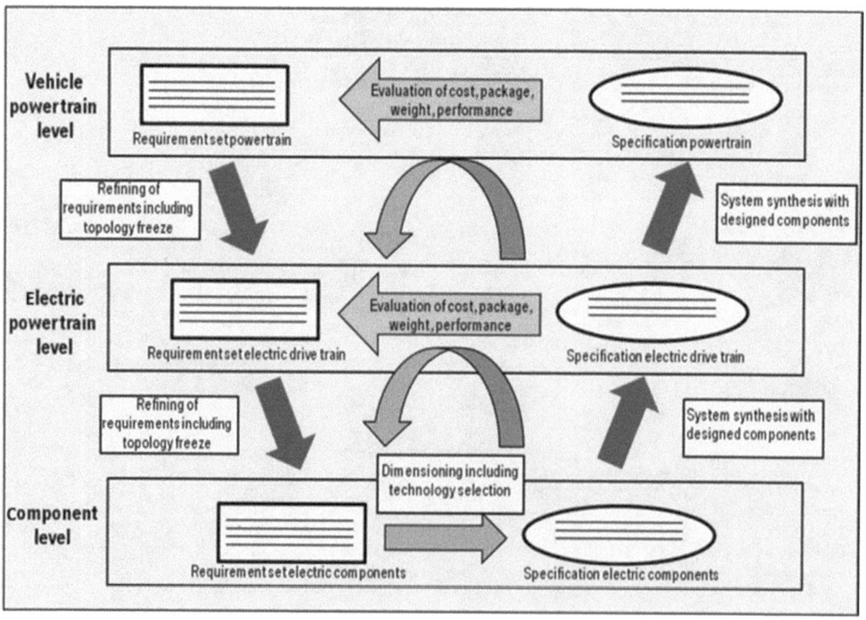

Fig. 31.4 Schematic diagram of the iterative development process for an electric power train

Usually, vehicle characteristics (especially range and performance) determine the technical specifications. Therefore, all power train components and especially their interaction and vehicle integration must be optimized in an iterative process until the desired functionality and other targets regarding installation space, weight, and costs are achieved. These targets also include safety, service life, and quality.

31.3 Vehicle and battery concept applications

Based on BMW's vehicle fleet, as an example, we will illustrate how state-of-the-art lithium-ion batteries are used.

One example of application in the hybrid vehicle class is the BMW Active Hybrid 5.

Battery requirements were derived from the objective to integrate an electric power train in a 535i with a six-cylinder twin-turbo engine in order to obtain the performance of a 540i with eight cylinders, and still have 15 % less fuel consumption than the 535i with "efficient dynamics". In order to achieve these goals, a certain percentage of electric driving is required in the reference driving cycle. The battery's capacity is determined based on this. In this case, the battery power is the result of the difference in performance between the eight-cylinder and six-cylinder vehicles which is compensated for by the electric drive. Fig. 31.5 shows all vehicle characteristics and technical data for the electric power train.

The battery used in the BMW Active Hybrid 5 is the first battery in series production which was developed and manufactured entirely by BMW. The lithium-ion cell and other sub-components are supplied by external companies. A cylindrical cell is used with lithium iron phosphate as the cathode material. Fig. 31.6 shows the battery and its technical data.

Another example of an application in electric vehicles is the so-called conversion vehicle, i.e., a conventional BMW 1 Series coupé converted into an electric vehicle. After the MINI E, the BMW Active E was the second electric vehicle on the market

Technical data	
Engine	6-cylinder gasoline engine
Displacement	2979 ccm
Max. power Engine	225 kW
Max. torque Engine	400 Nm
Max. power E-engine	40 kW
Max. torque E-engine	210 Nm
Acceleration (0-100 km/h)	5.9 s
Fuel consumption NEDC	6.4 ltr./100 km
Pure electric range	3 km

Fig. 31.5 The BMW Active Hybrid 5 and its technical data used for battery development

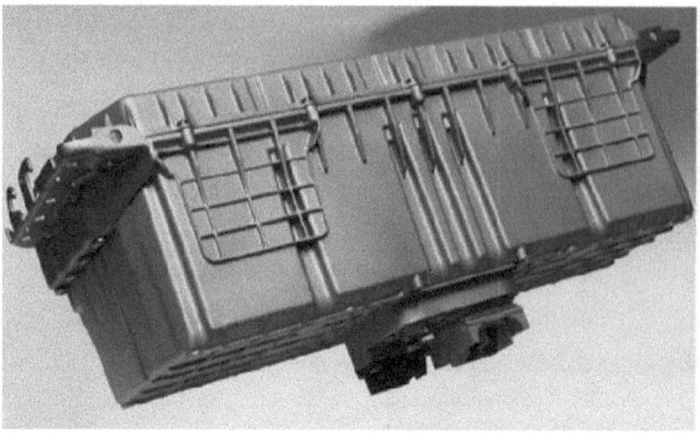

Technical data	
Total number of cells	96 (1p, 96s)
Nominal voltage	317 V
Voltage range	200 - 385 V
Nominal capacity	4 Ah
Nominal energy content	1.35 kWh
Usable energy content	0.6 kWh
Discharge power (20 s)	43 kW
Volume	40 l
Weight	46 kg
Cooling	Refrigerant

Fig. 31.6 Lithium-ion battery of the BMW Active Hybrid 5 and its technical data

Range (FTP72)	miles	150
Typ. customer range	miles	100
Seats	-	4
Trunk volume	litres	200
Acceleration	sec. (0-60 mph)	9.0
Speed	mph max.	90
Weight distribution	%	50:50
Drive type	-	Rear
Maturity	-	Concept ready for series production

Fig. 31.7 The BMW Active E and its technical data used for battery development

(with around 1,000 vehicles built). Like the MINI E, the BMW Active E was introduced to obtain customer feedback. However, in contrast to its predecessor, the MINI E, the BMW Active E was built to test technical components that would be used in the first series-produced electric vehicle, the BMW i3. Key battery data such as capacity and power are determined by the required range and performance (Fig. 31.7).

It is especially challenging to integrate the battery into an existing vehicle concept. The resulting solutions are usually a compromise and are rarely technically and economically sound (Fig. 31.8). However, this is acceptable for the BMW Active E as it is being used for an extended field test. The next step was the BMW i3, which was specifically designed as an electric vehicle (Fig. 31.9). The compact battery is integrated into the underbody. enabling a cost-efficient modular design consisting of standard cell modules. Also, the BMW i3's chassis consists of carbon fiber and features numerous highly innovative technologies which will be important for electric vehicles in the future.

These examples indicate that vehicles with a high customer value can be developed and marketed with the lithium-ion technology available today. For electric vehicles to substantially penetrate the market in the long term, battery technology needs to be continually developed further and optimized. The rest of this Chapter is devoted to today's technical state-of-the-art and the necessary aims for developing battery technology in the future from the vantage point of applying batteries in vehicles.

Technical data	
Nominal voltage	355 V
Max. current	400 A
Energy content	32 kWh
Discharge power cont. peak	78 kW 140 kW
Total number of cells	192 (2p, 96s)
Charging time (7.7 kW)	4-5 hours
Cooling and heating	Water

Fig. 31.8 The lithium-ion battery of the BMW Active E and its technical data

Fig. 31.9 The series-produced BMW i3 electric vehicle and battery concept

31.4 Battery requirements

The main required battery characteristics for usage especially in plug-in hybrid and electric vehicles are shown in Fig. 31.10. The generic target figures form the basis for all requirements for batteries used in electric vehicles. However, detailing these figures, i.e., deriving specific values, depends on the framework conditions of different automotive manufacturers and their corresponding model strategies. Nevertheless, this section will try to derive general battery requirements for the automotive industry from these figures and compare them, where possible, with the current technical state-of-the-art and identify the developments needed for the future. However, please note that these conclusions cannot be generally applied to the automotive industry.

31.4.1 General information

Today, lithium-ion cells are used in consumer electronics, such as computers, camcorders, and cell phones, and in battery-operated power tools. However, we must not forget that there are major differences between the above applications and the

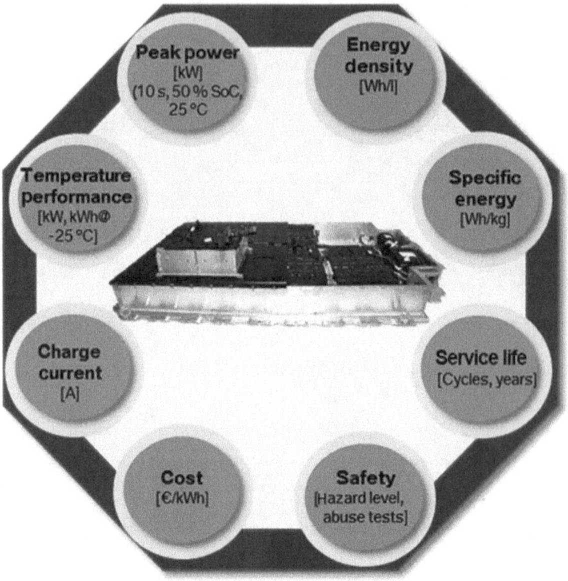

Fig. 31.10 Key performance parameters for electric vehicle batteries (targets for 2014 to 2020)

application of lithium-ion cells in vehicles (Fig. 31.11). These differences have resulted in new requirements, thus creating a need for new solutions.

Developing lithium-ion cells and batteries for use in vehicles goes beyond adapting a known technology to another application; it involves developing a specific product for automotive use from scratch, based on the lithium-ion technology know-how that is available today.

This article treats the battery as a single component of the electric power train. This solely refers to the battery as a high-voltage electrical energy storage system (typically around 400 V) and as the complete "battery" system, which in turn is made up of several components (Fig. 31.12).

We only refer to lithium-ion technology as the basis because it is currently the only technology in terms of energy and power density with the potential to fulfill all vehicle requirements.

The term lithium-ion technology is merely an umbrella term that comprises different active materials, cell designs, and types. These types are, for instance, cylindrical and prismatic cells with rigid steel or aluminum housing and/or cells with housings made of aluminum polymer composite film (so-called pouch cells). Today, carbon modifications are normally used as active materials for anodes. There is a greater variety of active materials for cathodes. They mainly consist of ceramic transition metal oxides such as lithium cobalt, lithium nickel manganese cobalt, or olivine structures such as lithium iron phosphate.

With regard to the goals shown in Fig. 31.10, the currently available materials are always a compromise because none of them fulfills all requirements, and all of them

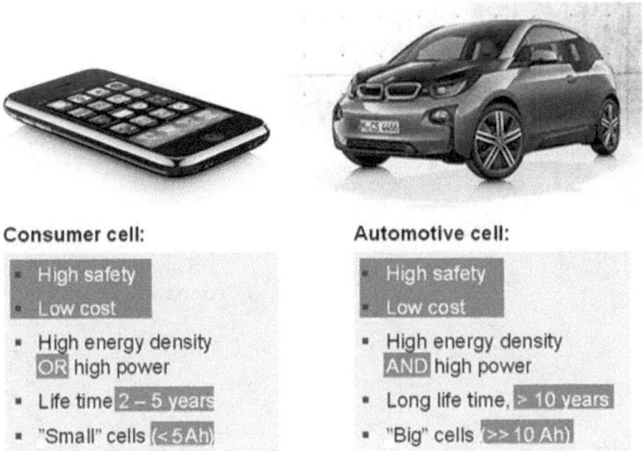

Fig. 31.11 Major differences between using lithium-ion technology in consumer products and in the automotive industry

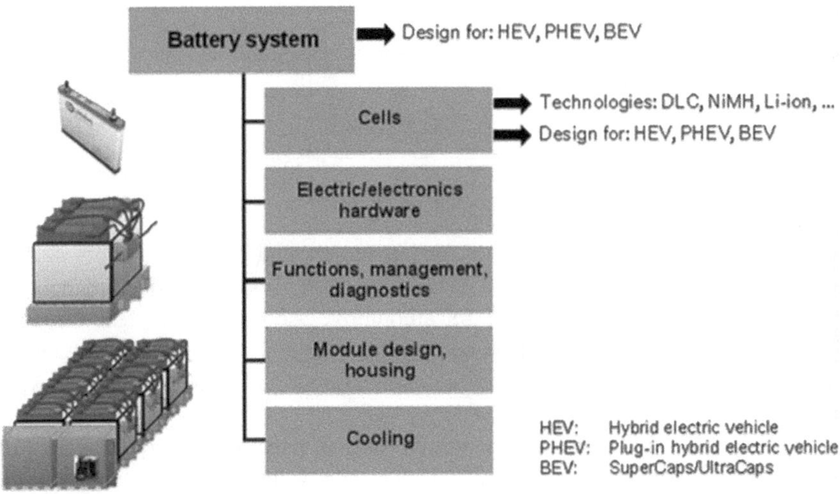

Fig. 31.12 The basic principle of a battery for automotive use as a complete system consisting of several sub-components, with the lithium-ion cell as the main component and technology carrier

have certain strengths and weaknesses (Fig. 31.13). Depending on the application and specific requirements, a specific material is considered best. In general, there is a need for further development and optimization in order to reach the goals depicted in Fig. 31.10. Lithium-ion technology is described in detail in Chapters 3 and 9.

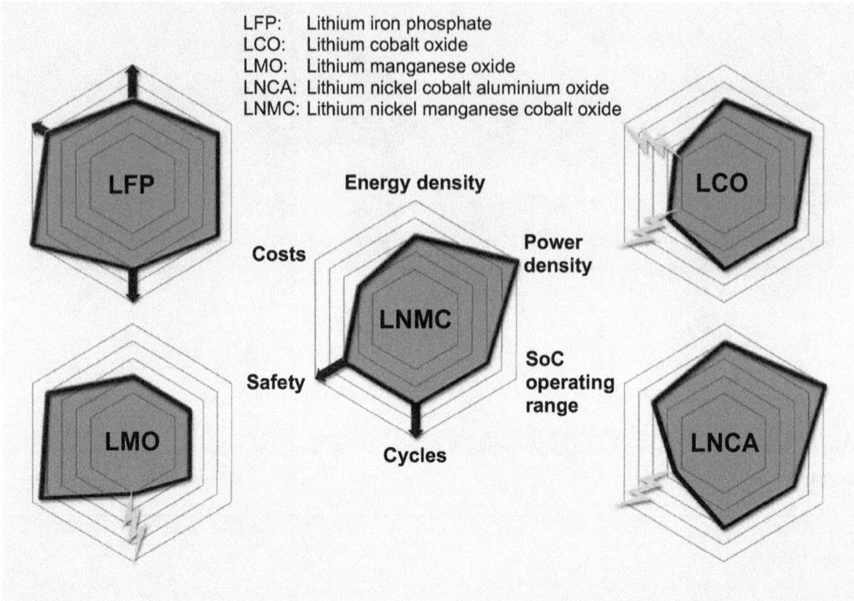

Fig. 31.13 The cathode active materials used in modern lithium-ion cells and their strengths and weaknesses for automotive use

31.4.2 Safety

In the automotive industry, hazards for passengers and other road users such as pedestrians must be avoided at all times. This is a top priority. The development and production of lithium-ion batteries for automotive use follow this basic principle. To this end, analyses are carried out and corresponding measures taken. Systems are checked according to recognized rules and regulations and corresponding standards requirements.

These include functional safety, which ensures safe battery system operation when using the vehicle, and applicable crash standards. In serious accidents that may damage the battery and in which the airbag and/or belt tensioner are triggered, the lithium-ion battery is automatically disconnected from the high-voltage, on-board electrical system. In very serious accidents that are not covered by legal regulations and tests, it is possible that lithium-ion batteries catch fire. This corresponds to the safety level of conventionally powered vehicles.

Fire can be caused by the chemical energy stored in the lithium-ion cell. The uncontrolled release of this energy, which ultimately causes the cell to combust, is called thermal runaway. It is a multistage process (Fig. 31.14) that is started by an external event (overcharging, overheating, etc.) and then continues with an exothermal reaction between the electrolyte and the anode's carbon material, causing the electrolyte and cathode material to degrade.

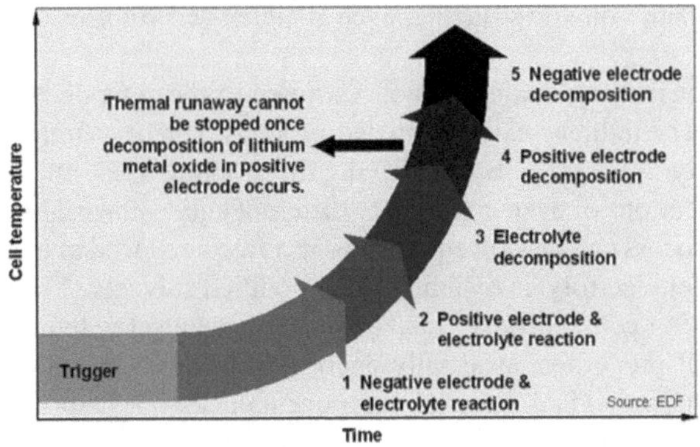

Fig. 31.14 Schematic diagram of the multistage process causing thermal runaway, i.e., causing the lithium-ion cells to catch fire

The amount of energy that is released depends on the materials used. For cathode materials, this is illustrated in Fig. 31.15. It is clear that lithium iron phosphate and manganese oxide are advantageous for the intrinsic safety of cathode materials. On the other hand, the former is inferior in terms of energy density and the latter in terms of service life (Fig. 31.13). The requirement is to link the high energy density characteristics and long service life with a low temperature energy release or with a release that only occurs at high temperatures.

Modifying lithium iron phosphate by replacing the iron with manganese, cobalt, or nickel is one possible direction of development. All these materials still possess

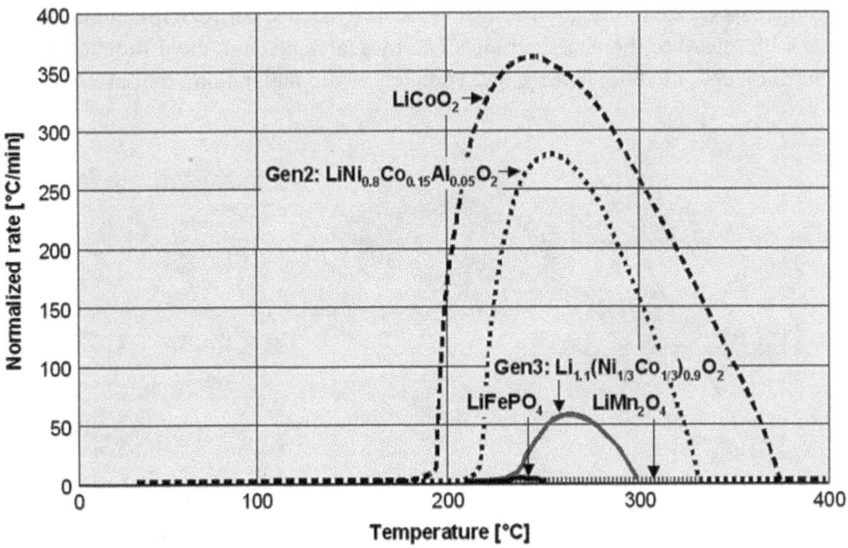

Fig. 31.15 Rate and starting point of energy release as a temperature function for different cathode materials

the iron phosphate's olivine structure, even at higher cell voltages and thus at higher energy densities.

However, intrinsic cell safety is not restricted to the cathode material alone. It can and should be influenced in the preceding steps. The best strategy is to stop the process as early as possible. By modifying the graphite used, utilizing alternative carbon modifications or even completely different anode materials such as lithium titanate, the process can be slowed down or its effects reduced at the anode. This is also possible for electrolytes by employing modified solvents or special additives.

For lithium-ion cells, intrinsic safety must be enhanced without compromising other important properties, especially energy density, power density, and service life. Optimizing individual material components does not help enhance intrinsic safety. The only solution is to enhance the entire lithium-ion cell system. In order to evaluate the lithium-ion cells' intrinsic safety, universally accepted tamper tests and other criteria such as ISO 12405-3 are currently used. Under test conditions, fires or even explosions should never occur.

The lithium-ion cells' intrinsic safety is only the starting point for ensuring the required product safety of the battery system. At battery level, typical measures include monitoring the cell condition by measuring voltage and temperature, reducing voltage and charging currents, and safely assembling the lithium-ion cells in the battery housing and vehicle (Fig. 31.16). All verifications take into account a larger number of safety aspects at component level (cell, module, battery system) and, of course, their behavior in the vehicle in the event of a crash (Fig. 31.16). All stress and strain tests are carried out in accordance with applicable standards, such as IEC, ISO, DIN, SAE, PVGAP, UN Transportation, or in-house regulations.

This guarantees that persons are not endangered. The efforts required to ensure safety at battery and vehicle level are reduced as the intrinsic safety of lithium-ion cells increases. One future aim must be to create a balance between functionality and costs with regard to the characteristics and measures taken at the different levels – lithium-ion cell, module, battery, and vehicle – while maintaining the safety level.

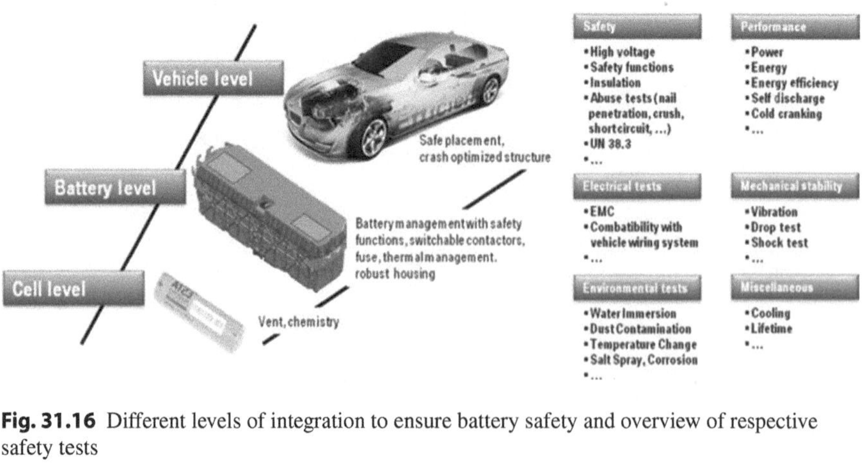

Fig. 31.16 Different levels of integration to ensure battery safety and overview of respective safety tests

31.4.3 Functional characteristics

The renaissance of electric mobility is largely due to the industrialization of lithium-ion technology in the consumer market and also the functional characteristics of this technology. Compared with all other modern battery technologies, lithium-ion technology has the highest energy density and power density.

However, this technology can also be specifically adapted to provide a certain ratio of power to energy, which is called P/E ratio. The P/E ratio is an important criterion for internal cell design and, when used in vehicles, it is mainly determined by the numerous requirements in hybrid, plug-in, and electric vehicle applications as well as by special requirements for individual vehicle models (Fig. 31.17).

The flexibility of lithium-ion technology enables it to be used in all types of electric vehicles. This is a major benefit because it creates synergies in development, purchasing, and production.

Based on the status quo of about 100 Wh/kg at battery level (for electric vehicles), lithium-ion technology can approximately double the energy density. Even if there is a potential to reduce weight and volume when designing cell modules, battery housings, and sub-components as well as when integrating them into the battery, the energy content will primarily be increased by optimizing the lithium-ion cell and

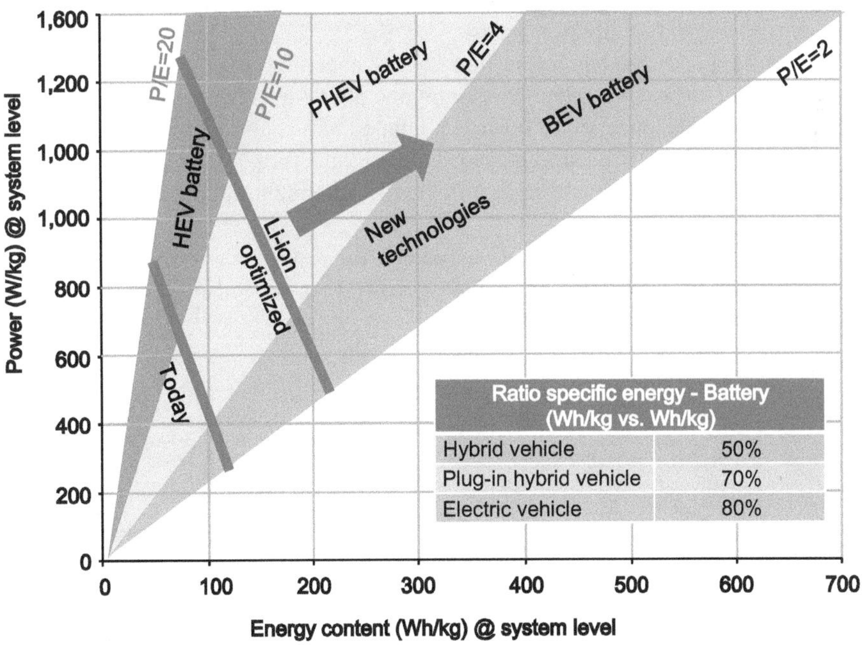

Fig. 31.17 The different battery requirements for use in hybrid, plug-in hybrid, and electric vehicles concerning power/energy content ratio (Areas and values indicated only serve as a rough guideline.)

the anode and cathode materials used, e.g., Si or Si-C for anodes, cathode materials with an open circuit voltage of 5 V instead of 4 V.

Increasing the energy density is imperative in order to fulfill market requirements concerning electric range and, thus, in order to establish electric mobility in the long term. To further penetrate the market, energy density and consequently electric range must be increased even more. Thus the development of novel storage technologies is essential.

The power densities that can be achieved today generally enable an acceptable level of driving performance for customers. This applies especially to the cells used in hybrid vehicles. If plug-in and electric vehicles are to succeed in the market, especially in the luxury car segment such as the BMW 7 Series or Mercedes-Benz S-Class, an additional increase in performance is desirable. Simultaneously achieving an increase in energy density and power density is a huge challenge in this area.

Simply designing the cell so that the power density is higher is insufficient. Its dependency on performance parameters such as temperature and state of charge must also be greatly reduced. The critical variable that determines the possible power output of lithium-ion cells is internal resistance, which depends on the temperature and state of charge (SOC). The internal resistance increases as SOC and temperature decrease, causing power to decrease as well (Fig. 31.18). This behavior must be taken into account in development activities, i.e., in the operational strategy. However, this does not fully compensate the situation. Usually, a certain oversizing of the battery is necessary in order to offer customers a largely steady road performance. This, in turn, increases costs.

Therefore, the dependencies on internal resistance in lithium-ion cells must be reduced as far as possible when developing the physical and chemical properties of these cells. The dependency of power on temperature is also a challenge; development activities urgently need to make progress in this area.

Notwithstanding performance requirements, minimizing the internal resistance is generally beneficial.

The lower the internal resistance, the higher the battery efficiency and the lower the heat generation and, thus, the lower the effort required to cool cells. This benefits costs, weight, and volume alike.

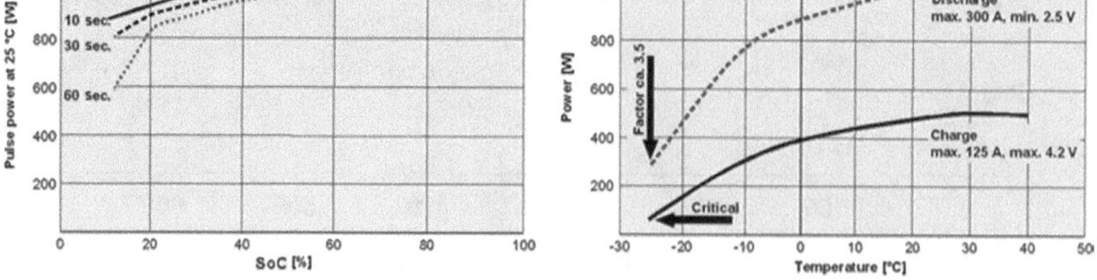

Fig. 31.18 Examples of lithium-ion cell power as a function of SoC and temperature. Development efforts must focus on reducing dependencies

The battery's thermal management, especially management of the lithium-ion cell temperature, is a crucial part of battery design and development. Efficient thermal management ensures that the cell's energy and power potential are largely exploited over a wide operating range and also makes sure that the desired service life is reached as aging of lithium-ion cells strongly depends on their temperature (Fig. 31.21). In accordance with the requirements for the respective vehicle model, air, coolants, or refrigerants may be used as a cooling medium. If performance requirements and consequently cooling demands are high, the use of refrigerants is the most technically demanding option. It is, however, the most efficient option in terms of weight, installation space, and costs. Efficient thermal management, which is the goal of development, involves dissipating heat from the lithium-ion cell's core to a cooling surface. It is also important that heat is homogeneously distributed within the cell, module, or battery. This must be taken into account when determining the outer dimensions of the lithium-ion cell, the materials used, and the wall thicknesses. The transversal cell dimensions or cell thickness are decisive for the cell's temperature gradients (Fig. 31.19).

31.4.4 Quality and service life

Other challenges in lithium-ion technology for automotive use and also highly complex subjects are quality and service life. Although they appear to be separate requirements, they are closely linked. The quality of the materials used and the production quality of lithium-ion cells strongly influence the possible occurrence of spontaneous defects as well as aging and degradation.

The battery's reliability – i.e., a low number of malfunctions and its capacity to maintain functionality over time – is a decisive factor that shapes customer satisfaction. It also influences costs if warranty claims arise. For electric mobility to be successful, reliability must be high.

Typically, quality is measured through the number of malfunctions over a given time period. Malfunction means the spontaneous failure of a battery caused by a defect of one or more components. These defects usually spread over the entire service life, with a larger number of malfunctions at the beginning caused by

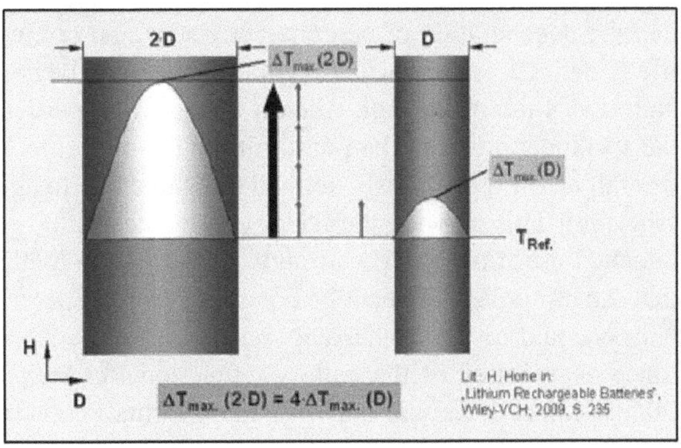

Fig. 31.19 Example of the internal temperature gradient's dependency on cell dimensions at a specified constant power and thus heat generation/dissipation

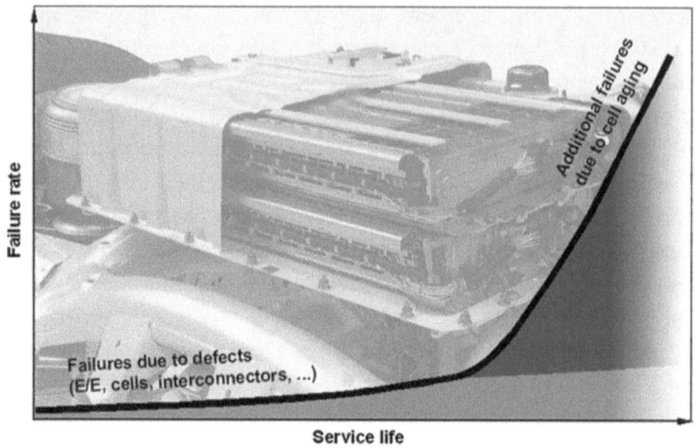

Fig. 31.20 An example of a battery's failure rate process. Toward the end of a battery's service life, this process can be dominated by the degradation of the lithium-ion cell's characteristics

manufacturing errors, for instance, that become noticeable early on. Wear and/or aging are factors toward the end of service life (Fig. 31.20).

Battery failure is caused not only by the lithium-ion cells, but also by other components such as the control unit, fuses, relays, connectors, or sensors. The failure rate of a battery is a function determined by the individual components' failure rates. The failure rates of lithium-ion cells should not be significantly different to those of other electric and electronic components. Even though there is hardly any market data available relating to lithium-ion cell failure rates and there are no meaningful statistics for automotive use yet, current experience shows that a certain reliability can be obtained through high manufacturing quality and quality assurance during and after production. Normally, a battery consists of many individual cells that are grouped in modules. Therefore, the connecting technology is very important. When designing lithium-ion cells and cell connections, laser welding should be implemented as a connecting technology. It ensures reliability and the high cycle times needed to meet cost objectives.

The production quality of lithium-ion cells strongly influences the aging process, i.e., the degeneration of cell characteristics and consequently service life. Service life is defined as the period of time between delivery (Begin-of-Life, BoL) of the battery or lithium-ion cell, when it meets all properties for new products set out in the specifications, and the point in time (End-of-Life, EoL) at which these properties fall below a previously defined threshold due to aging and/or degradation (EoL criterion). This is when the battery is deemed unfit – or faulty – for its originally intended use. EoL criteria strongly depend on individual use; there are no criteria that can be applied universally. For example, the EoL criterion for an electric vehicle battery could be that its capacity has fallen below 80 % in comparison to the BoL. Since degradation of the battery's functional characteristics mostly occurs in the lithium-ion cell, we will only provide information on the latter's service life below.

Even if the battery is not used, lithium-ion cells age due to chemical processes that occur within the battery (calendrical aging). The higher the temperature, the faster these processes take place (Fig. 31.21). Degradation also occurs when the battery is used, i.e., through charging and discharging (cyclical aging). Cyclical aging also depends on the temperature, but is mostly influenced by the energy throughput (Fig. 31.21). Another influencing factor is amperage during the charging process, i.e., the speed of charging and discharging. In addition to specifically developing and optimizing active materials in order to achieve the desired service life, operational strategies for batteries must also take into account the operating limits to avoid conditions that accelerate aging.

Any kind of contamination such as traces of water or water vapor can also speed up degradation. When choosing a cell technology, production quality and hermetically sealed housings are decisive factors.

Determining the desired service life is important. Today, the desired service life and EoL criteria are typically linked with the framework conditions that apply to the respective vehicle project. The most important framework conditions include the load profile determined by the driving cycle and the corresponding temperature histogram. The latter is a result of climatic conditions and self-heating of the lithium-ion cell determined by its temperature-dependent internal resistance and cooling characteristics (Fig. 31.22). Especially in batteries with high performance requirements and, consequently, higher internal heat generation, active cooling is needed to limit the temperature and, thus, slow down aging during operation.

The automotive industry typically requires a 15-year service life, which is quite long.

In contrast, the EoL criteria for lithium-ion technology used in consumer products are less stringent, requiring a service life of only two to five years. Currently, there is only limited knowledge about the long service life periods that the automotive industry requires. Against this background and due to the current capacity of lithium-ion technology, the required service life is sometimes reduced. The minimum requirement is ten years.

A service life of ten or 15 years includes calendrical and cyclic aging, i.e., a combination of the processes. The cycle numbers required during a lithium-ion cell's

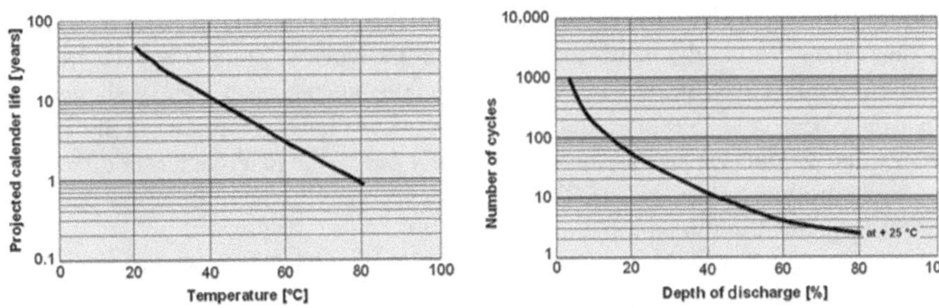

Fig. 31.21 Calendar life in years as a function of temperature and cyclic life as the number of cycles as a function of discharge depth

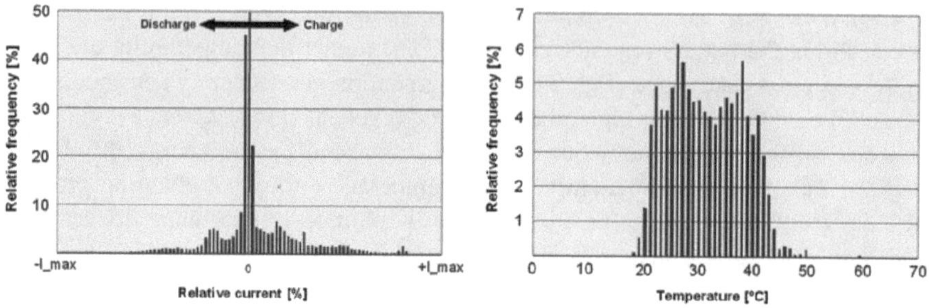

Fig. 31.22 Typical power and temperature histogram used to calculate service life in hybrid applications

service life are derived from the desired operational performance and energy consumption during electric operation, depending on the vehicle, driving profile, etc. For hybrid vehicles, the operational strategy also plays an important role. Therefore, there are hardly any universal criteria. It is safe to assume that around 2,000 cycles are realistic for electric vehicles and more than 4,000 cycles for plug-in hybrid vehicles.

Even accelerated aging tests such as storage at increased temperatures to determine calendrical aging or shortening cycles for cyclic aging take months or even years. Therefore, developing accelerated aging tests is both a necessity and tremendous challenge. Working with service life models is the only method for estimating the service life in early development phases, testing the cells' susceptibility to influencing factors, and exploring customer behavior. Currently, these models are not very reliable for long periods of time. The current service life models need to be further developed, optimized, and validated and must be able to correctly take into account all influences on the service life present in a vehicle, thus enabling reliable service life estimates.

31.4.5 Costs

The final and most important factor is costs. Today, it is possible but difficult to obtain a positive business case for electric vehicles equipped with lithium-ion batteries. The cost structure of an electrical power train shows that the majority of costs are caused by the battery. Within the battery, cell costs are the decisive factor (Fig. 31.23).

The price trend of lithium-ion cells in the consumer market clearly shows that costs have been considerably reduced over the years thanks to optimization of technology and production processes and especially due to the fact that larger quantities are being produced (Fig. 31.24). This applies especially to the lithium-ion cell type 18650, which is used in laptop computers.

By creating a geometric standard for the cells used in laptops, cost reduction potential has been fully exploited through economies of scale. In general, the costs

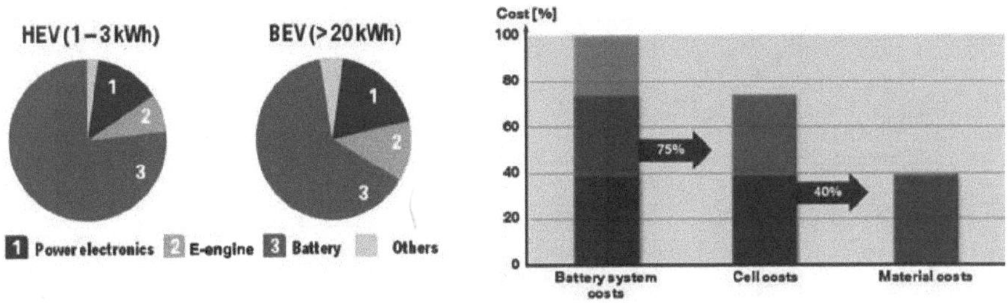

Fig. 31.23 Cost allocation for the electric power train of HEVs and BEVs (*left*), proportion of costs for material and lithium-ion cell in a BEV battery (*right*)

of lithium-ion cells used in automobiles are expected to drop (Fig. 31.24). The extent to which they decrease, however, will strongly depend on the development of standards, market volume, and competition.

Implementing a geometric standard for lithium-ion cells for automotive use is key for successfully developing and introducing electric vehicles in a cost-efficient manner. Development and especially quality assurance processes in the automotive industry do not enable cells, cell modules, or entire batteries to be developed from scratch and integrated into vehicles at short intervals. Implementing a phased, yet quick market introduction of new electric vehicles and simultaneously and instantly optimizing and further developing lithium-ion technology require standardized cell geometries and, in turn, standardized cell modules.

Against this background, the German automotive industry issued DIN specification DIN SPEC 91252:2011-01 (Electrically propelled road vehicles – Battery systems – Dimensions for lithium-ion-cells) at the beginning of 2011. Currently, a joint working group of ISO and IEC is formulating an international cell geometric standard (ISO/IEC PAS 16898: Electrically propelled road vehicles – Dimensions and designation of secondary lithium-ion cells). Even though discussion is more complicated at international level due to the countries' varying interests, there is a broad consensus that standardizing cell geometry for automotive use is indispensable.

Standardizing the cell geometry enables the creation of standardized cell modules that can then be configured in various ways to design batteries that suit different vehicle models and installation spaces. Synergies and cost efficiency potential can be fully exploited when, based on a standardized cell module, the same battery sub-components such as cell and module connectors, cell monitoring units, switch box, and the battery control unit are ideally used by all manufacturers as an industry standard (Fig. 31.25).

Combining the cell module and sub-components to make a battery must be a largely automated process. This must be taken into account when designing lithium-ion cells and all other components. It also means that not only the cell module, but

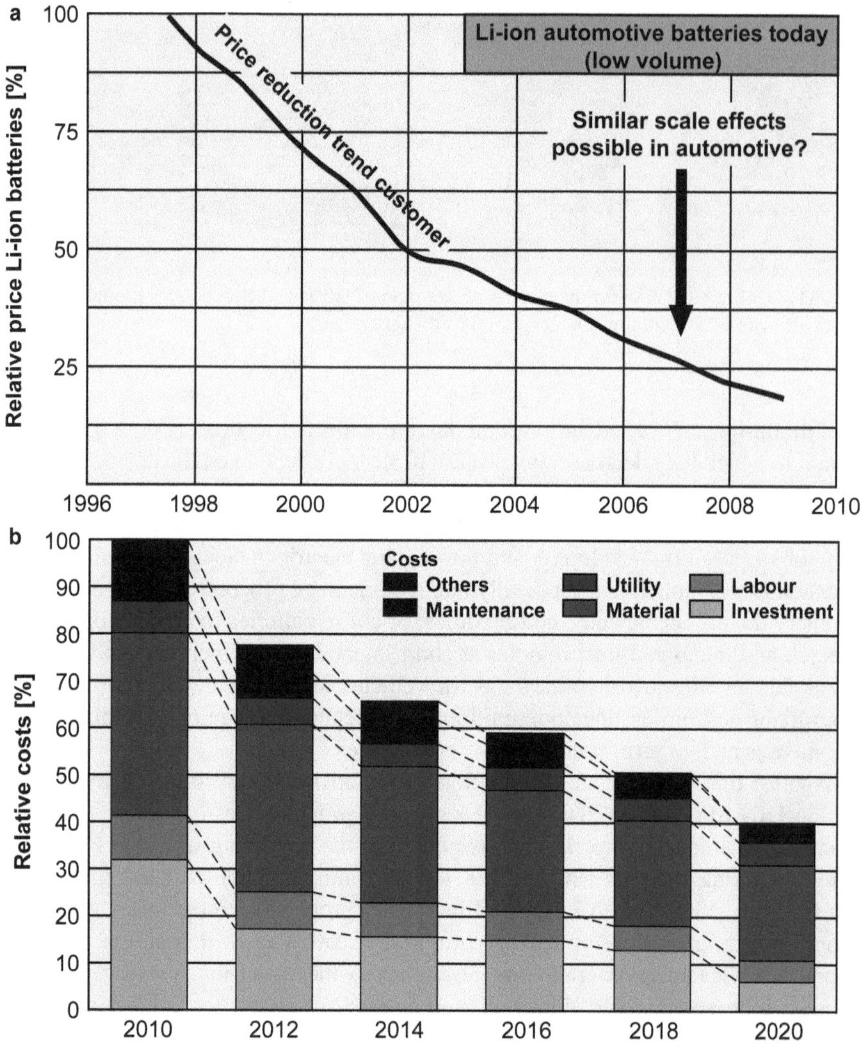

Fig. 31.24 Price trend history of lithium-ion cells for the consumer market (**a**), cost forecast for lithium-ion cells for automotive use (**b**) (*Source:* Roland Berger)

also a large number of individual components should be delivered as preassembled sub-systems for the final battery assembly. These measures reduce the costs of individual components and also enable manufacturers to reach the cost objective of no more than EUR 250/kWh (according to NPE) for an entire battery.

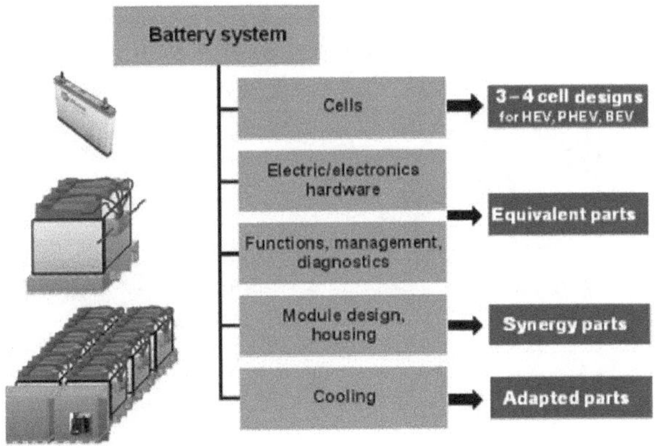

Fig. 31.25 Battery layout with sub-components (Fig. 31.12), modular principle, equivalent parts, and synergy parts

31.5 Outlook

While striving to reduce fuel requirements and, consequently, CO_2 emissions and to offer sustainable mobility in the long term, the entire automotive industry is devoted to not only constantly optimizing combustion engine vehicles based on fossil fuels, but especially to further developing electric mobility in all shapes and forms. Electric mobility is a key pillar of the industry's strategic orientation; large amounts of capital are being invested in it in spite of the market risks that still exist.

Consumers can already buy hybrid, plug-in hybrid, and electric vehicles today. The number of choices in electric mobility will increase over the next years. These types of vehicle will play a major part in reducing CO_2 and will be equally safe and beneficial for the customer. Energy storage systems used in such vehicles will be based on today's lithium-ion technology, with only very few exceptions.

For electric mobility to penetrate the market in the long term and for this trend to continue growing, lithium-ion technology must be further optimized. These optimizations will involve increasing a battery's specific energy content in terms of weight and volume to obtain an improved range. Also, costs must be reduced further by optimizing material use, standardizing components, and by fully utilizing the automation potential and corresponding economies of scale. To reach these goals, all industry partners involved in the entire value chain must collaborate.

Requirements for stationary application batteries

32

Bernhard Riegel

Contents

32.1 Introduction

Considerable structural changes are needed in public power grids to promote the use of renewable energies and to increase their share electricity production and introduce and implement climate protection measures. Decentralized energy conversion systems (e.g., photovoltaics, wind power) must be incorporated into the power grid. Another challenge in planning and management is the interaction of existing facilities with the decentralized production and storage units to be incorporated. Stationary energy storage systems will become more important over the coming years,

B. Riegel (✉)
HOPPECKE Batterien GmbH & Co. KG, Bontkirchener Strasse 1,
59929 Brilon, Germany
e-mail: bernhard.riegel@hoppecke.com

R. Korthauer (ed.), *Lithium-Ion Batteries: Basics and Applications*,
https://doi.org/10.1007/978-3-662-53071-9_32

because there will be a need to manage a great number of fluctuating and decentralized energy generating plants. Electrochemical storage systems, consisting of cells or batteries, will be essential in this respect.

These systems' requirement profiles hinge on different time perspectives: first of all, short-term fluctuations in the power grid need to be compensated; secondly, buffering and intermediate storage also need to be provided for several hours. Suitable storage capacities are thus required which feature a long service life and high cycling stability. Operating methods for PV storage systems connected to the grid and off-grid applications as well as industrial storage systems concepts up to 1 MW are key for future applications.

Long-established applications for industrial electrochemical energy storage systems are energy provision and safeguarding, telecommunication, IT and railway technology, emergency and safety technology, medical engineering, and logistics [1]. Industrial batteries must fulfill different requirements in terms of energy density and power, storage and discharge dynamics, reliability, maintenance, and costs, depending on the application type.

Stationary lead-acid energy storage systems such as uninterrupted power supply systems or solar power storage are already available and specially geared toward such applications. They are relatively inexpensive, but do not meet requirements for future storage systems since they have low specific energy and their aging depends considerably on their operational mode. Continuous operation in a partially charged state, for instance, causes sulfation and rapid loss of capacity. This is not the case for lithium-ion cells. Lithium-ion cells have favorable characteristics when compared to other electrochemical energy storage systems and therefore play an important role.

Electrochemical energy storage systems are a key technology. This is true for both electric mobility and renewable energy integration. Accelerated renewable energy implementation has a considerable effect on power grids. This is why it is important to understand the systemic interrelation between decentralized power production and available grid capacities. For example, capacity can be increased by using decentralized battery storage systems. This is only true, however, if operational strategies are developed that not only optimize the system from an operator point of view; requirements in grids also must be taken into consideration.

Several battery types for grid-connected PV battery systems can be used today: lithium-ion batteries, lead-acid batteries, and high-temperature or redox-flow systems. Because of their low costs compared to other technologies, lead batteries have been used in a wide variety of applications for many years. This includes numerous stationary and grid-independent applications. Lithium-ion batteries are an interesting alternative due to their longer service life and better cycling capability. However, they are currently mainly used in mobile applications. Their considerably higher costs are a disadvantage.

32.2 Requirements for industrial energy storage systems

Several characteristics determine battery system choice and suitability for specific applications. The most important factors in choosing an energy storage system are service life, reliability, power, availability, safety, and costs (Fig. 32.1).

Suitable energy storage capacities play an important role in the development of renewable energies and required grid capacity availabilities. Undoubtedly, lithium-ion batteries show great potential for electrochemical energy storage systems. The decisive factor for such energy storage systems are the load demands during different time scales. Firstly, they need to balance short-term electricity fluctuations; secondly, they need to work as buffers and intermediate storage for wind or solar power for up to several hours. These new requirement profiles call for customized lithium-ion cells, which are not available at the present time.

At this stage, lithium battery systems stand out thanks to their high energy and power density. These aspects can even be improved even further using a wide variety of active materials. Current developments also focus on increasing safety and reducing costs (life cycle costing). The many possible combinations of suitable materials are far from being fully exploited. New materials, nanocomposites, and new cell concepts offer promising potential for further improvements and new applications. Requirement profiles for stationary energy storage systems all contain the need for a high energy throughput, long calendar life, and high cycling stability.

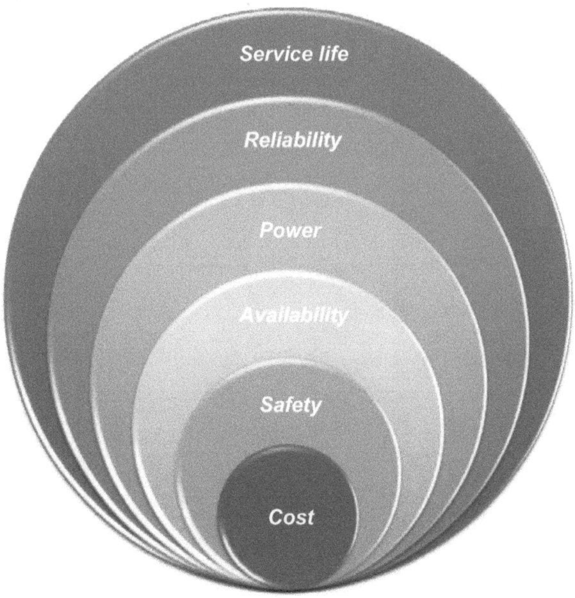

Fig. 32.1 Battery properties

32.3 Lithium-ion cells for stationary storage

Many automotive application manufacturers have developed large cells in the recent past. There are now cells available with a capacity between 10 and 400 Ah.

These cells have been customized specifically for the requirements based on the increase in demand for electric mobility battery systems. Example: Cells with a high specific power, referred to as high-power cells, have been developed for hybrid vehicles (HEVs) while cells with a high energy density, referred to as high-energy cells, have been designed for use in battery-electric vehicles (BEV).

As mentioned above, the focus is mainly on developing energy density and power density. Fig. 32.2 emphasizes this difference. It shows a weighted assessment of the different requirements for energy storage systems. In contrast, the requirement profile for stationary energy storage systems requires development of cells that are optimized either for short cycles with a low discharge depth or for cycles lasting several hours with a high discharge depth. The former are used to stabilize the grid, for instance, while the latter are used to store solar power, among other things. They also need to have a calendar life of over 10 (20) years.

In terms of standard stationary battery applications, cycling capability is of lesser importance compared to intermediate storage of renewable energies. Such standard applications include uninterrupted power supply (UPS), emergency power lighting, and diesel engine start-up. Other applications are telecommunications, the IT sector, power station coverage, and stationary applications for railroad crossings (Fig. 32.3).

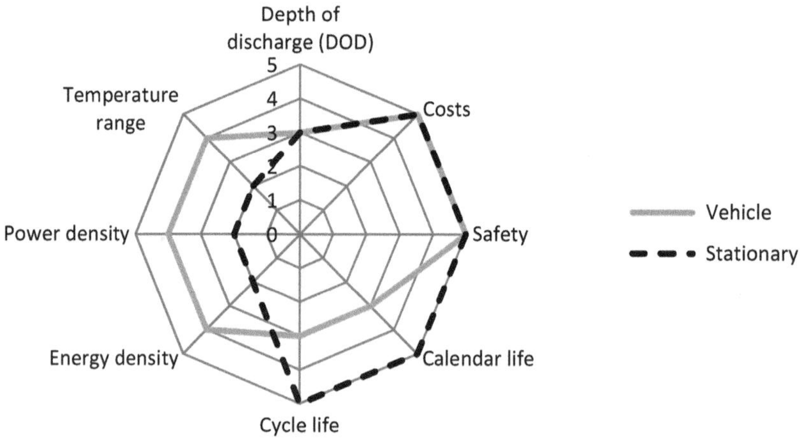

Fig. 32.2 Qualitative comparison of the energy storage requirement profile for vehicles and stationary applications (cycling operation) respectively. A distinction has been made between different states of vehicle electrification and different depths of discharge cycling for stationary applications

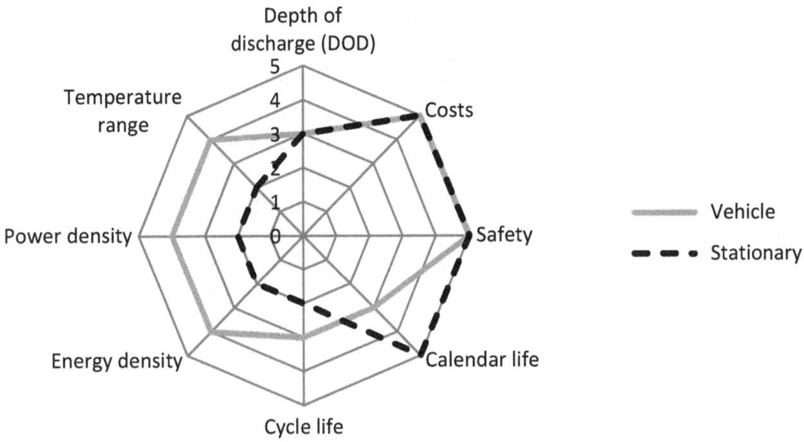

Fig. 32.3 Qualitative comparison of the energy storage requirement profile for vehicles and stationary applications (trickle charge, uninterrupted power supply)

32.4 Cathode materials for stationary lithium energy storage systems

Stationary storage systems require more attention to be paid to performance data, which are less important for the consumer market and mobile storage system applications. The most important factors are safety, cyclic and calendar life, cost, and storage efficiency. Several thousand cycles and a calendar life of over 10 (20) years are required to ensure economic viability for stationary lithium-ion batteries throughout their service life. There is thus a need for both improved operational expertise and customized development of cells for future stationary energy storage systems.

32.5 Trends in cathode material technology

Cathode materials play an important role in cell performance. The positive active material that has always been used in consumer cells is lithium cobalt dioxide ($LiCoO_2$). Its maximum usable specific capacity is only 130–150 Ah/kg. When charged, $LiCoO_2$ also exhibits thermally activated degradation (thermal runaway). Without suitable safety measures, this can cause complete cell destruction. Based on layered structure compounds such as $LiCoO_2$, transition metals such as nickel and manganese can be used to replace cobalt.

The compound $LiNi_{1/3}Mn_{1/3}Co_{1/3}O_2$ (NCM) achieves a specific capacity of 200 Ah/kg. Substituting manganese and nickel for cobalt also increases the temperature at which thermally activated degradation takes place, thus improving safety.

This material is currently often used in large lithium-ion cells for automotive applications. Other alternatives are phospho-olivines ($LiMPO_4$, M = Fe, Ni, Mn, Co). $LiFePO_4$ (Lithium iron phosphate, LFP) currently is the most promising in the short term and considered to be the most promising technology overall. John Goodenough of Texas University identified it as a cathode material for rechargeable lithium batteries back in 1969. LFP is very stable, preventing degradation reactions up to temperatures of 300 °C. It has a voltage of around 3.2 V (vs. graphite) and exhibits an achievable capacity of 160 to 170 Ah/kg. Moreover, iron, as opposed to cobalt, is an almost infinite resource. Test cells and smaller full cells have already demonstrated LFP's superior cycling stability. $LiFePO_4$ cells have a lower voltage and energy density than $LiCoO_2$ cells. However, this disadvantage is compensated by a lower capacity decrease throughout their calendar life in comparison to other lithium-ion batteries containing $LiCoO_2$ or $LiMn_2O_4$.

32.6 Trends in anode material technology

The most commonly used anode materials in lithium-ion cells nowadays are graphites with special morphologies. When compared with metallic lithium, they are shown to be considerably safer and feature higher reversibility in their electrochemical reaction. Intercalation up to the composition LiC_6 generates a theoretical specific charge of 372 Ah/kg. Having said all that, there is the danger of dendritic lithium particle formation, especially at high currents. This could cause an internal short circuit and, in the worst case, thermal cell degradation.

Lithium titanate $Li_4Ti_5O_{12}$ is a promising alternative. It is also known as zero-strain material because it exhibits minor changes in volume between charged and uncharged state. This characteristic, among others, ensures extremely high cycle numbers. Lithium titanate, however, has a disadvantageous low specific capacity of 175 Ah/kg and a relatively high intercalation reaction voltage of around 1.55 V that causes lower cell voltages [2]. Thanks to its favorable characteristics and the availability of raw materials, this substance could prove interesting for stationary electricity grid buffering or off-grid applications, which require a long cycle life.

32.7 The system lithium iron phosphate (LFP)/ lithium titanate (LTO)

Several different cell chemistries are available for lithium-ion energy storage systems. It is not possible to foresee which materials will come out on top for use in individual applications at the present time. Fig. 32.4 shows results for different materials. Fig. 32.4a shows charge-discharge curves for the cathode-anode

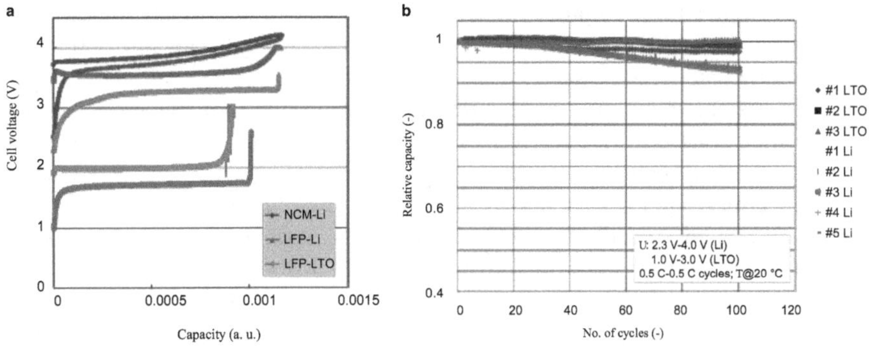

Fig. 32.4 **(a)** Comparison of charge-discharge curves for the cathode-anode composites NCM-Li, LFP-Li, and LFP titanate (LTO); **(b)** Capacity (standardized) in relation to the number of cycles for LFP-Li and LFP titanate

composites NCM-Li, LFP-Li, and LFP titanate (LTO). A new generation of lithium batteries using lithium titanate (LTO) in the anode and lithium iron phosphate (LFP) in the cathode may serve as an alternative for decentralized energy storage systems.

The LFP titanate composites, promising solutions for stationary energy storage systems, have a lower cell voltage, but their higher stability is expected to generate a considerably longer calendar life. Fig. 32.4b shows capacity (standardized) for LFP-Li and LFP titanate as a function of the number of cycles. It indicates the low capacity degradation of LFP titanate [3].

Electrochemists and electrical engineers involved in the "Langlebiges Energiespeichersystem" (durable energy storage system) project at the TU Muenchen (Technical University of Munich), Germany, are currently working on such an LTO-LFP battery. It is intended to withstand an extremely high number of charging cycles and could therefore be suitable as energy storage system for renewable energies. A service life of 20,000 cycles has already been achieved in prototype cells without a significant change in capacity. Today's lithium-ion batteries, for example, only achieve up to 4,000 cycles.

32.8 The complete energy storage system

A storage system's performance is defined by the available single lithium cells technologies. Advancements in innovative materials set out realistic cell level requirements regarding maximum cycle life for stationary battery systems. These are key to economical use of renewable energies.

However, the electrochemical cell is but one component in an overall energy storage system. This complex system consists of individual components such as battery management and monitoring, cooling, connecting technology, and housing. The design of an overall storage system is of paramount importance to ensure that the advantages of available battery technologies are fully exploited. The design is also relevant in terms of minimizing energy density and power density losses. Optimization and matching of individual components are a key factor in determining calendar life and overall system safety.

Large stationary lithium storage systems currently feature a modular design containing individual cells connected in series or parallel. This is necessary to meet energy and power requirements. These modules must also satisfy maximum electrical and thermal requirements. This is why module level battery monitoring is necessary. A system level management is also needed, incorporated into the large storage system as apparent redundant intelligence. Industrial storage systems require high voltages from the battery to the inverter with DC voltages up to 1,500 V possible. Special design measures need to be taken into account on a module and system level to provide isolation safety. The design also needs to take into account high electrical output in the MW range, which also poses a great challenge for power marshaling.

32.9 Examples of new applications

Following the introduction of the German law on renewable energies (*Erneuerbare-Energien-Gesetz, EEG*), grid-connected photovoltaic systems are a very interesting application in terms of electrochemical energy storage systems. Batteries are needed as energy storage systems to promote consumption of self-generated solar power in future smart grids. Energy self-sufficiency presents a future challenge for private households. Self-sufficiency refers to own electricity production and its usage. The energy storage system's purpose is to store surplus solar power until it is needed. Produced during midday peak periods, such solar power must be available for use at any other point in time (Fig. 32.5). This approach maximizes both on-site consumption and system efficiency.

Only surplus energy is fed into the grid and, depending on local legislation, the system owner will receive higher compensation for electricity that they feed into the grid during peak consumption periods.

It is a declared political goal to promote renewable energies that produce electricity. At this stage, the low-voltage grid does not always have the capacity to handle solar power, especially when many photovoltaic systems feed their power into the grid, causing bottlenecks. Such bottlenecks require increases in network capacity, otherwise generating systems need to be shut off temporarily. If more solar power systems are built, this problem will be exacerbated and have an impact on the transmission grid. Energy storage systems enable a great many photovoltaic systems to be incorporated into existing grids, allowing back-up power provided by conventional power plants to be reduced (Fig. 32.6).

Decentralized storage systems will play an important role in overcoming this problem. This is due to the stored energy's high economic value for the end customer and to the lower investment requirements compared to centralized storage systems.

Own consumption promotion under German renewable energy legislation also indicates a move toward increasing the proportion of own consumption from PV systems.

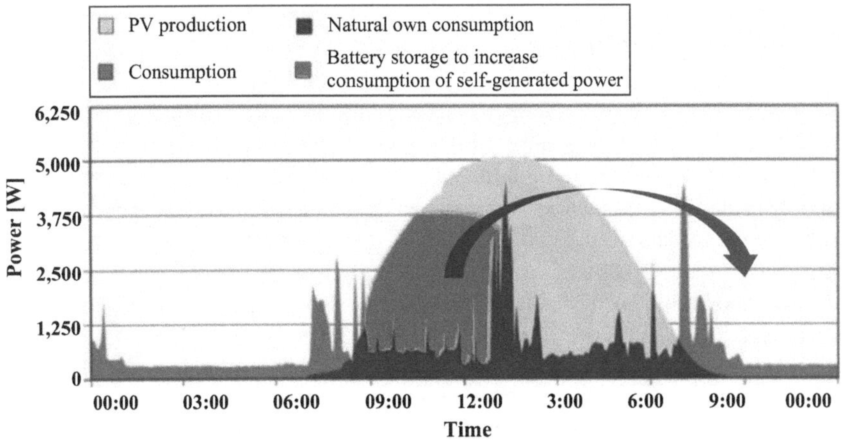

Fig. 32.5 Battery storage system to increase consumption of self-generated power [4]

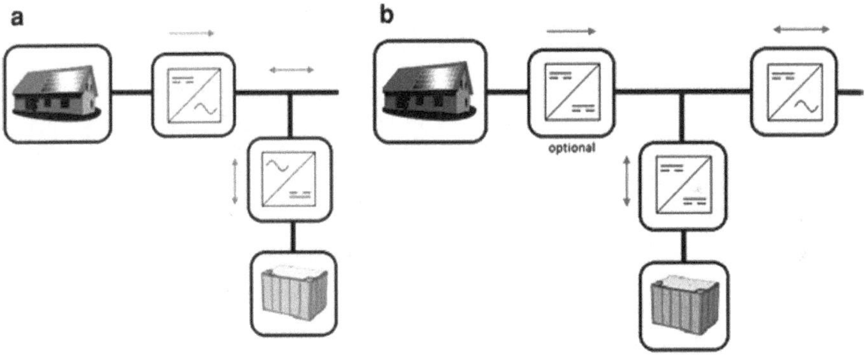

Fig. 32.6 PV battery system **(a)** with DC coupling, **(b)** with AC coupling

32.10 Stationary industrial storage systems

Stationary industrial storage systems will become more and more important in the coming years, because the German government has decided to increase the share of renewable energy in overall electricity production. Electrochemical storage systems are appealing, because they do not need to be installed in specific geographical locations, unlike compressed-air storage systems and pump storage stations. Other reasons include comparatively low planning outlay for such systems and their high scalability.

Other interesting applications are (a) renewable energy integration, such as feed-in power balancing for wind turbines, (b) decentralized provision of primary and secondary balancing power, especially when renewable energies provide a large share of production, (c) power trading, i.e., using electricity price differentials to create income, (d) reactive power compensation, and (e) harmonic component elimination.

32.11 Existing industrial storage systems

Lead-acid battery storage systems have been and are currently being installed to solve local energy supply problems. This includes network spur stabilization systems and frequency and voltage stability management systems. The largest plant built in Germany is the BEWAG plant in Berlin, which generates 17 MW. It was built in 1986 to stabilize the frequency and voltage in the West Berlin power grid and was operated as an isolated network at that time. The full storage capacity of 14 MWh was reached twice a day on average. For a lead-acid system, this plant achieved an unusually long service life with an overall nominal load volume of 7,000. German reunification meant that there was no longer the need to build other similar plants in Berlin, because the West Berlin power grid was then integrated into the European power grid.

Table 32.1 shows examples of other battery storage systems with up to 70 MW maximum power and 40 MWh storage capacity [5]. It should be noted that a battery storage system's peak power is not limited by the batteries used but by the power electronics installed. Lead-batteries can achieve high power values for short periods of time. There are currently three gigantic lead-battery storage systems operated outside Europe – in Puerto Rico (20 MW/14 MWh), Hawaii, USA (10 MW/15 MWh), and California, USA (10 MW/40 MWh).

Nickel-cadmium batteries require less maintenance than lead-acid batteries and have a longer service life. Consisting of 13,760 modular cells, the battery energy storage system in Fairbanks, Alaska, USA, is capable of achieving a possibly record-breaking 40 MW.

Other technologies tested and used for industrial storage systems include redox-flow and high-temperature systems such as sodium-sulfur batteries.

Table 32.1 List of operational battery energy storage systems (BESS)

Company	Location	Power [MW]	Energy [MWh]
BEWAG	Berlin, Germany	17	14
Kansal Electric Power Company	Tatsumi, Japan	1	4
Southern California Edison Company	Chino (California), USA	10	40
Vall Reefs	Godmine, South Africa	4	7.4
Hawaii Electric Light Company	Hawaii, USA	10	15
Puerto Rico Electric Power Authority	San Juan, Puerto Rico	20	14
Chugach Electric Assn	Anchorage (Alaska), USA	20	10
Golden Valley Electric Assn	Fairbanks (Alaska), USA	70	17

Lithium-ion-based industrial storage systems are also being installed. The Laurel Mountain wind farm (AES Corporation Arlington) was commissioned in October 2011. This wind farm consists of 61 wind turbines, distributed over 12 miles, each with an output of 1.6 MW, making a total of 97.6 MW, and an energy storage system with an output of 32 MW for 15 minutes. This wind farm can provide 260,000 MWh of emission-free energy per year. The storage system consists of several energy storage containers with a total of 1.3 million individual cells made by A123 Systems.

The latest project is the State Grid Corporation of China (SGCC) plant in Zhangbei, province of Heibei, China. It produces 140 MW from renewable energies with 100 MW generated from wind and 40 MW from solar power systems; it has a 36-MWh lithium storage system with an output of 20 MW for 1.45 hours.

32.12 Outlook

Renewable energies will play an increasingly important role in energy supply worldwide. In Germany, the share of renewable energies in energy supplies is currently 20 %. The intention is to increase the share to 50 % by 2030 and to 80 % by 2050. Solar and wind power fluctuate considerably, depending on the season and the time of day. Feed-in is increasingly decentralized on a low-voltage and medium-voltage level.

Lead-acid is currently the preferred electrochemical technology for integrating renewable energy into stationary applications. This is due to its ready availability, economic viability, and existing recycling provisions. Only lithium-ion technology can be used for electric mobility applications due to the required energy and power densities. Economies of scale in mass production associated with electric mobility can reduce costs in such a way that economic use of stationary lithium-ion battery applications is practical and competitive. Moreover, the lithium electrode material developments described above can play an important role in developing cost-optimized storage systems for new stationary applications.

Bibliography

1. Riegel B (2011) Elektrochemische Energiespeicher als Schlüsseltechnologie zur Erreichung der Klimaschutzziele. Solarzeitalter 1
2. Allen et al (2006) J Power Sources 159:1340–1345
3. Christensen et al (2006) J Electrochem Soc 153
4. Gewerblicher Eigenverbrauch von Solarstrom (2012) http://www.sma.de/de/produkte/knowledgebase/gewerblicher-eigenverbrauch-von-solarstrom.html. accessed on: 6 January 2012
5. Sauer DU (2006) Optionen zur Speicherung elektrischer Energie in Energieversorgungssystemen mit regenerativer Stromerzeugung. Solarzeitalter 4

Index

R. Korthauer (ed.), *Lithium-Ion Batteries: Basics and Applications*,
https://doi.org/10.1007/978-3-662-53071-9

T

U

V

W

Z

MIX
Papier | Fördert gute Waldnutzung
FSC® C083411

Kolibri 360 GmbH
Ferdinand-Jühlke-Straße 7
99095 Erfurt, Deutschland
produktsicherheit@kolibri360.de